Silva
The Tree In Britain

Silva
The Tree in Britain

Archie Miles

With contributions from: John White, Anne McIntyre FNIMH,
Stephen Daniels MA, MSc, PhD and Simon Brett

A Felix Dennis book

TED SMART

A TED SMART Publication 1999

First published in Great Britain in 1999

1 3 5 7 9 10 8 6 4 2

Editorial
Janet Visick – Editor (US)
Jon Goodchild – Designer (US)
Teresa Maughan – Editor (UK)
Jimmy Egerton – Designer (UK)
Sue Williams – Picture Researcher
Peter Gibbs – Production

Ebury Press
Random House, 20 Vauxhall Bridge Road, London SW1V 2SA

Random House Australia Pty Limited
20 Alfred Street, Milsons Point, Sydney, New South Wales 2061, Australia

Random House New Zealand Limited
18 Poland Road, Glenfield, Auckland 10, New Zealand

Random House South Africa (Pty) Limited
Endulini, 5A Jubilee Road, Parktown 2193, South Africa

The Random House Group Limited Reg. No. 954009

www.randomhouse.co.uk

Papers used by Ebury Press are natural, recyclable products made from wood grown in sustainable forests.

A CIP catalogue record for this book is available from the British Library.

ISBN 0 09 186788 6

Printed and bound in Italy by New Interlitho SpA.

Dedicated to the memory of

Jon Goodchild
(17.8.41 – 4.6.99)

*who sadly never lived to see the completion of this,
his last design project.*

The author's acknowedgements:
My partner Jan and daughters Rowan and Elly for their patience and encouragement. Felix Dennis for his faith and commitment, and without whom this book would never have happened. Contributing authors Stephen Daniels, Simon Brett, Anne McIntyre and John White.

All the wonderful people who have helped me over the last four years by generously giving their time, energy and expertise. I apologise to anyone who does not appear here, but you know who you are. Nancy Harrison, Colin and Daphne Gardiner, John Wyatt, Dr. Oliver Rackham, Tony D. Russell – Westonbirt Arboretum, Miles King – Plantlife, Mr and Mrs Bill Doolittle, Jane Kendall, Sue Clifford and Angela King – Common Ground, Karen Bradbury – Hereford Cider Museum, Jonathan Briggs – B.S.B.I, Guy Baxter and Barbera Holden – Rural History Centre, University of Reading, Paul Kendrick – Dept. of Paleontology, Natural History Museum, Roger Smith, Roy Finch – Arboriculturist, Professor John Birke, University of Bergen, Very Revd. David Leaning, Provost of Southwell Minster, Simon Cameron, Chris Robertson, Brian and Sal Shuel, Mike Abbott, Peter Faulkner, John W. Gittens, Dr. Andy Stevens, Plymouth City Council, J. S. Wright and Sons Ltd. – Willow Specialists and Merchants, Duncan Fearnley – Cricket bats and The Brasher Boot Company (who kept me well-shod!).

Contents

Preface

'They also serve who only stand and wait.'
Milton, Sonnet 16

Many years ago, very early on a Sunday morning in January, I quit my London flat in Kingly Court and wandered, aimless and miserable, through the back streets of Soho. Not a soul was abroad. My feet crunched through a thin layer of untouched pavement snow leaving a trail of scuffed footprints behind me in the desolate silence.

Eventually I found myself entering Golden Square, described by Dickens in *Nicholas Nickleby* as 'not exactly in anybody's way to or from anywhere...with a little wilderness of shrubs in the centre'. Little had changed since Dickens's time. A famous Ear, Nose & Throat Hospital was still situated in the northwest corner, lights blazing through its windows in stark contrast to the shuttered gloom of the surrounding buildings. On the roofs and bonnets of cars parked around the square the snow lay sparkling, reflecting the sulphurous glare of intermittent street lamps. I lifted my eyes idly to glance about and caught my breath...

Enormous, brooding and magnificent in their snow-laden outlines stood four sentinel trees, each flanking a pair of iron gates leading to the square's gardens. I must have passed them a hundred times before without so much as glancing at them. But somehow, on that morning, in that light, they seemed one of the most beautiful sights I had ever witnessed.

I know now that they were – and still are! – Pyramid Hornbeams (*Carpinus betulus* 'Fastigiata'). Like most other town dwellers I knew the common names of very few trees and still less of their habits and central importance to life on earth. To me trees were part of the scenery, something to climb as a child or shelter under, or to build houses or boats or furniture with.

All that changed on that early January morning. I do not know how long I stood transfixed, staring up at them, their branches reaching out to me, swaying slightly, the snow resting in every nook, cranny and fold of their skeleton, boughs dark beneath glittering, fantastic traceries.

At length I started as a hand pressed firmly onto my shoulder from behind. A policeman's helmeted face stared down at me, mildly quizzical.

'Are you all right, son?' he boomed.

Oddly, I realised that I was all right. My mood had undergone radical surgery, as the moods of youth are wont to do. Making mumbled excuses I strode back towards my home, pausing only to glance behind me once; the policeman was standing, rocking comfortably on his heels, looking not at me but at the ghostly silhouettes of the four guardians of Golden Square. What a sight they made!

From that day on I became obsessed by trees. I bought books. I walked and bicycled to seek trees out. I collected leaves and pressed them, often puzzling for weeks over their exact identification. (A narrow-leafed ash in Soho Square, I remember, caused a great deal of head scratching.) I began haunting Kew Gardens and nearby parks and arboreta.

Eventually I grew wealthy in commerce and acquired hundreds of acres in Warwickshire. With the help of a local conservationist farmer, Ralph Potter, I began planting in earnest. Nothing in the world gives me greater pleasure than to lay hands on the bole of a young sapling planted a few years back, imagining in my mind's-eye the day it will reach its maturity, its roots nourishing the earth, its leaves shading the ground, its fruits feeding wild life of all kinds and its beauty freezing the heart of humans yet to be born.

To those four hornbeams, standing yet in Golden Square, I owe all this. And to them my own small part in this present volume is affectionately dedicated.

Felix Dennis
Dorsington, Warwickshire
May 1999

OPPOSITE: *One of the four magnificent hornbeams in London's Golden Square.*

Silva More Precious Than Gold

You hold in your hand a very precious thing: it started its life as a tree growing free in one of the forests of the world.

Throughout its first life it soaked up carbon dioxide from an overheating global greenhouse, producing the purest of oxygen as a by-product of its growing pains. Evaporating water, it cooled, its ambience exhaling fluffy white sunshade clouds, clouds which move on the wind to protect other places and recycle the water they contain, nourishing landscapes further from the coast.

Its leaves, twigs and seeds fell to earth, feeding myriad worms, insects and arthropods. Fungi and bacteria who together make mor, (rich organic humus), binding the soil to hold minerals and water on the catchment and recycling nutrient, all in the service of generations to come.

Birds and mammals hunted, browsed and raised their young within the umbrage of strong branches, protected from but celebrating the changing of the seasons.

Shielding the earth from the icy winds of winter, the canopy kept the forest floor clear of snow so that frosts could penetrate deep, cleansing and rejuvenating the rich soils.

Spring and the rising of the sap saw leaves burst into bud as new roots made their way through the soil, binding it together and holding it ever safer from erosion, while the pulpit of branches held up to heaven gave vantage points for feathered choristers to sing their songs of love and hymns of praise.

The heat of summer ameliorated by the canopy witnessed the acrobatics of myriad inch worms dangling on silken threads, well-gorged acrobats spinning their cocoons of metamorphosis.

The rivers draining from the forest still ran sweet and cool even in the driest of summers, thanks to the magic of the forest which stores and releases last season's rain.

Lammas-tide is a very special season that pertains only to trees. This is the time when they produce a new crop of leaves as they prepare a harvest home for themselves and all that live among them or come to visit.

Autumn washes the strains of a job well done from the face of the earth, as leaves show their true colours, garnishing the forest with a host of fruits, nuts and seeds enough to see another winter through.

All this you hold in your hand; treasure it for its origins and the wisdom inscribed upon its pages; and when, like its authors, you have gained respect for the trees of the earth do all you can to safeguard the little that is left of the wildwoods of our planet.

Without those wildwoods and all the genetic stock they contain and conserve, there is no chance of helping those who are heaven-bent on rehabilitating this our planet, which is in a sorry state of repair.

It is a battle that must be won, for there are those hell-bent on destroying what is left with no regard for their fellow men and women, let alone the other helpmates of spaceship earth.

The cry of all these vandals is 'What about the workers, jobs come before environment.' They know this is untrue: most of the real jobs have already gone, thanks to mechanisation, while many of those that are left are as short-term as the shareholders' demands. The company moves on, taking the profits and the jobs with them, to continue their attack on the world's wild woods in someone else's domain.

If they had to pay the downstream environmental costs of just one percent of their rapine, they would go out of business. So on they go, all too often backed by backhanders to the local government.

What can you do to help put a stop to their greed and stupidity? Think when buying wood or wood products, and when you buy or trash paper. Contact your local conservation bodies for advice on recycling and for the names of both the goodies and the baddies in the timber business, and then campaign to ensure that your friends, colleagues and your local authority never again purchase from the latter.

This is a leviathan of a book, as majestic as the ents of the Lord of the Rings; this is your chance to share in the sense and sensibility of the trees, marching with Tolkien's army to reclaim the world.

David Bellamy
Bedburn, 1999

TREE SPIRIT

A crowded woodland of varied hues, ledge beyond ledge, climbs the hill's slow ascent, and in this dazzling dawn the sunlight plays upon the dewed leaves with gorgeous effect. Mellow limes contrast with sea-green chestnuts, now flaming with pinnacles of waxen bloom; the reddened foliage of the oaks seems to burn in the fierce light, while the pale tasselled birches are all a-quiver; and at the margin frail poplars change from grey to silver-whitening, as their leaves turn, with undersides uppermost, in the wind.

The Woodland Life *by Edward Thomas*

People love the company of trees. Woodlands stir up varying moods, whether springtime's natural rebirth, the raucous roar of blustery autumn, the dark cloak of a silent moonless night, or the still snow-settled white wood of winter. Clumps of trees form landmarks of fascination and mystery. Ancient individuals fire the imagination with their massive proportions and their contorted forms which may have evolved over several centuries. People build personal relationships with trees, making regular visits throughout the year to see how they are getting on, reassured by their renewed glory each spring, concerned by storm-torn boughs in winter, and soothed beneath the great canopies of summer.

Since childhood I have loved trees. As a small boy

OPPOSITE: *A weird tree face within an ancient oak in Richmond Park.*

RIGHT: *Although long dead, this ancient tree acts as guardian to a country cottage.*

I climbed hundreds, falling out of a good many of them, and even then I had my favourites. The best trees for building dens. Those with the densest foliage or hollowed trunks for hide 'n' seek. Golden pussy willows to squeeze through the back door and soften-up my mother if I'd been in trouble. Apple trees ripe for filling my face, as well as my pockets, for which I was frequently chased by an irate orchard owner. Elder stems to make wicked blowpipes. The great horse chestnut on the hill, above our house, where a lazy lower bough formed a great swing and my dad spent hours hurling great sticks far into the canopy to knock down the best conkers. I marvelled at their shiny, wet mahogany glow; took them home, strung up half a dozen to have a crack at the playground champion, and left two or three hundred in a bag to go mouldy.

Despite the rose-tinted memories of my early life, which seem sweeter as the years press on, I'm sure that, like any other small boy, I took all these elements of my country upbringing for granted. I was fortunate that I had parents who were both knowledgeable on countryside matters. Family walks always involved the identification of flowers, grasses, trees, butterflies and insects of all description. If it turned out to be something obscure then we either memorised it or took it home and ferreted through the numerous guidebooks. Fortunately, much of what I learnt stuck fast and, for that, I am eternally grateful to both of them.

With this background in mind I have, for several years, cherished the notion of producing a book upon trees in Britain. Not just a reference book, for many have trod this path before me, and there are numerous excellent works to be found, both in and out of print. Instead, I have attempted to produce a work which might loosely be described as a cultural overview of trees in Britain.

In some of the more remote corners of this island it is still possible to track down wild, untamed landscapes without the obvious imprints of man, but the reality is that such tracts of land are few and far between, particularly where trees are concerned. Man has cut them down when he needed to burn or build, swept them aside when he wished to farm; yet he built hedges to enclose his property and his stock and, when there seemed like a shortfall, he planted woodlands for his future needs. Although man has been a cardinal factor in the shaping of Britain's treescapes, the natural progressions and regenerations that have followed in

his wake have contributed to the wealth and diversity of today's British landscapes.

Trees in some shape or form are a basic necessity of life. Without them the world would eventually choke to death, and with this in mind the global implications of the loss of rain forest are of deep concern. In Britain they provide green lungs for our increasingly polluted towns and cities. Trees give our various landscapes the distinctive features which make them all such special places, whether it be willow-lined water meadows, gnarled old oaks of ancient forests, drooping boughs of a fruit-laden orchard, or majestic pines of Caledonia. Our houses are built and furnished with all manner of timber. Wood is used for making innumerable different implements, containers and craft. We burn it. We make it into paper. We need trees, and so we need to look after them, plant more of them, and then use them more wisely.

In order to promote a greater understanding of our tree heritage this book delves into several different avenues of association. Realising the cultural significance of trees in art, literature, medicine, folklore and industry will hopefully instil a reappraisal of all those hitherto barely noticed and widely unappreciated trees which have for so long just been taken for granted as part of the view. It is a daunting task to explore so many different aspects of trees within the covers of one volume, since each topic surely merits a substantial book in its own right, and yet the purpose of this work is to present a significant insight into the all-encompassing interrelationship between trees and man. If this book intrigues, inspires, surprises or even provokes, then it has achieved something.

The emphasis is principally upon the native trees and the most prolific of the naturalised species; for with some 2,500 different trees having the ability to thrive in at least some part of Britain, an encyclopedic format simply would not be possible. Instead, a wider vision of the more familiar trees and woodlands lies herein.

The leading players in this production appear in a multitude of different guises. They are not just the guidebook standards in mid-life respectability. They are shaped by the elements and the contours, as their costumes are wrought by the seasons. In infancy they are tiny yet hopeful, fighting for their place in the sun. In their dotage they writhe and grimace, tiring of the struggle, losing heart, crumbling gratefully back to the welcoming earth or crashing majestically in some grand finale.

Here is the mighty oak upon which a nation built its power and prosperity; the ash tree which burned brightly in many a hearth, or bore the blade of tool and weapon alike; the elm, whose demise was mourned, yet which lived to fight its round again; the hawthorn, the hedgerow stalwart and essence of May delights; the rowan, the darkest forces; yew, that enigmatic presence, the witness of death and the celebration of immortality.

During the three-year gestation period of this book I have pursued my own odyssey, making discoveries here, or photographs there, and all the while meeting fascinating people whose lives are enmeshed with trees and wood. The journey was always going to be too brief, however long it had endured. I have so much still to experience, and even more still to learn.

I will let John Wyatt have the last word here. I have met him but twice, and yet I feel that I know the man well. He writes with an enthusiasm and sensitivity as honest as it is beautiful. His two books devoted to his beloved Lake District, where he has spent a large part of his life, are gems. I quote from a passage in Reflections on the Lakes:

'...no writer could ever match the poetry expressed in the form of a single tree, for it speaks from its roots, through the fibres of its stem, the shape of the trunk, the turn and spread of the branches, the twisting and reaching of the twigs. A tree speaks. It speaks of a hundred summers and a hundred winters. Of storms and droughts and floods and snows, of plague and gales, of rocks and soils and hidden waters, of air and birds and pollinating insects. The whole of the message is contained in the way it has grown, precisely, to make use of what its environment has provided.*

It states in essence: "Here I am, and where I am is what I am."

The way of the saying can only be purely truthful. The natural laws make it impossible to lie. If a tree appears to be beautiful it is because beauty is truth and truth is beauty.'

Archie Miles, 1998

OPPOSITE: *A pool of sunlight catches tiny beech seedlings in a clearing on Selborne Hanger.*

BELOW: *An ancient yew with tortured visage grasps at the rocky slopes of the Wye valley.*

CHAPTER ONE

Evolution of Forest and Woodland

About 12,000 BC, long before mankind had made his mark on the land that is now Britain, the last ice age drew to a close. As the ice melted, the tree line migrated northwards from the warmer southern climes. Pollen analysis from earth core samples relates the story of the various species. The earliest arrivals were the pioneers: birch, aspen and sallow, still noted for their rampant colonising abilities. These were followed by pine and hazel, then oak, alder, lime and elm, which had all established major inroads across Britain by about 4000 BC. Other familiar species such as ash, beech, maple, yew and holly arrived later, as the climate grew warmer.

Countryside historian Dr. Oliver Rackham, has coined the description 'wildwood' for the prehistoric cover of woodland in Britain. For a long time this was widely assumed to consist of an almost blanket coverage of oak woods with a variety of under-storey trees struggling in their wake. However, pollen samples taken from 150 sites throughout Britain point more towards the existence of five mixed woodland types, though their precise proportions are masked because some species produce greater quantities of pollen than others.

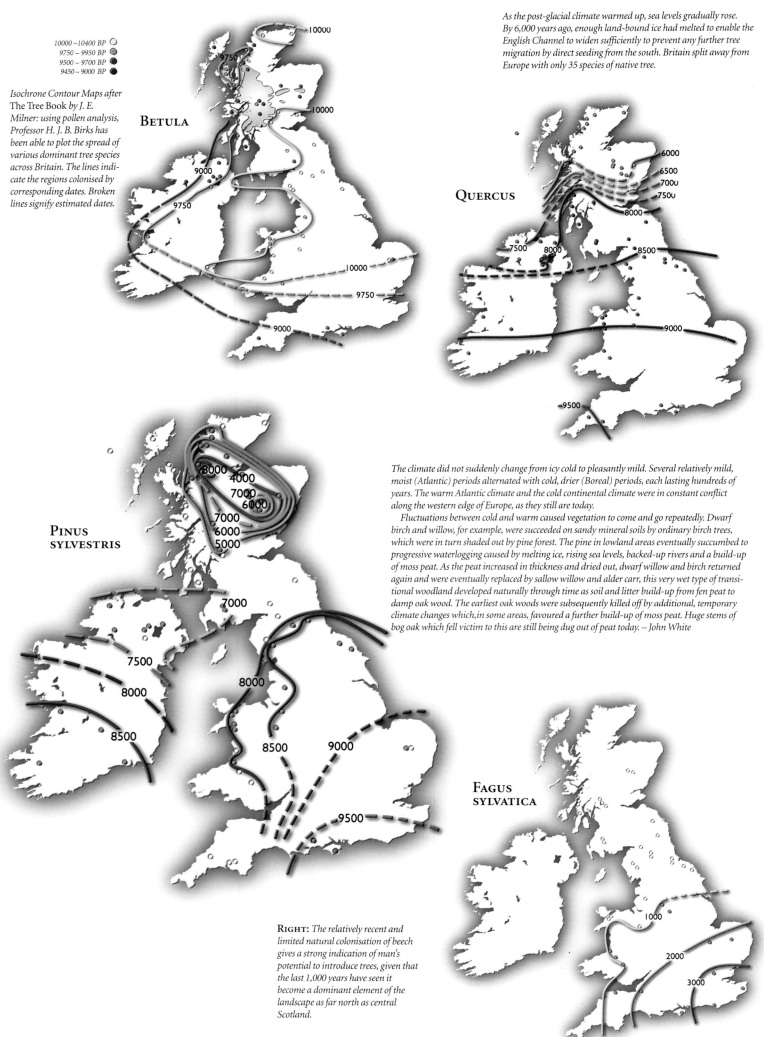

10000–10400 BP ○
9750 – 9950 BP ◔
9500 – 9700 BP ◑
9450 – 9000 BP ●

Isochrone Contour Maps after The Tree Book by J. E. Milner: using pollen analysis, Professor H. J. B. Birks has been able to plot the spread of various dominant tree species across Britain. The lines indicate the regions colonised by corresponding dates. Broken lines signify estimated dates.

BETULA

As the post-glacial climate warmed up, sea levels gradually rose. By 6,000 years ago, enough land-bound ice had melted to enable the English Channel to widen sufficiently to prevent any further tree migration by direct seeding from the south. Britain split away from Europe with only 35 species of native tree.

QUERCUS

PINUS SYLVESTRIS

The climate did not suddenly change from icy cold to pleasantly mild. Several relatively mild, moist (Atlantic) periods alternated with cold, drier (Boreal) periods, each lasting hundreds of years. The warm Atlantic climate and the cold continental climate were in constant conflict along the western edge of Europe, as they still are today.

Fluctuations between cold and warm caused vegetation to come and go repeatedly. Dwarf birch and willow, for example, were succeeded on sandy mineral soils by ordinary birch trees, which were in turn shaded out by pine forest. The pine in lowland areas eventually succumbed to progressive waterlogging caused by melting ice, rising sea levels, backed-up rivers and a build-up of moss peat. As the peat increased in thickness and dried out, dwarf willow and birch returned again and were eventually replaced by sallow willow and alder carr, this very wet type of transitional woodland developed naturally through time as soil and litter build-up from fen peat to damp oak wood. The earliest oak woods were subsequently killed off by additional, temporary climate changes which, in some areas, favoured a further build-up of moss peat. Huge stems of bog oak which fell victim to this are still being dug out of peat today. – John White

FAGUS SYLVATICA

RIGHT: *The relatively recent and limited natural colonisation of beech gives a strong indication of man's potential to introduce trees, given that the last 1,000 years have seen it become a dominant element of the landscape as far north as central Scotland.*

EARLY TREES

The dominant trees in early woodland across Britain were birch in the far north of Scotland, pine in central Scotland, oak and hazel from lowland Scotland down to the Midlands and through most of Wales and the West Country, a small area of hazel with elm in southwest Wales, and lime in most of central and southeast England. Palaeo-botanists have concluded from the pollen data that by 5,000 years ago most of the principal native species had finished migrating and colonising and settled in their respective provinces, remaining little changed to the present day, except for the widespread recession of small-leaved lime and the northerly migration of the pine woods. All the generalised provinces included wide varieties of woodland mixtures in localised areas, often featuring outliers from other regions, such as the small-leaved lime that still appears in isolated pockets of the Lake District and North Yorkshire.

Through dendrochronology (the study of annual ring data) a detailed portrait of an individual tree's age and growth history can sometimes be drawn. Ancient gnarled remnants of American bristlecone pines, of which the oldest living specimens are over 4,000 years old, have provided more than 9,000 consecutive years of annual rings from living and dead

LEFT: The earliest evidence of trees and their distribution on earth comes down as fossilised remains of bark and leaves, many millions of years old, while more recent information – from the last 10,000 years – can be gleaned from pollen grains preserved in peaty and marshy terrains. Some of the most beautiful fossils are those of leaves.

BELOW: The setting sun illuminates a massive old yew tree in Linton churchyard, Herefordshire. Measured at 33 inches in girth, it is estimated to be a stupendous 4,000 years old.

RIGHT: *Hackfall Wood in North Yorkshire is a strange and beautiful place on the north-facing slope of a gorge 100 yards deep cut by the river Ure. The wood is documented back to 1600, yet it may well be much older, for little if any agricultural grazing would ever have been feasible on such steep terrain. It was bought by John Aislabie, owner of the nearby Studley Royal estate and, in 1730, his son William set his hand to creating a romantic 'wild' woodland, complete with numerous follies, grottoes and cascades. Up until the 1930s, Hackfall was extremely popular with tourists, but when the wood was sold on to a timber merchant it was promptly plundered for its best trees and then left to decline.*

OPPOSITE: *In 1989 the Woodland Trust bought Hackfall and, by employing some sensitive management, they are resurrecting the woodland without losing the wonderful wilderness atmosphere that it has accrued since it was abandoned.*

BELOW: *The mid-nineteenth century etching presents the romantic vision of the forester's life in Hackfall.*

samples in the White Mountains of California; the information is being used to calibrate radiocarbon dating results. In Britain, there are very few trees with a vintage much in excess of a thousand years. However, semi-fossilised bog oak has been dug from peat bogs in Ireland, and a remarkable 7,200 years of oak history has been unravelled from its ring data. This information makes it possible to date oak timbers in buildings or artefacts on archaeological sites and provides evidence of environmental changes which are mirrored in anomalies among the ring patterns.

Britain's most obvious candidates for examination of ring data are its oldest living trees, the ancient yews still found, for the most part, in churchyards. Unfortunately, the most venerable examples tend to be either hollow or formed from a fusion of several younger stems, thus making ring analysis virtually impossible. Some amateur naturalists have investigated yew tree age through early documentation and folklore, while relatively young trees revealed cross-sections of a few hundred countable rings after being toppled by the gales of 1987 and 1990. Estimates of girth growth rates in yews up to about 500 years are roughly 7/16 inch per year, but this may well vary with growing conditions, and a deceleration of growth may have been a necessary life preserver for such ancients.

There can be no better place to experience an ancient yew woodland than beneath the chalk downs at Kingley Vale, in Sussex. All of Britain's primary woodland, or wildwood, disappeared long ago, so there are few other tracts of woodland of such enduring longevity. Richard Williamson, a former warden of this nature reserve, has estimated that the oldest of the trees may only be about 500 years old, and yet to move softly beneath the dark and looming forms, which writhe in the gloom and claw with arachnoid limbs at the bare earth, is an awesome and humbling experience. The oldest vegetation in

the great majority of woods is seldom more than a hundred years old, though its genetic lineage could well date back many hundreds of years. What, therefore, became of the original wildwood and how has it evolved? Woodland without evidence of man's intervention may still be seen in dense primary forest in South America, where regeneration and decay cycles continue in their own space and time. Prior to 4500 BC, Britain was probably almost completely covered by such wildwood, though traces of pollen from grasses and flower species which could not have tolerated woodland shade indicate that there must have been clearings in which such species could survive. Before early man got going with his axe, clearings could have been the result of trees falling, wild animals grazing or terrain that could not support trees. But for thousands of years man has had a continuing and irrefutable place within the landscape. As he evolved, so did the woodland.

It is generally assumed that the very earliest settlers in Britain would have made small incursions into the island's woodland cover, in order to build shelters, gather fuel and provide food for their livestock, but it was not until the Neolithic peoples came, around 4500 BC, that large areas of woodland were either grubbed out to provide tillage or managed to provide wood harvests. Rackham has postulated that half of the wildwood had gone by about 500 BC.

Shortly after the spread of 'farming man' a dramatic elm decline is shown in the pollen records from around 4000 BC. Several factors may have hampered the elms: human activities such as land clearance and the use of foliage for fodder, climatic changes, or an epidemic such as Dutch elm disease. R. H. Richens points out in his book, *Elm*, that it was the wych elm that suffered the decline, rather than field elms, which were still in the early stages of colonising. If Dutch elm disease infected the prehistoric elms, it would have been a strain different from the current bout, which is less inclined to attack wych elm than other members of the Ulmus family.

By the very nature of their biodegradability relatively few wooden artefacts have survived to provide a picture of wood culture in prehistoric life. Perhaps one of the most spectacular archaeological discoveries was the uncovering of an ancient trackway, known as the Sweet Track, beneath the Somerset Levels, dating from about 3900 BC. It was made from a variety of large and small poles to form a sturdy path across the marshy ground. Small-scale wood poles of various sizes, rather than mighty hunks of timber, were the mainstays for prehistoric wood use, both for construction and fuel. They were derived from the woodland management techniques of coppicing and pollarding.

By the time of the Roman invasion in AD 43, most of Britain had a well-developed Celtic agricultural

LEFT: *This sweet chestnut may be dead but the bleached and writhing boughs have an almost sculptural quality.*

system, with managed productive woodland as an integral element. The newcomers utilised the larger timber for building and boat construction, while the coppice wood fuelled their forges and warmed their villas. The full extent of Roman methods of woodland management is conjectural, although the writings of Columella in the first century prescribed various coppice rotations for sweet chestnut in Italy. Since the Romans introduced the chestnut to Britain in the first century AD it is fair to assume that similar practices were adopted in this country.

It is possible that the Romans cultivated the tree because their familiarity with it persuaded them that its growth rate, superior to the native oak, would provide a more productive coppicing regime. Both the nuts and flour ground from them formed part of the Roman diet.

OPPOSITE: *Winter sunlight seeps through a frosted elm. Elm has endured its own ecological winter for the last 30 years with the dreaded Dutch elm disease.*

BELOW: *The spidery claws of one of Kingley Vale's ancient yews, below the Sussex Downs.*

Sweet Chestnut

RIGHT: *By the time the foliage has taken on this mellow yellow, the spiky husks have fallen and burst to release their nuts. Savoured by squirrels, they are often too small for people to bother with in more northern climes, but with a little patience a few worthwhile nuts can usually be harvested.*

In spite of its visual similarity to oak timber, the sweet or Spanish chestnut does not form sound pieces of large timber. The tree grows throughout Britain, but it does best on deep, well-drained or sandy acidic loams. It has generally been grown for its coppice produce, most particularly in Kent and the Sussex Weald which, because it mainly comprises of tough heartwood, resists shrinkage or expansion even under the wettest conditions. It has thus been selected in the past for barrel staves, wine casks, coffins, and even underground water pipes, as an alternative to elm. Until the end of the last century, when wiring systems were introduced, huge quantities of chestnut poles were used for hop poles: in *The Wild Wood*, Peter Marren reports that as many as 2,000 poles, each about 14 feet long, were needed to support an acre of hops. Its principal use now is for fencing made of poles cleft from 10-12-year-old coppice wood and linked together by a twisted wire at top and bottom. H. L. Edlin, in *Woodland Crafts in Britain*, asserts that 25,000 poles could be harvested from one acre and would yield a mile of fencing. He also records that during World War II more than 1,500 miles of 'trackway fencing' (a scaled-up version) was used to enable military vehicles to travel over the soft, sandy beaches of Normandy.

The sweet chestnut is seldom allowed to mature to a large tree in the confines of woodland, where it would not find enough space and light. Most of the more impressive and ancient specimens are found in old deer parks, as massive individual landmark trees or in parkland landscapes, where they are sometimes grown in impressive avenues. Britain's sole triple avenue of the trees grows at Croft Castle in north Herefordshire. It is affectionately known locally as the Armada Avenue, due to the legend that homecoming victorious soldiers salvaged the nuts from the defeated Spanish Armada in 1588 and planted the trees in a fit of defiant patriotism. Sadly, these mighty chestnuts began to show signs of disease in 1990.

Without doubt, the most celebrated of all ancient chestnuts is the knobby old stager at Tortworth in Gloucestershire. There have been many accounts and endless debates upon the vintage of this remarkable tree. As far back as 1135, in the reign of King Stephen, it was known as the Great Chestnut of Tortworth. The approach to the tree, where it stands within an iron railing enclosure beside St. Leonard's church, fails to prepare the uninitiated for the monstrous writhing beast behind the bars. The main trunk has twisted, split and broken many times down the centuries, casting limbs which have dropped, layered and sprung anew, giving the impression of a small wood; yet the tree seems to be in good health despite its decrepit state.

OPPOSITE: *Sweet chestnut in its summery finery..*

BELOW: *Detail of one of the sweet chestnuts in Armada Avenue at Croft Castle, Herefordshire.*

THE MEDIEVAL WOODLAND

After the departure of the Romans in the fourth century Britain was thrown into the Dark Ages. It is evident that the agricultural infrastructure survived that era of upheaval and general decline, since neglect might have brought about a massive colonisation of secondary woodland in a very short time. Farming and woodmanship continued

OPPOSITE: *Ancient oak pollards in medieval wood pasture at Staverton Park in Suffolk.*

RIGHT: *The bronze plate placed in 1800 upon the fence which surrounds the Tortworth Chestnut, which almost certainly underestimates the true vintage of the tree by several centuries.*

BELOW: *The Tortworth Chestnut in Gloucestershire. Could this tree really be 1,200 years old?*

and became the fundamental basis for the Anglo-Saxon colonisers.

Anglo-Saxon charters form some of the earliest and most informative documents in respect of early rural life and the countryside in which it evolved. Several hundred have survived, dated from about AD 600 up to the Domesday Book of 1086. They provided detailed information for the conveyancing of parcels of land. Since there were no maps available for reference, the charters contained perambulations which described parcel boundaries, with many features that would still be familiar to this day, including references to hedges, hedgerow trees, woods and landmark trees. They also referred to the functions and produce of woodlands, with such details as quantities of swine that could be grazed upon the land (pannage) or the amount of coppice poles or rods that could be harvested.

Place names also contribute to an understanding

of Britain's ancient woodland. Rackham has listed several words which most likely referred to particular types of woodland: wudu (wood), graf (grove), scaga (shaw), hangr (hanger) and the more obscure bearu, holt and fyrhp. Settlements situated in clearings surrounded by woodland commonly bear the suffix ley, hurst or even the Norse thwaite, while field signifies an open space near to woodland. Study of any modern maps will reveal numerous such names, from tiny hamlets or farms to large towns.

A nationwide plotting of all such woodland related names suggests that, when the names arose, the principal areas of woodland were in the southeast quarter and the western Midlands, much as they are today. Such a distribution is consistent with the Domesday Book, which recorded woodland appertaining to the 12,580 listed settlements. About half had some measure of woodland and, although the areas were not quantified, the southeast appeared to be the most densely wooded area of Britain, where Kent and Sussex are still the most wooded counties today.

Many of the very earliest Saxon charters identify private ownership of woodland, with all the accompanying intricacies of custodianship and its attendant legal disputes; but from the coming of the Normans onwards the documentation of woodland, its extent, ownership, and function began to be recorded in even greater detail. Once William the Conqueror had installed himself as monarch in 1066, he set about the legal and physical organisation of his domain. Already it would seem that woodland was a hugely valued asset, over which measures for protection from thieves and vagabonds had become essential. Rigorous laws were laid down, and a hierarchy of officers was appointed to enforce them.

Nowhere else were such laws laid more heavily upon the common folk than in the New Forest. Forests were composed of different types of land, most of which contained a fair measure of woodland, although some, such as those on Dartmoor, Exmoor and the Pennines, were mostly moorland. They had legal bounds and were held principally for the king's deer, as defined in Manwood's Forest Laws of 1598:

'*A forrest is a certen territorie of wooddy grounds and fruitfull pastures, priviledged for wild beasts and foules of forrest, chase and warren, to rest and abide in, in the safe protection of the king for his princely delight and pleasure; while territorie of ground so priviledged is meered and bounded with irremoveable markes, meeres, and boundaries, either known by matter of record, or els by prescription. And also replenished with wilde beasts of venarie or chase, and with great coverts of vert for the succour of the said wilde beasts to have there abode in; for the perservacion and continuance of which said*

LEFT: *A bucolic scene of a boy attending hogs in the forest; an idyllic Victorian rendition of a simple woodlander's way of life.*

OPPOSITE: *Dense carpet of snowdrops in a Gloucestershire woodland.*

place, together with the vert and venison, there are certen particuler lawes, priviledges, and officers belonging to the same, meete for that purpose that are onely proper unto a forrest, and not to any other place.'

The king had authority to impose his forestal rights over any land he desired for hunting and to appoint courts to police the forest and levy fines. He also utilised the timber from his lands for construction and the underwood for revenue. Where such lands were governed by the nobility rather than the king, they were often known as chases rather than forests.

While the Domesday Book contains few details about woodland, it does provide a general picture of woodland distribution across most of England. Its data support the calculation that approximately 15 percent of England was wooded, all of which would have been managed by that time. By the end of the nineteenth century, some two-thirds of this woodland had been lost, half of it destroyed during the

TREE NAMES

Thousands of place names include specific trees, the most common being thorn (hawthorn or blackthorn), ash, willow and oak. Such a name may have indicated an area dominated by that species, or the presence of a small group or one magnificent specimen in a landscape.

BELOW: *Burnham Beeches, (J. G. Strutt).*

'How is it possible . . . to look upon its gigantic trunk, and widely spreading arms, without feelings of reverence! How many, not merely generations of men, but whole nations, have been swept from the face of the earth, whilst, winter after winter, it has defied the howling blasts with its bare branches, and spring after spring put forth its leaves again, a grateful shelter from the summer sun! Its tranquil existance, unlike that of the human race, stained by no guilt, chequered by no vicissitudes, is thus perpetually renewing itself.'
Jacob George Strutt, Sylva Britannica, 1830.

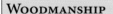
RIGHT: Bradfield woods in Suffolk is an excellent example of a long-established woodland once more being expertly managed to produce both coppice wood and mature timber. Even the coppice brushwood, which has no commercial value, has been woven into a deer fence to protect the regenerating stools.

WOODMANSHIP
Woodmanship is the logistic and physical management of woodland, incorporating the selective felling of large trees for timber, coppicing and pollarding, and understorey control to maximise regeneration. Thus the woodland remains vibrant and healthy, with readily accessible timber of all grades to supply the needs of wood crafts and industries. Charcoal burners, bodgers and hurdle makers, all of whom are beginning to appear once more in various woodlands, could be described as an intermediate stratum of woodmanship, for they all practise coppicing to produce their raw materials.

period between 1086 and the massive population cull wrought by the Black Death in 1349. Then, for a time, the conversion of woodland to farmland ground to a halt, allowing existing woods to recover their vigour and derelict agricultural land to regenerate with secondary woodland. Over the next 500 years relative stability of woodland cover would be established, due to its social and economic value.

From the mid-thirteenth century there are more frequent records of woodland surveys, legislation and transactions. One of the earliest, and most detailed, is the *Old Coucher Book of Ely*, which offers a comprehensive account of land uses within the bishopric in 1251. It recorded woodland acreages and estimates of the potential revenue to be derived from their wood harvest. The Ely document was a milestone in the enduring evolution of managed and valued woodland. The renewable nature of woodland means that, while the individuals living in the woodland may be of relatively recent vintage, the intrinsic character of a well-managed wood often remains unchanged. Thus, in the Ely area, names may have changed slightly down the centuries, but to all intents and purposes the distribution, shape and size of most of the woods in that area are largely the same today as in 1251.

From the earliest times of woodland exploitation in Britain, well-considered plans ensured renewal of a harvest so crucial to socio-economic well being.

Boundary banks and ditches, as well as fences and hedges, were constructed to keep animals from browsing the regenerating coppice growth. Accounts of such construction work exist from the thirteenth century, as do extant remnants to be found in many woods. The true vintage of woodbanks is usually indeterminable, although they are known to have existed prior to the Norman Conquest and, Rackham suspects, even as far back as the Roman occupation. The scale of some of these excavations and the toil they required surely bear witness to the value placed upon woodland.

In the Middle Ages, great quantities of wood were required for building, fencing and fuel. A mixture of mature timber, generally oak and elm, formed the framing in building construction, while pliable rods of sallow, hazel or elm were used for the wattle-and-daub infill. Fencing and fuel supplies were drawn from regular coppice rotations. The former used huge amounts of hazel and sweet chestnut while almost anything, including offcuts from felled timber, was hauled away for fuel. Such use was served by the format of most medieval woodland, which was coppice-with-standards, a style of management currently being encouraged once more throughout Britain. Well-spaced standards were left to mature into large timber trees, while the underwood was coppiced on short rotations between four and eight years, as revealed both by contemporary accounts and by the number of annual rings in the wattling of ancient buildings. Industries which required huge quantities of fuel also worked the coppice rotations with care to ensure a renewable resource, since there would be little future for industry without fuel.

Early concerns about availability of a plentiful supply of large construction timber were first officially addressed by Henry VIII. His Act of 1543 decreed that 12 standard trees per acre of woodland must be left to mature until they were 10 inches square at 3 feet from the ground. In 1559, Queen Elizabeth stipulated that only coppice wood could be used to make charcoal from trees which grew within 14 miles of the coast or navigable rivers.

But it was not until 1664 that a well-conceived and practical plan of action to regenerate the country's woodland was laid before the populace. In that year, the diarist and academic John Evelyn published his *Silva: or, a Discourse of Forest-Trees, and the Propagation of Timber in His Majesty's Dominions*. British woodland was showing the effects of overuse, notably in shipbuilding: records from a few decades later show that as many as 2,000 mature oaks were required to build one 64-gun warship. Evelyn's treatise encouraged positive political action, and exhorted the literate and learned members of the landowning classes to plant trees and regenerate their woodlands. *Silva* was published several times during the following century, while many other commentators expressed similar sentiments and contributed to a growing interest in the establishment of new plantations of both native and naturalised species.

As the seventeenth century slipped into the eighteenth, man's inevitable manipulation of British woodland became more pronounced. Until this time the native trees had largely been either continuously managed for their coppice wood, nurtured into mature timber trees or, when gaps appeared, planted with the same species. Woodland was therefore mainly 'natural' in its structure and appearance. But though beech, for example, was a native tree across the southern counties of England, particularly on chalk, it was now planted in huge quantities outside its natural range all the way up to Scotland. Its popularity with landowners and landscape designers alike caused it also to become manifest in numerous parks, gardens, avenues, hilltop clumps and hedgerows. Subsequent vogues saw wych elm, hornbeam and larch planted at particular periods in the nineteenth century, followed by Scots pine, Corsican pine, hybrid poplar and lodgepole pine: the fast-growing conifers were cultivated largely to service changing industrial requirements.

From Evelyn's time onwards, ever more explorers and plant collectors travelled to the farthest reaches of the world and returned with specimens of hitherto unknown trees. Most of these tree treasures borne home were planted in private parks, arboreta and botanical collections. With a few exceptions

they were mainly amenity trees, with little commercial value. After trial cultivation in Britain many of the trees that thrived and found favour were manipulated into an astonishing array of cultivars.

Of the hundreds of trees that were brought to Britain there have been very few which can truly be thought of as widely naturalised. Three trees in particular have colonised with remarkable vigour: the sweet chestnut; the sycamore, introduced in the sixteenth century (although there is evidence to suggest that a few specimens arrived even earlier); an individual now much pilloried for its weedlike, invasive ways and held in low esteem by some conservationists; and the larch, which came from central Europe prior to 1629. All three of these have established a significant foothold in British woodlands, where space on the woodland floor is fiercely competitive and only the strongest and most adaptable will succeed.

NEXT PAGE: *Typical dome-shaped canopy of the mature sycamore. As a field tree it provides excellent summer shade for livestock.*

BELOW: *Scene of foresters at work felling trees. From Prideaux John Selby's* History of British Forest Trees, *1842.*

BOTTOM: *Scene of timber hauling (Selby).*

Sycamore

ABOVE: *The earliest known representation of the sycamore in Britain is to be found in the Cathedral of Christchurch in Oxford, carved on the stone arches above the shrine of St. Frideswide, which dates back to 1289, long before the generally accepted date for the arrival of the tree in Britain. The carved stones which form the canopy include several different trees and plants with oak, hawthorn and field maple also among them. It had been presumed that the sycamore was either known to the carver, who might have travelled from Europe, or that a few specimens were established in Britain at this time.*

The whole stone canopy had been smashed up and 'lost' during the dissolution of the monasteries in 1538. The stones lay hidden until excavation work for new drains in 1889 brought them back to light. They were restored to the cathedral once again exactly 600 years after their original installation.

OPPOSITE:
A handsome sycamore in the Ribble Valley, North Yorkshire with Penyghent in the distance.

ABOVE RIGHT: *Evening light reveals the subtle contours of sycamore bark.*

RIGHT: *Portrait of the sycamore, which Prideaux John Selby considered was 'certainly a noble tree'.*

A*cer pseudoplatanus*, literally the false plane maple, was often referred to as the Great maple by early authorities and is still commonly known as a Plane tree in Scotland, due to the shape of its leaves. The name Sycamore, or originally Sycomore, derives from its early confusion with a Middle Eastern fig, *Ficus sycomorus*, again due to the similar shape of the leaves.

A somewhat sinister regional name for the tree hails form the west coast of Scotland, where it was known as a 'Dool' or 'Grief tree', since it was common practice for clan chiefs to use the tree as a gibbet to hang their enemies.

A native of Europe, the sycamore was probably first introduced in limited numbers prior to 1500. It has been intermittently revered and reviled by writers and naturalists ever since. Evelyn gave the tree short shrift, claiming that 'it is much more in reputation for its shade than it deserves; for the Honey-dew leaves, which fall early, like those of the Ash, turn to mucilage and noxious insects, and putrify with the first moisture of the season, so as they contaminate and marr our walks; and are therefore, by my consent, to be banished from all curious gardens and avenues'. The sycamore has been a prodigiously successful coloniser and will grow well in almost any situation and upon

all soil types; for over 200 years it has been one of the dominant trees of northern England and Scotland, spreading from its original domestic settings to feature in all aspects of the landscape. Some of the rampant upstart's vigorous opponents among modern conservationists across Britain spend their weekends systematically plucking its seedlings from woodlands, but the balance of opinion may be beginning to tilt back in its favour. In *Flora Britannica*, Richard Mabey has done his best to rehabilitate the sycamore and reports recent positive ecological revelations connected with the tree: those slimy fallen leaves, so abhorred by Evelyn, apparently decay quickly and, in doing so 'boost earthworm populations'. Many woodland flowers are also successful beneath sycamores. They encourage quantities of insect life (principally honey-dew producing aphids) which provide sustenance for birds, particularly in urban environments. The commercial value of the tree's timber and its protective capacity to shelter homesteads and younger trees from wind, are among the qualities John White notes in a pragmatic appraisal later in this book.

Larch

ABOVE: *Female larch flowers burst forth in the spring like bright pink jewels amongst the vivid green rosettes of new leaves. The flowers are often known as larch roses.*

BELOW: *Three images of the 'Mother' larch at Dunkeld in Perthshire, made at approximately 90-year intervals.*
BELOW LEFT: *In 1822 as depicted by J. G. Strutt.*
BELOW MIDDLE AND RIGHT: *In a 1907 photograph by Henry John Elwes and Augustine Henry; and in 1997. Recent high winds have ripped a couple of boughs from the 'old lady', otherwise she is in good heart, though now more hemmed in by neighbouring trees.*

*N*umber One: The Larch, decreed *Monty Python's Flying Circus* in its risible guide to British trees, back in the 1960s. Ironically, this bizarre comedic introduction to the larch, repeated several times over, was remarkably accurate. The European larch was the first of the larches to arrive on British soil around 1620. Initially, it was planted as a new and beautiful curiosity. Here was that strange beast amongst conifers, a deciduous tree with a wonderful seasonal repertoire of vivid emerald needles in spring, interspersed with the ruby red emerging cones, the foliage turning to deep gold in the autumn before falling to reveal the graceful poise of the skeletal tree.

Both herbalist John Parkinson, in 1629, and Evelyn, in 1664, refer to the tree, but only as a rarity, and yet Evelyn, while appreciating the tree's beauty, was also aware of its potential value for timber.

The timber of larch is remarkably hard and durable and this, combined with its quality of incombustibility and resilience in waterlogged conditions, has made it popular in building construction, civil engineering and boat building. By the middle of the nineteenth century, larch was one of the most widely grown and valuable timber trees in Britain.

In modern forestry, larch has proven ideal as a nurse tree for other species and as a partial firebreak on the edge of more combustible conifers. Its varying hues throughout the year make a welcome respite to the eye when planted amongst evergreen conifers. Larch seeds spread widely on the wind and, wherever they settle, particularly in sheltered hollows or along upland valleys, they spring anew, forming small, naturally set woodlands. Like any colonising plant, it has both protagonists and antagonists. At a time when tree lovers are busily promoting the resurgence of broadleaf woodland, the larch has become an unfashionable tree among those concerned with amenity values, but without it the landscape would be poorer.

Timber from the larch is remarkably hard and durable this, combined with its resistance to fire and its resilience in waterlogged conditions, has made it popular in building construction, civil engineering and shipbuilding. Timber from the Atholl estate in Perthshire was used to build a frigate named the *Athole,* in 1820, with keels, masts and yards made solely of larch. In 1842 Selby reported the continued success of the vessel, stating that 'its strength and durability were equal to one built of oak'. By the middle of the nineteenth century, larch was one of the most widely grown and valuable timber trees in Britain.

James, the 2nd Duke of Atholl, was the first to plant the tree for its commercial value. He planted two specimens of *Larix decidua* hard by the

cathedral at Dunkeld, probably in 1738. Jacob Strutt drew a delightful sketch of the trees during his Scottish travels and, almost a century later, Elwes and Henry made a splendid photograph. One of these trees, known as the 'Mother Tree', still thrives near the ruinous cathedral, its partner having been destroyed in 1906 when it was struck by lightning. Three more larches of the same vintage once grew by the castle at Blair Atholl. Before the 2nd Duke died, he had seen for himself the superior growth rate and quality of timber that larch offered, especially when compared with the native Scots pine.

The 3rd Duke continued in his father's footsteps, planting thousands more larches, but after his death his son in turn extended the cultivation of larch to hitherto unheard-of proportions. Not only did he plant millions of these trees (precise figures vary among authorities), but he also experimented with different mixtures of species and densities of cultivation,

determining that larch was even hardier than previously thought and could be planted at 1,000 feet or more above sea level. The 4th Duke established a trend which was doggedly followed by many a landowner, especially across Scotland and northern England, lured by the prospect of handsome returns from their forestry. Almost to the exclusion of other species in some regions, the larch forests marched relentlessly from one hill to the next.

In the midst of this runaway success story came disaster, in the form of disease. The future for larch looked grim until the arrival of the Japanese larch. In 1883 the Duke of Atholl brought his own seed from Japan and replanted great areas. By the early part of the twentieth century, it was established that the European and Japanese larches had crossed; the resulting hybrid was named the Dunkeld larch. The hybrid has proved to be remarkably disease-free, and it continues to be one of the most popular with foresters.

ABOVE: *Larch is a great coloniser and thrives well in exposed conditions such as this hilltop site in Wales.*

THE DECLINE OF THE ANCIENT WOODLAND

The nineteenth century was an important watershed for woodlands, largely due to the decline of woodmanship. Coal gradually replaced wood and charcoal as the principal fuel source. The demand for wooden implements and components fell away as steel began to dominate. Wooden construction gave way to concrete in buildings and bridges, and to iron and steel in railways and ships. In almost every sphere of life, durable substitutes for wood were introduced that required much lower maintenance and less frequent replacement, which effectively ended the demand for coppicing. The need for the products of woodland craftsmen diminished; mass-produced and machine-made components were cheaper and quicker to make in a factory. As rural woodland trades declined the woodland was in turn thrown into a state of neglect which continued into the early years of the twentieth century.

The history of British woodland is an intriguing and complex story, which contributes to a comprehension of the nation's woodland legacy as it stands at the end of the twentieth century. A survey of the trees illustrates the wealth and beauty of Britain's woodlands.

Most woodlands can be divided into large timber trees and the smaller trees of the underwood or shrub layer. The former occur either as maiden trees (these are the 'standards', which are allowed to grow untouched to produce top grade mature timber), as coppiced stools, or as pollards. The pollards are the rarest in continuous woodland as this type of management is far more relevant in wood pasture or hedgerows, where the trees are at risk from browsing beasts. The underwood trees may be permanent features, such as holly, hazel, and even yew, which grow well beneath their larger brethren, or transient colonisers such as birch, which will eventually be crowded out by larger trees.

How ancient or natural any tract of British woodland might be depends on the interpretation of these terms. Since human beings have affected the structure of all woodlands, whether by direct intervention or indirectly through neglect, the description of woodland as natural may be applied to those woodlands containing the species that would have coincided with the original postglacial development of wildwood. Although there are innumerable shades of 'ancient', Peter Marren, in his *Wild Wood*, adopts a dateline of 1600, at which time the earliest accurate maps were drawn. Since tree planting prior to that date was generally restricted to orchards, parks and the gardens of the gentry, the woods depicted on the early maps were almost certainly of natural origin. If man had carried out planting within existing woodland it would have been by way of adding to and nurturing the naturally occurring species, to infill gaps caused by damage from storms, animals or disease.

In 1984 the Nature Conservancy Council estimated that about one quarter of Britain's 4.9 million acres of woodland, which cover about one tenth of the land, were, as Marren phrases it, 'of probable ancient origin', while a little over half of the total acreage has now been replanted to some degree with conifers. A map of England and Wales with individual county quotas as percentages of national ancient and natural woodland shows that, while every county can boast at least some measure, Kent, Sussex, Surrey and Hampshire, along with Gloucestershire, Herefordshire and Gwent, support the lion's share. The most striking statistic is that woodlands within areas of urban sprawl have survived particularly well.

The layman can identify ancient woodland in Britain and begin to understand its evolution and ecology by training the eye to distinguish the pieces

RIGHT: *'Chair bodgers at work in a beechwood', a wood engraving by Clare Leighton, from* Country Matters *(Gollancz, 1937).*
As Clare watched these men at their labours she sensed the impending demise of their way of life: 'I thought how calm and satisfied they looked, with none of the harassed strain of the factory-worker on their faces. Was this, then, the way to live and work? Did the beauty of the woods around them, and the sense of direct craftsmanship, balance the long hours at their trade? And how long can they survive? I wondered. The old man will carry his values and convictions to the grave; but his helper is young and soft and the impress of the town will not be long in stamping him. I looked from the old man as he stood at the primitive lathe, working the foot-treadle that vibrated the wooden poles stretching out from the hut, to the younger one as he sat on the shaving-horse, shaping the chair legs in readiness for the lathe. Neither flagged at his work, but I wondered for how long the old man would keep his companion.'

of the jigsaw and build the whole picture. The location of specialised species reflects their respective provinces; and many woods will provide clues to the site's antiquity through evidence of human influence upon the landscape. Centuries-long activity by man can be seen in earthworks such as wood banks, originally built as boundaries or defences to exclude livestock; old sunken tracks or hollow ways; and the remnants of practical forestry manifesting in old pollards and massive coppice stools; even long-disused saw pits or charcoal hearths.

Certain parts of Britain will be more likely to contain ancient woodland than others. Great tracts of land dominated by arable farming will contain much less ancient woodland than the boggy or mountainous regions where wheat will not grow and livestock cannot nibble. For many centuries the necessity for productive woods ensured their harmonious coexistence with crops and pastures. Indeed, wood pasture was once a common feature of the landscape; now it is most often found as a relic of a once functional agricultural system, in open semi-wild conservation areas or within private land where it survives as deer parks or the residue of a landscape design from the eighteenth or nineteenth century. Wood pasture is most easily distinguished by ancient pollards randomly spread across rolling grassland, often with recently added maiden trees displaying

distinctive flat-bottomed browse lines.

The stability of the greater majority of woodland, wood pasture included, altered radically at the beginning of the twentieth century. The demise of many industries once reliant on wood products and the adoption of alternative fuel sources put many woodlands on the margins. They were either grubbed up to supply extra farm land or neglected and allowed to decline into tangled dereliction. As the pace of modern farming gathered momentum wood pasture in particular was given little credence by those with profit in mind. The ancient trees had no timber value, and in many cases they were simply extra obstacles around which to plough. If trees were to be

ABOVE: *A young birch sapling thrusts bravely forth from the rotting root plate of a long-toppled oak in Staverton Park, Suffolk. Such natural colonisation is typical of woodland devoid of man's intervention.*

LEFT: *Foresters at work in the Kielder Forest in 1959, by which time horses were beginning to be eclipsed by machinery for timber extraction. Forty years on horses increasingly have a role in forestry once again, particularly where damage to the woodland floor must be minimised in the interests of conservation.*

NEXT PAGE: *Once a main thoroughfare for foresters, this ancient track through an Oxfordshire wood now serves merely as a recreational footpath.*

NATIVE TREE SPECIES

Trees in bold are the fifteen principal native species contained within the 'natural' or 'ancient' woodlands of Britain. The other native species can also still be found, but predominantly in habitats other than woodland with the exception of the strawberry tree, which no longer grows wild in Britain; Southern Ireland is its last stronghold.

Alder
Crab Apple
Ash
Aspen
Beech
Silver Birch
Downy Birch
Blackthorn
Box
Bird Cherry
Wild Cherry
Wych Elm
Hazel
Hawthorn
Holly
Hornbeam
Juniper
Small-leaved Lime
Large-leaved Lime
Field Maple
Pedunculate Oak
Sessile Oak
Wild Pear
Scots Pine
Black Poplar
Rowan
Wild Service Tree
Strawberry Tree
Whitebeam
Almond Willow
Bay Willow
Crack Willow
Goat Willow
White Willow
Grey Sallow
Yew

considered as a crop then it was conifers that would provide the speediest and safest returns. Had it not been for a few romantics and nature lovers far more of our ancient tree heritage would have disappeared in the early part of this century.

Consternation has been expressed since the time of John Evelyn about the status of British trees and woodland. In the past the best management endeavoured, through the practice of skilled woodmanship, to make woodlands profitable from their timber and coppice wood resources, while also nurturing cover for game. Modern conservation efforts have often focused on non-intervention, on arresting the disappearance of a unique landscape or providing a secure habitat for an endangered species.

With changes of attitude to woodland and forestry in recent years a gradual yet positive change is coming about. Commercial forestry is experimenting with a greater variety of trees, in different types of habitat, and also showing much more sensitivity to the landscape which it utilises. Farmers and landowners are being encouraged to grow broadleaf trees, revitalise their woodlands, cherish their hedgerows and veteran trees, or even to plant poplars or willows for biomass production. In the agricultural world, where profitability is paramount, incentives to plant or conserve must be sufficiently tempting to encourage implementation. There will always be a few philistines who will bulldoze ancient hedgerows or poison splendid old trees which get in their way. However, the cheering prospect is that there are far more individuals out there quietly getting on with planting trees or tending their woodlands and hedgerows.

Conservationists appear to have found agreement upon methods of caring for trees and woodlands, although differing sites and species present their own peculiar needs. A broad policy of sensitive manage-

BELOW: *Shipwrights working on a massive piece of oak timber (Selby).*

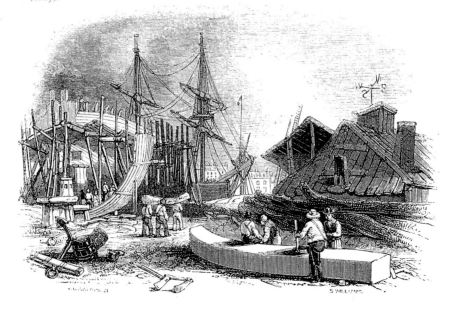

ment in close harmony with the indigenous species and the specific ecologies of each site is generally adopted countrywide. In some of the larger woodlands where several different types of habitat prevail, measures are taken to maintain their distinctive qualities. This may result in some areas of non-intervention, while others are managed to some lesser or greater degree.

The future for woodland may be uncertain, but it was always that way. Who, when they were planting woodlands in the nineteenth century, could have foreseen the supremacy of oil and gas or the radical decline of coal mining, which were to eclipse the use of wood? Some comfort may be drawn in the 1990s from the knowledge that there are currently more individuals and organisations than ever before dedicated to ensuring the sustained vigour and continuity of British trees and woodland.

LEGACY OF THE WOODLAND

Prior to the nineteenth century, the quantity of wood used in Britain was rigidly controlled by laws of ownership and daunting physical difficulties. The total amount cut each year was more or less equal to the amount being produced. There was strict legislation to protect trees from improper unauthorised use or removal. Convicted offenders were harshly punished. Hand-operated tools were primitive and time-consuming to use. Great skill and strength was required to wield a seven-pound axe, and cross-cut saws hammered out and ground by the village blacksmith lacked the precision needed for smooth operation. Transport depended upon what a horse or ox could pull or a man could carry. The work was seasonal, and for periods of the year trees were not cut at all. It was recognised that cutting coppice or pollarding out of the proper season would have a detrimental effect on subsequent growth. It was equally

well known that the quality of timber would be profoundly affected by the time of year when it was cut. In winter, muddy forest tracks and poorly surfaced roads would become impassable so the extraction of large tree trunks would have to stop, except of course when the ground was frozen hard. Generally, everything to do with tree cutting was done during the most appropriate season. The work was carried out to a very high standard but at a slow pace. New saplings were nurtured and protected from damage, even when larger trees were cut out around them.

As populations increased and became more prosperous, the human way of life became more turbulent and demanding, and the requirement for wood continually increased. In Britain, it eventually could not be sustained by home-grown produce. The need for oak to build warships, for example, far outstripped availability. New oak trees were planted for the navy but

new ships were required before the trees had enough time to grow. Timber was imported to overcome this dilemma, or ships were simply built overseas, until eventually they were made from iron. World trade in timber, however, has continued to flourish, fuelled by vast resources of natural tropical, temperate and boreal forests. Huge quantities of timber have been exported from these, and extensive additional deforestation by burning to provide space for agriculture has become a common practice.

In the 1950s, a new kind of precision saw was perfected. Its teeth were set on an endless chain and was driven by a petrol engine. The introduction of the 'chainsaw' heralded the exploitation of forests on a vast scale and at a furious rate. Forests could not withstand the lethal combination of chainsaws and fire. In a relatively short time deforestation could render some places uninhabitable. – John White

ABOVE: A crusty old hornbeam pollard in the ancient wood pasture of Hatfield Forest, Essex. Even though this tree has hollowed out and split into two halves, in similar fashion to several others of its tribe in Hatfield, it is still flourishing. The witches' brooms which clutter the branches are caused by the fungus Taphrina carpini which causes the tree to throw out these galls composed of densely packed shoots, often mistaken for birds' nests. They do not appear to harm the tree, and related fungi cause similar galls on birch.

CHAPTER TWO

Trees as Habitats

In a world which moves through changes and progressions at seemingly gathering speed, one of the constants is trees. Many may appear quite similar to the untrained eye, but like any other living entity each species is unique. Each has its own likes and dislikes when it comes to putting down roots. All creatures have the capacity to move into the conditions which suit them best. Plants and trees too, will only thrive in optimum conditions suited specifically to their own species. An important aspect of trees is that as they flourish, they in turn encourage all manner of flora and fauna around them, playing a vital role in the food chain and as cover and shelter. Throughout the life cycle of trees, an ever-changing habitat linked to the developmental stages evolves, bringing with it a continually varying biodiversity. Trees are the very linchpin of the ecosystem, creating habitat to suit all manner of flora and fauna, including human beings.

The Woodland Circle of Life

A walk in the woods may seem to be a peaceful diversion, with nothing but the sound of the twitter of birds far overhead, the drone of insect life and the occasional flurry of some small mammal disappearing into the undergrowth. However, millions of wildlife dramas are constantly unfolding beyond the senses of the casual passer-by, and it is the trees of the woodland which are the pivotal influence upon them all. Although any given ecosystem is essentially a cyclic structure built out of multi-layered interdependencies, it is trees that are the instigators of the food chain in woodlands. The ecosystem is a natural habitat viewed in its entirety, frequently incorporating human influences, where mutual relationships between the flora and fauna fund the biodiversity and give each and every habitat its own distinctive profile.

The hierarchy of animal life is largely dependent upon the plant layer structure which can be divided into four principal tiers: mosses and lichens, herbaceous plants, shrubs and small trees of the under-storey, and the mature tree canopy. It is these trees and plants, functioning as the prime producers of organic matter by using water, solar energy and minerals in the earth, which supply the energy source so vital to herbivorous animals and insects. The carnivores prey upon the herbivores and they in turn eat vegetation and deposit waste matter and ultimately their own decomposing carcasses, when they die, back into the soil. Invertebrates, fungi and bacteria use this fuel resource as well as rotting vegetable matter and in turn provide a continuity of nourishment for the trees and plants once more.

Man's intervention into this process may often change the delicate balance of the ecosystem and while in Britain this seldom has lasting or far-reaching effects, the devastating impact of human greed upon the world's rain forests paints an altogether gloomier picture. Intervention in woodlands may be for commercial exploitation or, at the other end of the spectrum, by way of conservation initiatives, but whatever the reason there is a risk of compromising the biodiversity equilibrium. The desire to 'improve' woodland so that certain species can proliferate may prove equally as detrimental to the ecology as it would in those woodlands where trees are simply extracted.

There seems to be a constant three-cornered conflict between commercial demands, conservation ideals and nature simply doing its own thing. The consequence of ecological imbalance has only been a high profile issue for the last thirty or forty years. Increased mechanisation of the forestry industry has brought about wide-scale degradation of woodland habitat during timber extraction and the general demands upon land for building and road construction, along with changes of agricultural use, has resulted in the wholesale destruction of woodlands. It must be conceded that the forestry industry now has a much more sensitive approach to the wildlife aspects of

the woodlands in which it operates, to the extent that in some difficult terrain there has been a resurgence in the use of horse-power to extract timber, thus minimising the damage to the woodland floor, and some of the larger machines are even able to 'walk' instead of barge their way through the under-storey. Similarly, the previously unhampered march of progress which frequently took little notice of trees and woodland has been tempered by the presence of numerous conservation bodies, increased legislation, a growing public awareness and even direct action by concerned individuals.

Historically the state of British woods has been moulded by the rigours of commercial necessity, with scant regard for the broader ecological consequences, although the less intensive level of operations gave time for recovery. Woodland as a sustainable resource was vital. Human beings have been culpable for some of the species imbalance from which woodlands now suffer, witness the grey squirrel, increases in wild deer and the introduction of rampant colonisers such as rhododendron and sycamore.

During the nineteenth century many landowners began planting great tracts of coniferous forests, a process which was pushed into yet a higher gear by the Forestry Commission during the early part of the twentieth century. Although the bleak prospects of coniferous monocultures upsets many people, there remains a steady demand for the timber and the

land on which they grow is seldom suitable for any other types of trees.

Roughly half of Britain's woodland is coniferous and of this the great majority is made up of imported species, principally the firs and spruces from Europe and North America. Sitka spruce is the most frequently planted, due to its remarkable tolerance to poor and often damp soils as well as extremely exposed conditions,

ABOVE: *Pigs at Pannage fulfilling the commoners' rights of mast in the New Forest.*

BELOW: *Fungi colonise an old beech stump on the woodland floor.*

moreover it will return the greatest yield of timber. The wood is of a high quality and suitable for many interior construction purposes as well as making excellent pulp for paper making. Although such plantations were for many years grown in very tightly packed linear tracts, recent forestry policy has increased more random planting patterns, with both regular thinning and intermittent inclusion of alternative tree species, often including broadleaves. This not only improves the effect on the landscape, but also encourages biodiversity. However, even the coniferous monocultures are not quite as barren in terms of wildlife as popular perceptions would claim. It seems strange to consider that these imported trees were introduced to Britain without the complementary species array of their native habitats, and so the British habitats which have evolved around them over the last two hundred years have undergone a steady process of colonisation and adaptation.

Different species are able to benefit from the developing coniferous wood. Initially, various insects and spiders colonised the grassy woodland floor around the young trees. These provide food for shrews, foxes and various insectivorous birds, while the abundance of voles also makes good hunting for raptors. As the trees grow they provide excellent nesting sites for a variety of birds such as wrens, willow warblers and sparrows and later, as the trees grow taller, thrushes, blackbirds and chaffinches move in. When the trees reach maturity the canopy becomes denser and different birds such as jays, crows and magpies set up home.

At this point the appearance of the woodland floor may seem somewhat stark, since little light filters down and the diversity of plant life is limited to a few grasses, ferns and brambles. However, this is still a habitat admirably suited to numerous types of fungi; organisms which provide a crucial role in breaking down organic matter in acidic soils which is frequently lacking in invertebrates. Dense dark forest does provide good cover for deer, foxes, hares and owls, but in turn creates poor cover and a paucity of food resources for smaller birds and mammals. Some of Britain's rarer denizens of coniferous woods, such as pine martens, wildcats and red squirrels, prefer the more rugged and remote remnants of natural pine forests. The secret to successful coniferous habitat management in commercial forests is to create variety, by providing many small blocks of woodland, preferably using a variety of species, planted at different times, regularly thinned and interspersed by ridges and clearings so that the biodiversity stands the best chance for the future.

Woodland management of any type must have a constantly evolving methodology which can adapt to changing circumstances. There have long-been accepted problems and much debate on how best to manage trees and woodland, but rather than adopting a blanket policy it seems more prudent to consider each site and its associated habitats as individual cases.

Frequently the dominance of certain species to the exclusion of others or recouping the natural balance of woodland after individual species have been eradicated or appear to be at risk are topics which arouse concern. Man has been shaping most of Britain's woodlands for centuries, but it is misguided to believe that nature struggles to maintain or regenerate a natural balance. By their very nature, and that of the climate, some British trees have gradually declined in number

OPPOSITE: *A natural woodland habitat is continually changing – a constant flux of regeneration and decomposition.*

LEFT: *The roots of the alder most often provide a safe haven for the otter* (Lutra lutra) *holt.*

BELOW AND BOTTOM: *The red squirrel (* Sclurus vulgaris) *and pine marten (*Martea martes)*, two of Britains' rarer woodland mammals.*

RIGHT: *To spot the vivd blue plummage of a kingfisher on a woodland river-bank is a rare treat.*

OPPOSITE: *The small-leaved lime is no longer a common tree in Britain.*

RIGHT: *To spot the vivd blue plummage of a kingfisher on a woodland river-bank is a rare treat.*

OPPOSITE: *The small-leaved lime is no longer a common tree in Britain.*

RIGHT: *Stag beetles (Lucanlis cervus) come face-to-face on an oak bough. Sadly, like many other woodland creatures, the stag beetle is now becoming increasingly rare in Britain.*

BELOW: *Ancient oak pollard, recognised by entomologists as one of the finest habitats for some of Britain's rarest invertebrates.*

over thousands of years, witness small-leaved and large-leaved lime, which now seldom sets fertile seed in Britain. Others, such as black poplar, have disappeared as man drained their preferred floodplain habitats to suit his agricultural needs. While trees such as wild service and some of the more obscure whitebeams have probably always been relatively thin on the ground, some people delight in planting as many of these rarities as they can, while others decry the idea in the belief that part of these trees' intrinsic allure is their very scarcity. Frankly, it doesn't really matter one jot. Man's influence on tree numbers and distribution pale into insignificance when set against the forces of nature.

British ecosystems are subject to inevitable events, from the gradual changes wrought by global warming with melting ice-caps and the shifting Gulf Stream to the cataclysmic effects of hurricane force winds. Nature, and most particularly trees, will adapt and survive. In the wake of the storms of 1987 and 1990 much gloom and doom was expressed about the long-term devastation of trees and woodlands. In some places people rushed around tidying up the mess and carefully replanting young trees to replace those losses. In many cases these trees failed while the naturally regenerating seedlings which sprang forth in the light-flooded clearings created by the storms flourished. Some trees that fell took root and then sent up new

RIGHT: *Bluebells are one of the most familiar woodland flowers, growing here in a ash wood on the Malvern Hills.*

RIGHT: *Bluebells are one of the most familiar woodland flowers, growing here in a ash wood on the Malvern Hills.*

BELOW LEFT AND MIDDLE: *Two rare flowers of the woodland – herb Paris in Bear's Wood, Herefordshire, and oxlips in Hayley Wood, Cambridgeshire.*

BELOW RIGHT: *The ghost orchid is a strange plant of the beechwood floor. Devoid of chlorophyll, its colouring is highly unusual.*

stems. Others died, and their decaying carcasses have provided fine habitats for all manner of fungi and invertebrates.

The issue of dead wood, whether to leave or to clear, has been keenly debated in recent years. Some authorities believed that dead wood, which may have once been trees which died through disease, presented a threat to the health of living trees. Foresters also considered that it was good management to tidy up the woodland floor, thus improving accessibility (particularly for forestry work) and creating an altogether neater aspect in those woods used for public amenity. However, there is now a strong dead wood lobby which recognises the ecological benefits this brings to woodland ecosystem. Across Britain around a thousand species of creatures, including beetles, flies, slugs, snails, woodlice and centipedes, as well as hundreds of different fungi, live on dead wood and,

in turn, many other species of birds and mammals are reliant upon these colonies. Dead or dying wood is not necessarily part of the woodland floor, for it can also be in the form of veteran trees. These old-timers with their hollowed trunks provide habitats for another range of creatures and, in particular, great larders, nest sites and roosts for many different birds and bats.

The species-rich habitats in ancient broadleaf woodlands are a delight to naturalists and recreational visitors alike. The different native broadleaves influence the ground flora, which in turn attract different insects and animals. Woodland flowers are a very special element of our native flora; many of them being specific to certain regions or woodland types. They in turn provide good indicators as to the age of woodland, many species being poor colonisers and seldom found in more recent woods, and similarly their presence outside existing woods often revealing the sites of

cleared woodland. Such common plants as bluebells, wood anemones and dog's mercury in a hedge bank almost certainly indicate the remnant edge of an old wood. The density of the woodland canopy is critical for the ground flora. Woodlands of predominantly ash, with their light feathery leaves, tend to allow more light to filter through and thus the flowers are profuse. Managed woods comprising of other large trees such as oak, chestnut and lime promote flower growth in early spring before the leaves are fully formed or when partially cleared in coppicing with standards. It is at these times that seeds which may have lain dormant in the soil for many years finally manage to spring forth. One of the most striking displays can be dense stands of foxgloves, which have seeds that can wait for decades to germinate. Beech is the one broadleaf which tends to limit the ground flora with its dense foliage, but even though a beechwood floor may appear rather barren it still doesn't receive such bad press as the conifers. Shade-tolerant flowers such as wood-sorrel and woodruff thrive beneath beech, and among the leaf litter grow the rather strange bird's nest orchids which do not require sunlight and take their nutrition from the leaf mould. This is also a most productive habitat for a large range of fungi.

Certain woodland flowers are now restricted to very limited areas usually as a result of agricultural improvement and to a lesser degree the thoughtlessness of people picking them. Typically, wild daffodils still exist in a few enclaves in south and west England and Wales, while the oxlip is another flower now restricted to a few localised woodlands in Suffolk, Essex and Cambridgeshire. A few other woodland flowers may still be relatively widespread in Britain, although their distribution is again localised to certain specific woods with exactly the right conditions. Herb Paris is one such plant, prefering damp calcareous soils, with its poor colonising abilities denoting ancient woodland.

Climatic conditions are of particular significance, not only to woodland flowers, but also mosses, lichens and ferns. Damp conditions with plenty of unpolluted air suit these plants best. Mosses, lichens and liverworts find purchase principally on the rough bark of older trees, and flourish most luxuriantly in the more humid parts of western Britain. Since these epiphytes, as they are collectively known, draw their nutrition from the air and water around them they are highly sensitive to any atmospheric pollution, and thus thrive in the damp sheltered valleys far from roads and industry, creating landscapes of Tolkienesque appearance.

There are dozens of different woodland ferns, but visually these are divided into those with feathery fronds, which include shield ferns, lady ferns, male ferns and buckler ferns, and the many varieties of the hart's-tongue ferns with glossy spike-like fronds and prominent ribbed sori (groups of spore capsules) on the under-sides. Polypody is a small feathery shaped fern which is exclusive to the more humid western woodlands, where it grows in large colonies, often taking root in the moss colonies on oak.

The presence of the many hundreds of different species of fungi in woodland provide a crucial link in the ecology. Surprisingly, although most people associate these with dead or dying material they are quite definetely implicated in the very life-cycles of other plants and animals. Several fungi have mycorrhizal relationships with specific living tree hosts; a symbiosis which improves the tree's take-up of mineral nutrients, particularly nitrogen and phosphorus, while providing the fungi with a source of carbohydrates for their own nutrition. Other fungi are part of the diet of certain insects and animals, not to mention human beings.

LEFT: *Honey fungus is a common species of British woodlands. Though the mushrooms are good to eat when they first appear, they often signal the death knell of the trees on which they grow.*

LEFT: *The branching formation of the fungus mycelium beneath the leaf litter is a remarkable mirror image of a tree's roots and branches.*

BELOW: *Upland oak wood in Devon – the perfect habitat for lichens and mosses.*

As even the dedicated hunter of mushrooms (the fruiting bodies of fungi) will know most species are very particular about the conditions in which they will grow, again illustrating the precarious balance required for species proliferation.

Trees, and in particular those of the deciduous woodlands, provide food, nest sites and cover for a dazzling array of birdlife, some resident, and some migratory incomers. Tits, wrens and warblers haunt the under-storey; woodpeckers, nuthatches and treecreepers both live in and dine upon the delights of the tree trunks; thrushes, robins, finches and jays seek fruits and invertebrates on the woodland floor; the perfectly camouflaged woodcock finds cover and nest site amongst the leaf litter; at the top of the tree goshawks and sparrowhawks hold sway.

Woodland is home to many different mammals, from the tiny wood mouse to the rarely seen, but plentiful, deer population. Woodland provides marvellous cover for timid creatures such as deer and yet, because they no longer have any natural predators since bears and wolves are long gone, they are able to multiply unchecked other than by the hand of man. A large herd of deer have the capacity to do extreme damage to trees by eating the bark, which when completely ringed will kill them. Commercial foresters become infuriated by this and the public outcry when moves are made to selectively cull. The 'Bambi' association is hard for many people to overcome. The cull does stop populations overstretching the habitats in which they live however, while also protecting the interests of foresters and landowners. Even the grey squirrel which wreaks horrendous damage on beech and sycamore with its bark-ripping exploits is held in great affection by the masses, but if they realised the extent of squirrel despoliation this might be reassessed. It is partly due to human intervention and in turn to commercial demands that these creatures are now perceived as pests.

Other animals which find homes in the woods don't arouse quite such vehement maledictions. Both foxes and badgers may not be to everyone's liking, but they find safe harbour in the depths of many a wood, where they bring up their families and forage for smaller mammals, insects, beetles and fruits. The masses of tree roots make ideal foundations for their networks of chambers and tunnels.

Much of the beauty and interest of the British countryside lies in its diversity, and within this landscape trees have a highly influential effect upon the different wildlife habitats and, in return, that wildlife moulds the treescape in which it lives. The coniferous woodland scene has already been discussed, but there are several other general types of habitat with their own particular tree complement.

The water-loving trees are described in Chapter Six, but their role in the general scheme of the lowland countryside is crucial. Rivers, canals and streams are most typically lined with willows and alders, although poplars, and even oaks, are also happy with their 'feet' in the water. Once established riverside trees form extensive root networks along the banks, which provide perfect homes for otters, water voles, water shrews and even the villainous escapee, the mink. The root systems also provide the necessary structure to hold the river bank together, thus preventing erosion and loss of the mammal habitat. Many birds populate the river bank, from the diminutive but vibrantly coloured kingfisher, which darts back and forth, to the stately grey heron; slender elegance at rest transformed into a wind-tossed mackintosh as it rises, only to reclaim its dignity once more as its leisurely measured wingbeat sweeps it away on another fish

LEFT: *Trees provide superb bird habitat: woodpeckers, tree creepers and little owls are but three species who depend on them for shelter and food. Top, great spotted woodpecker* (Dendrocopos major)*, middle, treecreeper* (Certhia familiaris) *at its nest site and below, a jay* (Garrulus glandarius) *flies home with a tasty morsel firmly in its beak.*

OPPOSITE: *A little owl* (Athene noctua) *peaks out of its haven of a hollow log.*

ABOVE: *Rare whitebeams have only survived on these crags above the Usk valley because they were innaccessible to browsing animals.*

BELOW AND BOTTOM: *Deer are persistent feeders in the woodland. Often they cause serious damage to trees which may even eventually kill them. Deer attack has exposed the fibrous bast on this lime.*

patrol. The combination of nest sites, cover and food attract the large array of waterside birds. Willows, which are the most widespread of water-loving trees form the most productive habitats when they have been pollarded. The broad open crowns provide excellent nest sites for the larger birds, and as the elements and trapped detritus in these crowns cause the heartwood to decay smaller birds move in and are also able to feast upon the increased invertebrate populations. Many invertebrates, including aphids, beetles, weevils, sawflies and gall wasps, depend upon the willows, in fact roughly 450 different species, more than any other tree, which in turn provide plentiful repast for the resident birdlife.

The other most common tree of wet locations is the alder, a tree of great versatility, equally at home beside a large river, a fast-flowing upland stream or in stagnant swampy situations, known as alder carr, where the tree dominates. Alder carr is a relatively scarce habitat today since many such marshy areas have been drained for agricultural purposes, but where it does still survive, predominantly in the fens of East Anglia, it harbours rich collections of mosses, ferns and wild flowers. The little woody cones of alder produce great quantities of winged seeds which are attractive to siskins and redpolls, as well as mallard and teal when they fall into flowing water. In this way the tree maximises its reproductive potential with seeds that can be distributed by air, water or bird. It is also the roots of alder which most often provide a safe haven for an otter holt, since the root ball which often extends far beneath the river's surface provides a perfect underwater concealed entrance. Alder thrives in even the poorest of soils, and so it may be found in quite exposed and barren upland situations, where it provides shelter for both wild animals and sheep. Due to their remarkable ability to fix atmospheric nitrogen by way of the root nodules all alders are able to improve the fertility of the soil in which they

grow, thus making them useful trees for the reclamation of derelict industrial land.

Other successful trees of the moor and mountain are rowan and hawthorn. Both of these throw great clusters of musky scented creamy flowers in the spring which attract beetles, flies, bees and wasps, giving much needed sustenance, while in return they provide the duties of pollination. Come the winter and the red berries of both trees are avidly sought by birds and small mammals as a highly nutritious food, their bright red colour making them conspicuous. Even yew berries, with their toxic seeds hidden within the bright pink aril, prove appetising to some birds; and while most will either discard or disgorge these seeds the green finch and the great tit have been known to eat them with no ill effect. These creatures in turn carry out seed dispersal for the trees, and thus a seed which killed the very mechanism of dispersal would be rather counter-productive. Why some animals are affected by toxins, but others are not, is a confusing issue, but one which evolution seems to have sorted out rather well.

The fruits and seeds of many trees are critical for the survival of certain species, and in some years when there are poor sets of fruits the respective animals that depend upon them, particularly migratory birds, will struggle to survive. Even trees such as oak and beech are prone to either a dearth of fruits, or in certain years, known as 'mast years', excessive fruit production. The domino effect of nature's bounty is not to be underestimated, for if part of the food chain is missing then the following tiers of dependants will have insufficient food sources.

The foliage of some trees is also particularly palatable to certain animals and this can have detrimental consequences for their very survival. Both deer and sheep have prodigious appetites and sometimes it is only the severity of the terrain that prevents them from wiping out whole tracts of trees. An interesting example is the scattered colonies of rare and site-specific whitebeams which are mainly to be found in highly inaccessible locations in southwest Britain. It begs the question as to whether or not they once had a much broader distribution and were simply pinned back to their rocky crags and ledges by the over zealous attentions of hungry beasts. It is possible that animal predation has taken a fair toll on small-leaved and large-leaved limes, trees which were once even stripped for winter fodder by farmers, since the many coppiced trees in woodland have become increasingly accessible to deer as a result of widespread relaxation of management regimes. Although long before this lime distribution was hampered by long-term climatic changes, which saw a gradual reduction of mean summer temperatures from about 3000 BC , resulting in the lime's inability to produce fertile seed. However, some trees seem to be able to handle browsing by animals.

Hawthorns and yews are two trees which are so

tenacious that they can still thrive even when reduced to bonzai proportions. In exposed situations adverse weather conditions also contribute to these forms. Various species specific insects, usually moth larvae, will have a good attempt at total defoliation of a tree, but as long as the tree is big enough and healthy enough it is usually able to bounce back and produce another flush of leaves. Here again a balance must be found, since for the insect species to kill the very trees upon which they rely would be a pointless exercise.

Although the animals and plants which live both upon and around trees will influence the distinctive ecosystems in which they play a part, the most potent factors of all are climate and light. The most basic requirements for success of a tree are water, sunlight and a soil type most suitably balanced for the particular species. Life expectancies of different species are mirrored in their ecological strategies which correlate with the prevailing climatic conditions and habitat types. Much of this information was gathered and interpreted by Professor Philip Grime of Sheffield University. His team's conclusions expressed simplistically were represented in a triangular model. At the three extreme corners were: maximum environmental stress (poor soil and extreme climatic conditions – stress tolerant, slow growing, hardy trees such as yew and juniper); maximum environmental disturbance (due to destruction by man or weather extremes – colonising trees, which produce large quantities of seed, such as birch, poplars and willows); maximum nutrients and most favourable conditions (typically the highly competitive trees which thrive upon good soil with ample shelter, such as oak, ash, beech and lime). Obviously some trees perform well even when they are not in their ecologically most ideal situation, but the basic tenets of this research provide a sound platform upon which to understand British trees in their place. Thus, without the effects of human activities it can safely be assumed that certain trees would have specific roles in the national ecology.

However, human beings will remain on this globe for a little while longer and, after climate and soil, they are the biggest influence on the success and distribution of trees. This may not always be a direct or even a conscious influence, but merely the knock-on effect of human demands and progressions. Forestry and its associated woodcrafts and management have shaped British woodland, along with the ebb and flow of agriculture and, to some lesser degree, the parkland manipulation of the landed classes. Even so, there are innumerable other trees in parks, gardens and streets which all have their part to play.

In the urban environment trees are frequently described as the 'lungs of the cities', as indeed they are, but they also form a vital network for wildlife, so that without them towns and cities would be much

BELOW: Countryside management reveals laid hedges and pollarded trees on the Worcestershire Plain. Both customs will prolong the lives of the hedgerow and the willows.

the poorer. Prior to the 1950s, when air pollution in cities was particularly bad, certain trees were selected for planting because of their ability to thrive in even the most adverse of conditions. Of the larger broadleaves London plane and common lime proved the most successful, and even with today's formidable pollution levels from the internal combustion engine they are still doing very well. Lime foliage provides food for several different moth larvae and the blossom attracts bees and other insects. The infestations of aphids with their resultant excrement, confusingly known as honeydew, frequently coats cars beneath with a sticky deposit – perhaps nature having its own little riposte at the motor car. Birches, rowans, and their huge array of cultivars are also popular urban trees; good for insects as well as birdlife with their abundant seeds and fruits. Many a gathering of larger trees also provide roosts for magpies, rooks and pigeons; there even

being a substantial heronry in Regent's Park in London. However, the down-side for trees in the city, especially in the street, is the increasing excavation activity for cable installations. Workmen with productivity levels to maintain and deadlines to fulfil usually think little about the ramifications of hacking through large root systems. The result can be whole streets full of dead and dying trees, which are unsightly, dangerous and eventually gone. Monitoring and legislation against this evil must be strengthened. Even the salt we scatter upon icy winter roads causes some trees to struggle. Some unusual suggestions for urban tree planting can be found in Chapter Ten.

It is not only the ecological functions that trees fulfil in the urban environment, but also their amenity value which makes them so special. Most people need a bit of greenery to relieve the hard-edged colourless monotony of buildings and thoroughfares. They delight in the light and shade created by trees in park, garden and street. Children clamber about their boughs, whack conkers off the horse chestnut and run full tilt through great drifts of rustly autumn leaves. Lovers tryst, families picnic and octogenarians doze and remember days long gone. It was people's fascination with trees from afar that sent the plant hunters to the ends of the earth and when they came home set them to furnishing every estate from the grandiose mansions to the humbler town villas with trees to amaze. Whether it was cedars from the Middle East, ginkgo from China, eucalyptus from Australia, Wellingtonia from America or monkey puzzle from Chile, they all had their moments of fashionability. It is perhaps slightly sad to observe today that they were not always planted in the most appropriate places. With the benefit of hindsight it is now patently obvious how all these trees will mature, and whether or not they will crowd out lesser species, overshadow the homestead, or disrupt foundations with their vast root systems. Plenty of books are out there to advise on the subject, although few as entertaining and accessible as Hugh Johnson's excellent *International Book of Trees*. Find it and enjoy his very personal accounts of selecting trees for the domestic setting, taking into account their form, colour, size and texture.

Ultimately people must live with the trees that they plant, nurturing them, as well as the habitats which develop around them. Without trees in the urban environment the diversity of flora and fauna would be a pale remnant, so it is encouraging to find that increased awareness of the urban natural world is making positive moves to conserve what is there.

Even though there are frequent rumblings about loss of habitat and species, particularly in the wild, it is safe to assume that, because there are so many different kinds of habitat, many of which have overlapping ecosystems, with a very few exceptions most of the native flora and fauna associated with trees and woodland have a secure future.

CHAPTER THREE

The Principal Woodland Broadleaves

A great variety of trees may be found in British woodlands and the types of woodlands themselves may be equally diverse. Indeed a whole volume could be written about the broadleaves alone.

The intention in tracing the manifestation and distribution of some of Britain's principal woodland species is to expand the appreciation of trees which pass for common. To this end there follows a small collection of portraits of the more familiar trees of which the majority of native woodland is comprised.

From a distance, most woods have a similar appearance: dense collections of stems and boughs in winter or undulating canopies of foliage of an incalculable array of greens in spring and golden hues of autumn. A foray into the depths of any woodland will unveil the splendour of these individuals, with their own distinctive shapes and textures.

Oak

The oak has always seemed particularly British, perhaps because of its long association with the country's history. Oak sprigs have traditionally been a feature on British coinage and a common device in heraldic designs. The oak has stood for qualities the British respect in their countrymen –strength, pride and steadfastness; they talk about 'hearts of oak'. – Anne McIntyre

ABOVE: *Oak leaves touched by early autumn frost.*

BELOW: *A windswept sessile oak that has been blown up the hillside over many years on Moughton Fell, above the village of Wharfe in North Yorkshire. Its size probably belies its antiquity, since oaks will grow very slowly in such extreme habitats.*

THE ROYAL OAK

The oak tree was adopted as the symbol of the annual May celebrations of Oak Apple Day, or Royal Oak Day, to commemorate Charles IIs concealment in the oak tree at Boscobel House in Shropshire, while on the run from the Parliamentarian army after the battle of Worcester in 1651. The diarist Samuel Pepys, in 1680, recorded details from Charles' personal account of the escape. The original tree no longer exists: by the early eighteenth century so many souvenir hunters had plundered bits of it that it had virtually disappeared; another oak on the site has taken on the mantle of 'Royal Oak'. A prodigious number of Royal Oak public houses commemorate the Restoration. Many of them feature the monarch in the Boscobel Oak upon their inn signs.

The oak is undisputedly Britain's greatest and most definitive woodland tree. Whether it be sessile or pedunculate, in their many diverse forms, or even some of the numerous foreign genotypes which have infiltrated from eastern and central Europe, the oak is the symbol of strength and longevity and is the quintessential element of many a pastoral vision.

Pedunculate oak (*Quercus robur*), alternatively known as English oak or common oak, is by far the more dominant of the two native species and grows throughout Britain, particularly in lowland woods, but also along hedgerows and as a field tree. Its timber was the most sought after for shipbuilding, as its naturally shorter-stemmed and bushier configuration was well suited for the ribs and knees. The tree was often planted, or thinned, to a fairly wide spacing in order to encourage this natural form to proliferate. The pedunculate was the oak most favoured by the planters of great forests, who were first enjoined to regenerate Britain's 'wooden walls' by the diarist and academic John Evelyn in his discourse to the Royal Society in 1662.

The sessile oak (*Quercus petraea*) is typically found in upland woodlands of the north and west of Britain, thriving in wet and rocky locations which are frequently too windswept and inhospitable for its pedunculate cousin. Traditionally, it has been considered to produce both slightly inferior timber and poorer crops of acorns, a concern to pig farmers, and has thus been relegated to coppicing, although in the best growing conditions the tree is capable of developing long straight trunks in maturity. From the sessile oak woods have come vast quantities of fuel wood, charcoal and oak bark. The two types are often found together in lowland woodlands, accompanied by a healthy understorey and ground flora. One easy method of distinguishing between them is to observe how the acorns are borne. On the sessile, they are hard up against the twig (the Latin *sessilis* means stalkless), with leaves on short petioles; the pedunculate bears acorns upon stems (peduncles), with leaves that have little or no petiole.

Oak colonises well, though not usually within its own shade. As an individual species, it supports the greatest variety of insect and other invertebrate life of all British trees – more than 500 different species. A large oak tree can usually handle all these dependants, even when partially defoliated by the larvae of such moths as the green oak tortrix or the winter moth. Young seedlings, however, cannot cope quite so well and usually perish or are eaten by squirrels before they so much as push forth from the acorn. The best chance for a potential oak tree is for the acorn to be taken by a bird and buried in open ground. Indeed, according to John White, the prime agents of natural

regeneration of British oaks, particularly during the earlier part of this century when there was relatively little planting, have been jays. The jay buries acorns for its winter larder without damaging the seed as a squirrel might do. Rooks also eat or bury acorns: Robinson's *Natural History of Westmorland and Cumberland* (1709) recorded a flock near Rose Castle industriously digging holes, dropping in acorns, and covering them over with moss. These acorns germinated and, 25 years later, were large enough to provide nesting sites for yet more rooks.

Man has also been an ardent seed dispersal mechanism, from the little child who puts a single acorn in a pot and marvels at its awakening to the crusty old landowner who plants by the thousand in memory of those 'wooden walls'. In *Silva*, Evelyn provides exhaustive instructions for the cultivation of oaks and eulogises the many virtues and applications of the timber. Hunter's 1776 edition of the book summarises the ideal manner of growing fine oaks:

'*The Oak will grow and thrive upon almost any soil, provided the trees be properly planted, though we cannot suppose that their growth will be equal in all the sorts: A rich deep loamy earth is what Oaks most delight in, and in such they will arrive at the greatest magnitude in the fewest years; tho' they grow exceedingly well in clays of all kinds and on sandy soils, in which last the finest-grained timber is produced.*'

The last point is slightly misleading as poorer, free-draining soils produce timber that is more inclined to shake and crack, and many plantations of oak, particularly in the nineteenth century, were set on the most inappropriate sites, which resulted in a widespread supply of oak timber of an indifferent quality. Admiral Collingwood, one of the great naval commanders in the Battle of Trafalgar in 1805, was reputed to have always carried a pocket full of acorns which he would carefully plant wherever he went on his travels. At his estate in Northumberland, he planted many oaks with which he hoped to build warships: a supposed vestige of them may still be found upon a hill called Hethpool Bell, though poor soil and fierce winds have stunted and deformed the few survivors. The British love their oaks, not just as the providers of glorious universal timber, but also for their splendid amenity value in all manner of landscapes.

Large standard forest oaks or free standing field oaks, a hundred years or more of age, may be found flourishing countrywide. Some of the largest expanses of oaks are in the New Forest and the Forest of Dean. The tallest oaks reach 140 feet, though this is exceptional. The oldest, some of which are 800 to 1,000 years old, are the ancient pollards, usually found in medieval parkland. The pollarding process has created a particular style of short stumpy trees with massive trunks and varying degrees of bushy foliage on relatively short boughs. This practical manage-

The Oak Tree

LEFT: *Botanical illustration of the oak from the First Hunter edition of Evelyn's* Silva. *A. Hunter, MD, FRS, was an eighteenth-century medical doctor who had a great interest in trees and their cultivation.* Silva *had already gone to five editions by 1776 when Hunter decided that it was time to again raise the issue of planting fine timber trees. He added his own annotations to Evelyn's text and commissioned a splendid series of engravings for his edition. Except for Evelyn's original, the First Hunter edition is the most desirable edition.*

PREVIOUS PAGE: *Autumn oak, Berkhamsted, Hertfordshire.*

OPPOSITE: *Single skyline oak on a crisp winter's day in Hampshire.*

ment has caused the trees to survive: their relatively short trunks are of little interest to foresters seeking construction timber, and as they decline through their natural lifespan there are no long boughs to be blown off, no tall tree structure to put increased stress on a weakening root system, and less foliage for the tree to maintain. For a non-pollarded tree, 300 to 400 years is a more realistic life expectancy.

BELOW: *Veteran oak pollard in Savernake Forest, Wiltshire; resembling some shambling great beast of prehistoric times. Completely hollow, the tree is still most definitely alive. Such trees make a strange sight in Savernake set, as they are amidst great stands of commercial beech and conifers.*

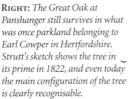

PREVIOUS PAGE: *Field oaks amid oilseed rape probably remnants of a defunct hedgerow.*

BELOW AND BOTTOM: *Wistman's Wood in the spring and autumn. The mixture of English oak and sessile oak in western Britain testifies to the effect of climate fluctuations on the postglacial woodland migration through Britain. English oak, which thrives best in a boreal climate, preceded sessile oak, which prefers Atlantic conditions. Wistman's Wood in the centre of Dartmoor is English oak, part of a boreal colonisation, but nearly all of the more recent oaks round the edge of the moor are sessile. – John White*

RIGHT: *The Great Oak at Panshanger still survives in what was once parkland belonging to Earl Cowper in Hertfordshire. Strutt's sketch shows the tree in its prime in 1822, and even today the main configuration of the tree is clearly recognisable.*

The Great Oak at Panshanger.

Some extreme habitats in Britain have generated their own peculiar forms of oak communities. In Devon, virtually all oak woods were once coppiced, though more recently a few stands have been allowed to bring forth standards, some by design and some by neglect. Most of the Dartmoor oak woods are to be found in the deep river valleys, or cleaves, which run off the moor and are almost exclusively composed of sessile oaks. By contrast, on the windswept upper reaches of Dartmoor lie three strange colonies of pedunculate oak. These are Black Tor Copse, Piles Copse and, the most weird and wonderful of all, Wistman's Wood.

If ever a place was meant to be experienced, rather than photographed or written about, then this is it. From a distance, Wistman's Wood is somewhat unprepossessing in appearance, a rather scrubby little bunch of oaks clinging to the valley side of the babbling upper reaches of the West Dart river. The inside of the wood is altogether a different proposition. The name is widely considered to be derived from the Devonian dialect word 'wisht', meaning melancholy, uncanny, wraithlike.

The prospect inside Wistman's Wood is one of silent turmoil, of plant life untamed, rampant, primordial. Amid giant moss-clad granite boulders, ancient contorted oaks festooned with lichens, ferns and more moss, claw agonisingly this way and that. The oaks are barely 15 feet high, but many are estimated to be over 1,000 years old. In 1868, a Mr. Wentworth Buller was given permission by the Prince of Wales to fell one tree in order to ascertain its age. The tree, a mere seven inches in diameter, was found to contain 163 annual rings, an amazing 40 years' growth in a single inch. All the older trees in Wistman's Wood, then as now, are neglected coppice growth, which partially masks the extreme age of the site. For centuries the nooks and crannies of the hillside clitter have provided shelter from the elements and

Above: *Young sessile oaks by Ober Water in the New Forest.*

protection from munching sheep, thus allowing young trees to establish themselves, but the perennial battering of the biting moortop winds has sculpted the larger trees into fantastic writhing forms. Wistman's Wood is as historically and culturally significant as ancient stone circles or the finest castles and cathedrals. Fortunately, however, it is far less likely to be trampled by tourists' boots, as there are no signposts and no visitor centre to betray the wood's whereabouts; perhaps it will remain a secret place to all but the most diligent.

Before leaving the West Country it's worth noting the remarkable oaks that thrive within the creek woods around the Helford and Fal rivers. Dense, almost impenetrable, oak woods flourish there, often with their roots embedded in the salty mud flats while their lower boughs brush the tidal waters. Although the woods present a remarkably natural vista today, they have long been intensively managed and harvested. The main reason they have survived was industry: great quantities of coppice wood fuelled the region's lead and tin smelters, particularly from the Middle Ages to about 1700, when coal began to replace wood as a fuel.

Further east, on the southernmost reaches of the New Forest, lies a similar area of oak woods along the coast. Poet naturalist W. H. Hudson (1846–1922) was much taken with the vista that he beheld in the Exe valley, particularly in May or June:

'...*when the oak is in its glad light grene, for that is the most vivid and beautiful of all vegetable greens, and the prospect is the greenest and most soul-refreshing to be found in England. The valley is all wooded and the wood is all oak – a continuous oak-wood stretching away on the right, mile on mile, to the sea. The sensation experienced at the sight of this prospect is like that of the traveller in a dry desert when he comes to a clear running stream and drinks his fill of water and is refreshed. The river is tidal, grown round with oaks – old trees that stretch their horizontal branches far out and wet their lower leaves in the salt water.*'

In Gwynedd, northwest Wales, several oakwoods of great age have survived in the upper reaches of the

'Holding the power of a giant
Within the green glazed dome of promise
Astounding energy
Coiled within
Waiting to burst forth
Waiting to touch the startling startled earth
Mother Earth that waits knowing not quite what
Some day
Some hundred thousand days
A mighty oak will set its seed
Within her safe dark sanctuary
Where through time beyond belief
And all that wildest elements might usurp
The proud oak stands resolute.'
Archie Miles, Oakspan.

WISTMAN'S WOOD ON DARTMOOR

'...*it is the silence, the waitingness of the place, that is so haunting; a quality all woods will have on occasion, but which is overwhelming here – a drama, but of a timespan humanity cannot conceive. A pastness, a presentness, a skill with tenses the writer in me knows he will never know; partly out of his own inadequacies, partly because there are tenses human language has yet to invent.*'
John Fowles, The Tree.

RIGHT: *The Major Oak in Sherwood Forest has a girth of 33 feet and an estimated weight of approximately 23 tons. It is presumed to be about 800 years old. Originally, the tree was known as the Cockpen Oak, as it was the custom to store baskets of fighting cocks in the hollow trunk. In 1790 the tree was described, along with many other nearby veterans (such as the Greendale Oak), by local historian Major Hayman Rooke; as a result, shortly afterwards, the tree became known as the Major's Oak.*

This is probably the most famous oak tree extant in Britain. It is a pedunculate oak and its shape suggests that it is an outgrown pollard, although some think that it could be several smaller trees which fused together centuries ago. The various stays and props began to be installed at the beginning of the twentieth century, and now the tree's welfare is regularly attended to by specialist tree surgeons. Recent developments include the application of arboricultural paint to reduce the effects of both fungal decay and the persistent penetration of water and frost upon the exposed wood.

In the 1970s it was still possible to clamber all over the tree, but it was fenced off because increasing visitor pressure (there are currently about 800,000 visitors each year) was causing severe compaction around the tree, reducing the flow of water and nutrients to the root system. In 1994, even the grass was removed to conserve nutrients which might otherwise benefit the tree, and the ground was covered with an inert mulch to prevent the earth from drying out.

NEXT PAGE:
Splendid mature oaks in Moccas Park, early on a winter's morn.
Oaks gather strength for ages;
and when at last
They wane, so beauteous in decrepitude,
So grand in weakness!
E'en in their decay
So venerable!
'Twere sacrilege t'escape
The consecrating touch of time.
From an anonymous poem quoted by J. G. Strutt, in Sylva Britannica.

Vales of Ffestiniog and Maentwrog in Snowdonia and the Artro valley near Harlech. On the precipitous rocky sides of the valleys the trees have escaped the regular attentions of both foresters and sheep, thus, with an abundance of shelter and water, they have created woodland not dissimilar to the inner depths of the Dartmoor woods. The trees are less stunted, but the wealth of mosses, lichens and liverworts is even more intense.

The oak wood theme continues, if a little more spartan in ground flora, mosses and lichens, further north in Cumbria. The woodland there has been valuable for fuel, tan bark, charcoal, and also an array of wood-based manufacturing industries, including the forges of the Furness district. Though the oak woods of central Scotland saw their most vibrant period as a commercial resource between 1750 and 1820, they have not, contrary to popular myth, all been plundered and burnt to fuel Scottish industry. The Scots were canny managers of their oak woods and employed the coppice system into the late

nineteenth century. Scottish oaks in general do not tend to mature into fine timber trees so, while their coppice wood was most useful to industry, the period also saw an astronomic development of conifers in plantations north of the border.

Throughout England there are many exciting communities of veteran oaks. Sherwood Forest, to the north of Nottingham, may appear superficially to be a manicured tourist attraction, but within the wood much of its natural antiquity has been preserved. At Domesday, notes Dr. Rackham, Sherwood was heathland, with about one-third of the forest given over to woodland. A sense of this can be experienced in Birkland. Much of the ground cover there is still of typical heathland character, and the ancient oaks, many of medieval origin, are either long-dead monuments, hosting important colonies of insect and bird life, or marvellous stag-headed individuals. As the trees age and decline, they gradually shut off nutrition to certain boughs, which then die back, causing the stag-headed appearance. This self-limitation of the canopy does not signal imminent demise but seems to be the oak's way of retaining maximum vigour by reducing its biological demands.

While most of the Sherwood oaks have managed to endure as ancient maiden trees, most of the veteran oaks elsewhere have survived as pollarded specimens. Some of the most familiar and beloved examples are the magical old monsters of Windsor

'*...through all the stages thou hast pushed*
Of treeship – first a seedling, hid in grass;
Then twig; then sapling; and, as century rolled
Slow after century, a giant bulk
Of girth enormous. . . .
Time made thee what thou wast, king of the woods;
And Time hath made thee what thou art – a cave
For owls to roost in.'
William Cowper, Yardley Oak (1791)

Great Park, Berkshire: though ancient oaks may be found at numerous sites throughout Britain, Windsor can boast the greatest varieties of form in both sessile and pedunculate oaks. Over 1,000 years of tree population are represented on the one site, where man and beast have taken their toll, as have the ravages of thunderstorms, howling gales, insects and infestations of numerous fungi. Single landmark oaks quite often occur in the public domain, because they were once sited as boundary markers or meeting places.

Larger gatherings of ancient oaks are most often to be encountered upon private land, in deer parks or wood pasture, where public access is not always available. Two amazing sites on opposite sides of the country are Moccas Park in Herefordshire, a medieval deer park complete with authentic oaken park palings, and the ancient wood-pasture of Staverton Park, which lies in Suffolk. Although both parks originated as private wood-pasture they have survived in two radically different ways. Staverton has become a relict, a wilderness, almost certainly a close approximation to the wildwood from which it was once enclosed, while Moccas has retained a decidedly managed and landscaped appearance, with a resident herd of fallow deer along with cattle and sheep. Here the grass is cropped short, the old trees are tended, and the young trees are protected. English Nature has been given special permission to manage the Moccas site and carry out research. Sadly, it is deemed too risky to allow the public to experience these old massive-boughed veterans, described by Reverend Francis Kilvert in his diary entry for the 22 April 1876, after a ramble through the park with friends:

'I fear those grey old men of Moccas, those grey, gnarled, low-browed, knock-kneed, bowed, bent, huge, strange, long-armed, deformed, hunchbacked misshapen oak men that stand waiting and watching century after century biding God's time with both feet in the grave and yet tiring down and seeing out generation after generation, with such tales to tell, as when they whisper them to each other in the midsummer nights, make the silver birches weep and the poplars and aspens shiver and the long ears of the hares and rabbits stand on end. No human hand set those oaks. They are "the trees which the Lord hath planted". They look as if they had been at the beginning and making of the world, and they will probably see its end.'

Historically, oak woods have delighted and excited naturalists, historians, diarists and poets who appreciate the ancient and picturesque prospects of the wild. The eighteenth and nineteenth centuries saw a burgeoning interest in veteran trees, especially oaks: poets eulogised them, artists painted and sketched them, while many self-styled authorities wrote exhaustive accounts of the trees that they discovered upon their travels. Jacob George Strutt's *Sylva Britannica* is a celebration of selected British trees that he describes with acutely observed sketches and lyrical prose. He devoted more than a third of his book to the oaks 'distinguished for their antiquity, magnitude or beauty'. At the time, many of Strutt's celebrated veterans were fast approaching the limit of their natural lifespan and the majority of these trees are now long gone. A few remain, including the Great Oak of Panshanger in Hertfordshire, which Strutt described as 'a fine specimen of the Oak in its prime'; the tree, still identical to Strutt's engraving, continues to grace the landscape. Sadly, another of Strutt's heroes, Sir Philip Sidney's Oak at Penshurst in Kent, has not fared quite so well. All that remains is a ragged remnant. Until recent years, many venerable oaks had only survived due to chance, or because of local superstition or some farmer's sentimental whim which left them untroubled amid productive farmland. Today most ancient oaks are treasured by the landowners or communities fortunate enough to have such trees in their safekeeping; and their future seems assured in these more enlightened times.

BELOW: *A 1905 postcard of the Major Oak. The Edwardian postcard boom brought forth depictions of all manner of local scenes and landmarks. Though pictures of individual trees are relatively scarce, the frequency with which various renditions of the Major Oak are still found today reflects an enduring fame. The whole myth of Robin Hood, the wild greenwood and the mighty oak where he was rumoured to have hidden from the wicked Sheriff of Nottingham are deeply ingrained in folklore tradition. The Major Oak would have been but a mere sapling in his time.*

BOTTOM: *One of the most impressive veteran oak pollards in Windsor Great Park, Berkshire which, although stag-headed, still appears to be in resoundingly good health.*

Ash

The name ash is derived from *aesc*, the Anglo-Saxon word for spear. Its Latin name is *Fraxinus excelsior*; excelsior means higher, a suitable name for a tree which frequently achieves 110–130 feet in height. The very sound of excelsior suggests vigorous towering strength. The ash, much like the oak, is a great coloniser and adapts to a wide variety of situations. Although it prefers heavy and calcareous soils, it has a remarkable ability to thrive in the most adverse conditions, and it has evolved a reliable and productive seed manufacturing system. It bears both male and female flowers on a single tree, and often develops hermaphrodite flowers as well. The countless millions of winged seeds which ashes produce each year can germinate in weed-like proportions so, that instead of being admired for its reproductive success it is quite often despised. However, since the foliage of ash has proved tasteful to animals and has been used in the past as fodder for sheep and deer, the tree must indeed produce prodigious amounts of fertile seeds where predation occurs. Many ancient woods have evidence of coppiced ash stools, but this form of forestry could only prosper in woods which were enclosed.

Ash may be found in the northwest highlands of Scotland, sometimes as pure ash woodland, it also grows prodigiously in the English Midlands where, until the latest bout of Dutch elm disease, ash and elm were the two dominant large trees. As the elms now lie in abeyance, the ash has proved a great restorer of the hedgerow landscape. Young saplings are able to thrive amid the protective cover of dense hedge growths of hazel, thorn or holly; as they emerge above the hedge line, the young ashes have often been pollarded, both for their rapidly produced poles and to protect them from hungry animals.

A mature ash tree will develop an incredibly large and widespread root system which makes high demands on available water and nutrients and can operate to the detriment of lesser trees and plants in its immediate vicinity. But its relatively light and airy canopy affords plenty of encouragement for ground flora, so that many predominantly ash woods will contain impressive quantities of wild flowers which, while not usually of any great variety, often appear as amazing carpets of snowdrops, wood anemones, dog's mercury, ramsons and bluebells. Since ash requires plenty of light, it does not fare so well in the shade of other large woodland trees.

Nineteenth-century Wiltshire writer Richard Jefferies found an ash copse in his *Wild Life In A Southern County* which provided a myriad natural delights in a seemingly distant realm, recording that:

'*The green sprays momentarily pushed aside close immediately behind, shutting out the vision, and with it the thought of civilization. These boughs are the gates of another world. Under trees and leaves – it is so, too, sometimes even in an avenue – where the direct rays of the sun do not penetrate, there is ever a subdued light; it is not shadow, but a light toned with green.*

In the spring the ground here is hidden by a verdant growth, out of which presently the anemone lifts its chaste flower. Then the wild hyacinths hang their blue bells so thickly that, glancing between the poles, it is hazy with colour; and in the evening, if the level beams of the red sun can reach them, here and there a streak of imperial purple plays upon the azure. Woodbine coils round the tall, straight poles, and wild hops, whose bloom emits a pleasant smell if crushed in the fingers. On the upper and clearer branches of the hawthorn the nightingale sings – more sweetly, I think, in the freshness of the spring morning than at night. Resting quietly on an ash stole, with the scent of flowers, and the odour of green buds and leaves, a ray of sunlight yonder lighting up the lichen and the moss on the oak trunk, a gentle air stirring in the branches above, giving glimpses of fleecy clouds sailing in the ether, there comes into the

Opposite: Ash trees of the Derbyshire Dales, a part of Britain in which this tree is pre-eminent.

Below: A splendid mature ash on the shores of Wastwater in Cumbria.

ABOVE: *An infant ash, tucked out of harm's way in a limestone pavement gryke in North Yorkshire.*

BELOW: *Fine silhouette of an ash against a dramatic windswept sky in Shropshire.*

mind a feeling of intense joy in the simple fact of living.'

Though it has been valued as an excellent utilitarian, the ash has rarely been considered in the same aesthetic frame as oak or beech. The 'Venus of the Forest' is one of the last to bring forth foliage. She is also one of the first to shed her attire in autumn, and at the fall there is seldom a golden show.

Edward Thomas was still moved to record the autumn metamorphosis of an ash in his memorable countryside prose from the early years of this century:

'As the sun rose I watched a proud ash tree shedding its leaves after a night of frost. It let them go by threes and tens and twenties; very rarely, with little intervals, only one at a time; once or twice a hundred in one flight. Leaflet – for they fall by leaflets – and stalk twirled through the windless air as if they would have liked to fall not quite so rapidly as their companions to that brown and shining and oblivious carpet below. A gentle wind arose from the north and the leaves all went sloping in larger companies to the ground – falling, falling, whispering as they joined the fallen, they fell for a longer time than a poppy spends in opening and shedding its husk in June. But soon only two leaves were left vibrating. In a little while they also, both together, made the leap, twinkling for a short space and then shadowed and lastly bright and silent on the grass. Then

the tree stood up entirely bereaved and without a voice, in the silver light of the morning that was still young, and wrote once more its grief in complicated scribble upon a sky of intolerably lustrous pearl.'

William Gilpin, an eighteenth-century arbiter of the picturesque, appreciated the qualities of the ash, noting that ...

'...its chief beauty and Vicor of Boldre in the new forest, consists in the lightness of its whole appearance. Its branches at first keep close to the trunk and form acute angles with it: but as they begin to lengthen, they generally take an easy sweep; and the looseness of the leaves corresponding with the lightness of the spray, the whole forms an elegant depending foliage. Nothing can have a better effect than an old ash, hanging from a corner of a wood and bringing off the heaviness of the other foliage, with its loose pendent branches'.

In the nineteenth century, Sir T. Dick Lauder, who published a new edition of Gilpin's writings, expressed caution that, 'Ash trees should be sparingly planted around a gentleman's residence, to avoid the risk of their giving to it a cold, late appearance, at a season when all nature should smile'.

Today, the virtues of ash seem to have been forgotten. Many of the larger pollards have been neglected for a long time, having now developed into bulbous wrinkled old bystanders. Most are found in close proximity to habitation or, intriguingly, to

what might once have been vibrant rural communities. For centuries, almost all communities or farmsteads relied on ash trees for poles and firewood, either from coppiced ash trees in woodlands, the stools of which are still found in abundance, or from pollards, most often found along hedgerows. As with oak, the pollarding of the ashes has resulted in the salvation of the trees. Ash does not normally attain as great an age as oak, and yet a few truly venerable specimens have endured. Some ancient ashes which have been in wood pasture for many years develop wonderfully knobby bases to their boles, as if the tree was lodged on a plinth of concrete. The knobs are actually scar tissue, which the trees have built up over years of being kicked or nibbled by livestock. Very likely the huge conical base of the ancient Clapton Ash, in Somerset, has evolved in this way.

No tree has tougher, more elastic and flexible wood than the ash. The ancients knew this well; as the name implies, it was a first choice for spears, arrows and later pike-shafts. Although the average ash may live for about 200 years it has usually attained its prime before it is half that age. It is often referred to as the 'husbandman's tree': the larger timber was traditionally used for such items as wagons or furniture, while smaller poles were ideal for hop-poles, wheels, ladders, oars, walking sticks and all manner of shafts for tools. Seasoned ash has a remarkable capacity to be steamed and bent into acute forms and still retain its strength. The chassis for the Morgan motor car is made from ash. The original little three-wheelers were formed from imported Tulip wood; soon thereafter a reliable source was discovered for native ash, and today Morgan actually owns and manages its own ash woods.

As a fuel wood ash is remarkably useful as it will burn with equal intensity when green as when dry.

'*Burn ash-wood green, 'Tis fit for a queen.*' John Evelyn was well acquainted with the value of ash and recommends that 'so useful and profitable is this tree, next to Oak, that every prudent Lord of a Manor should employ one acre of ground with Ash to every other twenty acres of land, since in as many years it would be more worth than the land itself.'

Strutt was brief yet complimentary in his appreciation of ash.

'*It is in mountain scenery that the ash appears to peculiar advantage; waving its slender branches over some precipice which just affords it soil sufficient for its footing, or springing between crevices of rock, a happy emblem of the hardy spirit which will not be subdued by fortune's scantiness. It is likewise a lovely object by the side of some crystal stream, in which it views its elegant pendent foliage, bending, Narcissuslike, over its own charms.*'

Ash is a major component in two particularly dramatic landscapes: the valleys around the White Peak in Derbyshire and the much more rugged and

LEFT: *As if moulded in concrete, an old ash pollard stands in pasture near a Herefordshire farmstead. A healthy new crop of poles thrusts forth yet again creating a glorious mixture of rugged antiquity and productive vigour.*

windswept limestone pavements above the Yorkshire Dales. In both locations the ash succeeds best in areas that sheep cannot reach. In the Derbyshire Dales the trees cling to the steep valley sides, or form dense stands where land has been fenced off from the predations of livestock. Above Ribblesdale, in North Yorkshire, the Ling Gill ashwood grows in a steep ravine, while Colt Park Wood runs along an outcrop of seemingly inhospitable limestone pavement some 10 to 15 feet above the neighbouring pastures on its eastern edge. Sheep would need springs on their feet to get at this wood. Colt Park lies on a tracery of grykes (the hollows wrought by the action of rainwater over millions of years) among clints (the remaining limestone boulders, often resembling some giant Henry Moore sculpture). Not surprisingly, many of the ashes that thrive in this harsh habitat are gnarled and stunted. The rock is covered in mosses, lichens and rich carpets of woodland flowers. So dense is the ground cover that it is easy to suddenly disappear down a hole: because of the need for caution, and the fact that it is a rather sensitive National Nature Reserve, special permission is required to visit this wood.

Naturalists often talk of 'climax woodland', which is considered to be the final stage of the natural evolution of a woodland type. Because man has been busy for thousands of years manipulating virtually every aspect of the landscape, either directly or with the help of his livestock, such sites are few and far between. However, Colt Park may be an indicator of what could have been climax ashwood in the upper reaches of the Pennines, before sheep-farming, wood-burning man appeared on the scene.

Logs to burn, logs to burn;
Logs to save the coal a turn.
Here's a word to make you wise
When you hear the wood-man's cries.
Never heed his usual tale
That he has good logs for sale,
But read these lines and really learn
The proper kind of logs to burn.
Oak logs will warm you well
If they're old and dry,
Larch logs of pinewood smell,
But the sparks will fly.
Beech logs for Christmas time;
Yew logs heat well;
'Scots' logs it is a crime
For anyone to sell.
Birch logs will burn too fast,
Chestnut scarce at all;
Hawthorn logs are good to last
If cut in the fall.
Holly logs will burn like wax:
You should burn them green.
Elm logs like smouldering flax
No flame to be seen.
Pear logs and apple logs,
They will scent your room.
Cherry logs across the dogs
Smell like flowers in bloom.
But ash logs, all smooth and grey,
Burn them green or old;
Buy up all that comes your way;
They're worth their weight in gold.
Anonymous

Beech

Now summer blinks on flowery braes,
And o'er the crystal streamlet plays,
Come let us spend the lightsome days
In the birks of Aberfeldy.
Bonnie lassie, will ye go,
To the birks of Aberfeldy?
Robbie Burns

Above: *Emerging foliage of beech with male and female flowers.*

There is a ring of beeches on the hill
Lifting to the quiet evening sky
Layered tracery of green and grey,
Voicing reed – like in the hill – wind's sigh
Tales of long ago and far away.
Beneath them lies the valley wide and still.

There is forgetfulness within this ring,
Where memory's steeped in calm and thought is stilled.
For they the secrets of old earth enfold
Within their shadowed forms whose crowns are filled
With lustre of the west and evening gold,
And in the listening silence seem to sing.

Margery Lea, On a Hill in Shropshire from These Days *(1975)*

A third large tree most often associated with British woodland is the statuesque beech (*Fagus sylvatica*). Though its native distribution is largely confined to the southeastern quarter of the kingdom, it has also naturalised broadly and with extreme success. The name beech is derived from the Anglo-Saxon *boc* and/or the early German *buche,* which in turn gave rise to the word 'book'. It is said that early books were either carved on thin tablets of beech wood or written on vellum leaves which were bound between beech boards. The word 'buck', usually as part of another name, occurs often within the native beech kingdom, witness Buckinghamshire, or Buckholt (beechwoods) Wood in the Cotswolds.

The latter portion of the Latin name for beech refers to woodland, and the beech is most often found in woodland settings, a habitat which it may dominate to the exclusion of virtually all other species. The continuous foliage of its huge outspreading canopy becomes so dense that little or no light reaches the woodland floor beneath mature beeches. Some people find this bleak vista, where a shining floral face seldom rises to the sun, a distinctly sour and forlorn landscape, while others take delight in the drama of massive and graceful beeches – be they silver-grey winter skeletons, vivid emerald spring bursts, or mellow autumnal giants of innumerable golden hues. Hugh Johnson, author of the *International Book of Trees* (1973), was deeply moved by the beechwoods that he knew as a child: 'An old beechwood has the longest echo of any woodland: a sound so eerie and disturbing that I remember it vividly from when I was a boy of four living on the chalk hills of Buckinghamshire.' He remarks upon the barren nature of the woodland floor, deducing that his childhood fear lay in 'the long-answering echo; the echo of an empty room'.

The Chiltern Hills have long been a stronghold of beech, and the tree is most successful on such chalky soils, although it does thrive on an acidic surface layer. Handsome mature beeches in close proximity will grow tall and straight, sometimes reaching a height of 140 feet. For such large trees their lifespan is relatively short, seldom more than 150 years. This is in part because of their vulnerability to high winds, one of beech's main adversaries. The tree's root system, although large and spreading, is generally quite shallow; thousands of fine old beeches succumbed to the gales of 1987 and 1990. As many of them grew close together, the domino effect was catastrophic. Most commercially planted and regularly managed beechwoods are no longer allowed to grow big enough to become this hazardous. In the past the trees have often been planted in avenues and as hilltop clumps: it is even possible to come across ancient burial mounds marked with stands of beech trees and, near Wallingford a splendid group surmounts the Iron Age hill-fort at Wittenham Clumps, a landmark often depicted by the artist Paul Nash. An avenue of beeches marred by decay and storm runs up the remote and exposed Linley Hill, in Shropshire, where neatly protected saplings suggest that restoration moves are afoot. Chanctonbury Ring on the crest of the South Downs in Sussex was planted in 1760 and cherished as a distinctive local landmark for over 200 years, but one night's raging storm

Right: *Two stems of this unusual specimen have fused together, and since it would appear to have grown from an old coppice stool, this crossing of stems might have been the result of some forester's foolery at the turn of the century.*

Opposite: *A view along the top of Selborne Hanger, Hampshire, early on a May morning. Beyond the towering beeches a distant deer watchfully nibbles at the understorey.*

tore down most of the trees in 1987. Replanting has begun, and the extra light reaching the ground now encourages the natural seedlings to thrive, so that with luck the clump will arise again.

Since the Romans occupied Britain, beech has been valued mainly as a fuel source, but traditionally its wood has also been used for furniture, carriage panels, tool handles, buckets and bowls. Its ability to remain sound when submerged in water suited it for the keels of ships and for floodgates and sluices. It is said that the Normans built Winchester Cathedral upon piles of beechwood set into the local peaty marshland. At one time, the beech nuts, known as mast, were fed to cattle, but since good mast years are somewhat unreliable this must always have been viewed as a bonus rather than a staple feed.

In the Chilterns, beech became the mainstay of a part of the furniture industry known as bodging, or the turning of chair legs. H. L. Edlin wrote, just after World War II, that the industry was in decline, and it proceeded to disappear. It has tentatively resurfaced, albeit in a limited way, as a few craftsmen begin to use the woodland once again. Turned products find favour with smaller-scale furniture makers, and the wood is also used for parquet flooring.

Large tracts of native beechwood lie in the Cotswolds and on the downs, with smaller fragments to be found within the New Forest and even the southeast corner of Wales. In Hampshire, one of the most evocative and best-documented stands of beech clothes the steep hills behind the village of Selborne where, in the late eighteenth century, Gilbert White, one of the greatest amateur naturalists and diarists, wrote in minute detail about local wildlife. A climb alone through Selborne Hanger at the break of day evokes White's belief that beech was 'the most lovely of all

ABOVE: *A clump of beeches surmounts an ancient Cotswold burial mound.*

OPPOSITE: *Massive old beech pollard in Epping Forest.*

BELOW: *Exposed beech avenue on Linley Hill in Shropshire.*

Portrait of beech from Selby's History of British Forest-Trees, 1842

The tallest hedge in the world, at Meikleour in Perthshire, pictured in 1907 (below) and in 1997 (bottom). The beeches are trimmed and remeasured every 10 years, using a hydraulic platform and hand-held equipment, an operation which takes four men approximately six weeks to complete.

Elwes & Henry

A violent enemy of the beech is the grey squirrel, which has a great passion for stripping away the bark, a process which eventually kills the tree. In the past it was assumed that squirrels used the bark to add variety to their diet; however it now appears that male squirrels wreak this destruction as a display of machismo to impress the females and ward off rival males on their territory. Such frolics are beginning to seriously threaten large stands of beech in areas such as the New Forest. Programmes of squirrel sterilisation are becoming a real prospect. Cowper might today reassess the charm of these little horrors which he saw as:

'Flippant, pert, and full of play;
He sees me, and at once, swift as a bird,
Ascends the neighbouring beech.'

forest trees, whether we consider its smooth rind or bark, its glossy foliage, or graceful pendulous boughs'.

William Gilpin, also a resident of Hampshire, was not so greatly taken by the beech. Of its winter skeleton he opined: 'We rarely see a beech well ramified. In full leaf it is equally unpleasing; it has the appearance of an overgrown bush. This bushiness gives a great heaviness to the tree; which is always a deformity. What lightness it has, disgusts.'

Some of the most beautiful and unusual beeches are found in areas of ancient wood pasture, such as Epping Forest and Burnham Beeches, where an abundance of aged beech pollards have survived for several centuries. In Epping, it is possible to find massive multi-stemmed trees, which clearly began life as pollards, but which have not been lopped for well over a century. In Burnham Beeches, about 700 gnarled and twisted individuals are carefully accounted for by little numbered plates and cared for by teams of conservationists and foresters working for the Corporation of London, who purchased the site back in 1880, since when it has been a public open space. Recent programmes of pollarding are aimed at protracting the lives of the veterans and forming new pollards alongside, thus preserving the character of the site. Similar activity is underway in Epping Forest with

beech, oak and the famous hornbeams.

The eighteenth-century poet Thomas Gray was much given to dalliance amongst the ancient beeches at Burnham, being moved to write of the trees in a letter to Horace Walpole in 1737, where he reports that 'both vale and hill are covered with most venerable beeches, and other very reverend vegetables, that, like most other ancient people, are always dreaming out their old stories to the winds'.

In the west a different set of beechwood habitats can be seen. The steep slopes of the Wye Valley are home to many dramatic stands of beech, often found clinging to the steep valley sides in configurations similar to the Hampshire beech hangers; the trees become long and leggy in their search for light among the rival dense undergrowth. The protected nature of the valley also affords a footing to such rarities as large-leaved lime, wild service and some obscure whitebeam varieties.

The valleys of southeast Wales harbour some of the most visually exciting beechwoods in Britain, such as the group of woods on the west side of the Usk valley, overlooking Abergavenny. As with other ancient woodlands, they have survived remarkably intact, because of their value to the coal and iron industries. Though the commercial need for them has now disappeared, their relative inaccessibility and their more recently perceived importance as historic wildlife environments would seem to assure their future. The beeches in and around these woods present virtually every different ramification imaginable. There are pollards and coppice stools, mostly wildly overgrown; great maiden trees with serpentine root systems which loom above the hollow ways and occasional crags; and beyond the woods, on the wild hilltops, wind-bent trees stand in dutiful lines on long-forgotten field boundaries.

Beech has been well regarded as a hedging tree; it keeps some of its old leaves throughout the winter and is often visually more appealing than a bare and twiggy hedge. Though large beeches in the path of a gale often sustain severe damage, smaller trees which have perennially been exposed to high winds tend to develop glorious windswept shapes, as seen in many of the beech hedgerows typical of Dartmoor and, in particular, Exmoor in Devon. Here, even when the trees are sitting on top of a bank and the wind and rain has winkled out most of the soil from around their roots, the beech hedgerows stand fast.

Perhaps the most dramatic garden hedge of all is at Meikleour, near Perth. The men who were planting it in 1745 were called away to fight at the battle of Culloden; they hid their tools beneath the hedge and never returned to reclaim them. The mighty hedge stands as a monument to those who fell, presently a third of a mile long and 100 feet high, the highest hedge in the world.

Lime

The native lime appears in Britain as both the small-leaved lime (*Tilia cordata*) and the much rarer large-leaved lime (*Tilia platyphyllos*). Pollen records show that both limes were major components of primeval woodland, although it would take some effort to track down either of them today. Common lime (*Tilia* x *vulgaris*), also known as European lime (*Tilia* x *europacea*), is the hybrid of the two and is much more prolific, most often found in formal settings such as in avenues, in parks and along roadsides, and seldom as a woodland tree. It can occur as a natural hybrid where both small-leaved and large-leaved varieties are present, but as the flowering periods of the two rarely coincide, such natural crosses are uncommon. European lime is generally believed to have been introduced from Holland or Germany in the seventeenth century, whereupon it became an instant success as a fashion tree.

The common lime has proliferated to a phenomenal degree, while the small-leaved and large-leaved varieties remain in virtual obscurity. This is largely due to two factors – climate and commercial value.

Since the arrival of small-leaved and large-leaved limes approximately 8,000 years ago, the climate has, on balance, become cooler. While native limes have survived in various isolated sites, usually quite warm and sheltered, and generally in the southern half of Britain, they set much less fertile seed. The lime's success from seedlings has been further hindered by the palatability of its foliage to sheep, cattle and rabbits. The hybrid newcomer was much less susceptible to the cooler climate and was able to replicate with vigour wherever it grew.

The other contribution to the native varieties' retreat was their limited commercial value. While lime has served as general coppice wood to be harvested for poles, as well as ropes and nets which have been woven from the fibrous bast or bass of the underbark (limes in America are known as 'basswoods'), it is inadequate as a major construction timber and as fuel wood; it actually smells rather unpleasant when burnt. Its most valued property is that its soft white timber has the ideal qualities for wood turners and carvers, notably the great Grinling Gibbons (1648–1721), whose breathtaking carvings still adorn many churches and stately homes. In contrast, common lime became very popular as an amenity tree because it grew easily, and was resilient to heavy pruning and pleaching; when in flower it is also a joy to the senses, beautiful to behold, with a heady perfume and the soporific drone of thousands of attendant bees. Perhaps its only drawback is the sticky honeydew generated by lime-loving aphids.

Most small-leaved and large-leaved limes which still grow in British woodlands are of considerable antiquity, which in many cases has been preserved through coppicing and pollarding. In addition, these trees can proliferate by layering themselves: when a large bough reaches down to the ground, it generates a new root system at the point of contact. This may then remain as an overlying extension to the parent tree or become separated and flourish in its own right. During recent years the lime expert Dr. Donald Pigott has extensively researched the locations and antiquity of ancient coppice stools, particularly of small-leaved lime deep within the confines of long-established woods across England, and has arrived at some staggering conclusions regarding their age.

BELOW: *Shrawley Wood in Worcestershire is a rather special 500-acre site, much of which is predominantly small-leaved lime coppice. At the beginning of the twentieth century, the owner of this wood saw little value in the limes and considered replacing them with larch as a more commercially viable proposition. Although some parts of Shrawley succumbed to coniferisation, for some reason this was not carried out to the exclusion of the lime.*

ABOVE: *An ancient small-leaved lime coppice stool which may well be 1,000 years old. This one is at the northernmost limit of its range in Britain, hidden away in sheltered woodland above the east side of Coniston Water, Cumbria.*

BELOW RIGHT: *Evening light filters through a huge large-leaved lime in the Wye Valley, illuminating a stool which is once more ready for coppicing.*

One of the most celebrated stands of old lime was recognised in Silk Wood, at Westonbirt, Gloucestershire, in 1964. The Forestry Commission had set about removing all the hardwoods with a view to planting an unusual mixture of Norway spruce and elm. The Conservator of Forests, Morley Penistan, who happened to be on hand at the time, realised only hours before chemical eradication that a circle of lime trees some 48 feet in diameter might be worth saving. This eleventh-hour action turned out to be even more crucial than first suspected: DNA finger-printing subsequently showed that what appeared to be a ring of

about 60 trees was actually one tree. Dr. Pigott and Dr. Rackham have postulated that this lime tree could actually be as old as 6,000 years, back to the dawn of the species' arrival in this country: older in all probability than even the very oldest yews. The tree was known not to have been coppiced since 1939; in 1994, then-curator John White, after much consultation, instigated a new coppicing regime. This would seem to assure the survival of one of Britain's oldest trees for posterity. It is sobering to realise that, but for one man's vision, it would have been annihilated.

That limes were once a principal element of Britain's wildwood is perfectly mirrored in the great evidence of lime-related place-names across the country, though no limes may still be present in these locations. Linde, the Anglo-Saxon name for lime, and variations of it (lynd, line, lin, lyne) occur in the names of many settlements and woods; Lyndhurst, for example, in the New Forest, literally translates as an isolated wood (possibly on a hill) of lime trees. Linden is also the alternative name for lime trees in the United States. Geoffrey Grigson mentions that Whitewood in Worcestershire also refers to lime, while Oliver Rackham notes that Pry is the old Essex name for the tree. Pollen records indicate that the small-leaved lime has been the more widespread of the two natives. The large-leaved lime now occurs only in a few isolated enclaves, some of which, such as those within the woods along the edge of the South Downs, have only been identified within the last few years. Undoubtedly a few other secret specimens await the diligent searcher.

For the novice lime hunter the differences between the types can be confusing, especially where natural hybrids have established. On small-leaved limes the leaves are generally 1¼–3½ inches long while on large-leaved trees they are 2¼–3½ inches long. Careful examination of the leaves reveals that the former have leaves

LIMES IN LITERATURE

Many early writers failed to draw attention to the native limes. If they knew of the small-leaved and large-leaved limes they typically mentioned them in passing before presenting a lengthy dissertation upon the Common Lime. This is hardly surprising; John Evelyn was preparing his Silva at precisely the same time that European limes began to flood into Britain.

As the introduced tree became more familiar the two natives were all but forgotten: allusions to small-leaved and large-leaved lime in texts, particularly poetic and literary appreciations, are well nigh absent. Lyrical writers almost always referred to common limes, for it was the sheer presence of long avenues of splendid trees dripping with their rich scent and loaded with bees which sparked the muse. In W. H. Hudson's Hampshire Days, he described a drowsy

summer between Winchester and Alresford, near a formal lime avenue which had reverted to a semi-wild character:

'The old limes were now in their fullest bloom; and the hotter the day the greater the fragrance, the flower, unlike the woodbine and sweetbriar, needing no dew nor rain to bring out its deliciousness. To me, sitting there, it was at the same time a bath and atmosphere of sweetness, but it was very much more than that to all the honey-eating insects in the neighbourhood. Their murmur was loud all day till dark, and from the lowest branches that touched the grass with leaf and flower to their very tops the trees were peopled with tens and with hundreds of thousands of bees. Where they all came from was a mystery; somewhere there should be a great harvest of honey and wax as a result of all this noise and activity. It was a soothing noise, according with an idle man's mood in the July weather; and it harmonised with, forming, so

to speak, an appropriate background to, the various distinct and individual sounds of bird life.

The birds were many, and the tree under which I sat was their favourite resting-place; for not only was it the largest of the limes, but it was the last of the row, and overlooked the valley, so that when they flew across from the wood on the other side they mostly came to it. It was a very noble tree, eighteen feet in circumference near the ground; at about twenty feet from the root, the trunk divided into two central boles and several of lesser size, and these all threw out long horizontal and drooping branches, the lowest of which feathered down to the grass. One sat as in a vast pavilion, and looked up to heights of sixty or seventy feet through wide spaces of shadow and green sunlight, and sunlit golden-green foliage and honey-coloured blossom, contrasting with brown branches and with masses of darkest mistletoe.'

ABOVE: *Large-leaved lime in Rook Clift, near Treyford, at the edge of the South Downs in Sussex. These limes were only discovered in this wood as recently as 1989 by Dr. Francis Rose.*

RIGHT: *This stand is in fact one tree, a very old small-leaved lime coppice stool with about a dozen main stems. The circle which they form is about 13–16 feet across and has evolved over many centuries of coppicing. As each rotation was cut, the new growth slowly moved further out to form the circular clump. The bank upon which this lime grows was woodland until about 40 years ago, but the farmer, who cleared it to provide more grazing land, realised the significance of such a special tree. However, recent unwelcome attention from cattle has damaged bark and compacted the soil, and the present owner will soon be obliged to fence off the tree and coppice once again to assure its continued survival.*

RIGHT: *In Gloucestershire, small-leaved lime occurs in several woodlands across the county, yet Collin Park Wood, near Newent, has some of the most impressive stands. Until the mid-nineteenth century the coppice wood was used to make charcoal for the iron industry, but wood extraction gradually declined here until the last coppicing was carried out in the 1920s. The result of such close confines was that the limes and the accompanying oaks grew into incredibly tall and slender trees. A dense canopy was formed and much of the ground flora was shaded out. The Gloucestershire Trust for Nature Conservation is now actively managing the site, creating a multi-featured woodland with coppiced areas. These have opened up spaces for flower and butterfly resurgence, high forest which will become a future timber source, and non-intervention areas where the natural progression can be studied. These amazingly tall limes are only visible in this way because the surrounding trees were recently coppiced.*

PREVIOUS PAGE: *Fine example of a mature common lime. Though not strictly a native, it is a remarkably successful introduction.*

with a dark and glabrous (hairless) upper surface with a paler underside and tufts of little red-brown hairs in the vein-axils, while the latter have leaves that are hairy on both surfaces, particularly underneath, feeling velvety to the touch.

Shrawley Wood in Worcestershire is composed in large parts almost purely of small-leaved lime. Now in the care of English Nature, this wood is in remarkably good shape and contains large stands of coppice stools and a few pollards of lime. While good stands of small-leaved lime will be found in widely distributed lowland woods, its largest concentrations are in Essex, Suffolk and Lincolnshire. As with many other trees growing in close confinement, the woodland lime tends to be a long, leggy affair, which is ideal for foresters, yet seldom allows the tree to mature to its finer natural form.

One of the best sites to see small-leaved limes in their full magnificence is at Turville Heath in Buckinghamshire. In the 1740s, William Perry, the local landowner, planted an avenue along the drive to his home at Turville Park. The small-leaved lime was an unusual choice of tree for this purpose at that time, when common lime was the rage. Thirty-five of the original planting still stand, though gaps from storm blow and natural decay have been filled in less enlightened times with common lime and nursery-sourced small-leaved lime. But it's still an awesome experience to walk beneath these towering limes.

Hornbeam

LEFT: *Ancient hornbeam coppice in Hatfield Forest, Essex.*

ABOVE: *The leaves of the hornbeam which are often confused with those of the beech.*

The hornbeam is a relatively unfamiliar tree which forms a major element of some woodlands in England. Its Latin name, *Carpinus betulus*, comes from the Celtic *carr*, meaning wood, and *pen*, which means head; it is popularly assumed to be an allusion to the use of the wood for harness yokes for oxen. The sixteenth-century herbalist John Gerard added his perspective to the name in his *Herball*: 'In time it waxeth so hard that the toughnesse and hardnesse of it may be rather compared to horn than unto wood, and therefore it was called hornbeame, or hardbeam.' In parts of East Anglia the name hardbeam still persists.

The tree is most often encountered as a pollard or a lesser tree of woodland broadleaf mixture. Its native distribution is the southeast quarter of England with small inroads into the West Country and South Wales. However, it has been successfully planted and grown as far north as Scotland, although in the cooler northern areas it prefers some degree of shelter from the worst winters. It is most often associated with beech woods, though it is not quite so forgiving of poor and stony soils, preferring good clay or deep loam, and, like the beech, often retains some of its dead foliage through the winter. Though from a distance its canopy may resemble a beech, hornbeam can be recognised by its sharply toothed leaves and distinctively fluted trunk. Exceptional specimens have been known to attain 90–100 feet in height, while some ancient pollards achieve girths of up to 15 feet. Hatfield Forest in Essex has some of the oldest and most visually interesting pollards, many of which have been split right down the middle of the bole yet still burst forth anew each year. It's hard to determine their precise age, but it most likely exceeds 300 years. The inclusion of the tree in wood pasture has long been a safe choice since the foliage is not particularly attractive to grazing animals, and, even if they decide to take a nibble of the bark, hornbeam withstands the damage sturdily. Its other common manifestation is

as a hedging tree, either in old country hedgerows or as a dense alternative to beech in domestic settings.

Hornbeam's extremely hard wood has limited its traditional uses since it tends to blunt the tools of any craftsman who tries to work it. The best-known use of the wood has been for items where exceptional durability was required, such as the cogs in the mechanisms of windmills and watermills. It has also been prized as excellent fuel wood; the hornbeam woods around north London and Essex were once regularly harvested for the hearths of the city. The pollards of Epping Forest are a particular reminder of this tradition.

Unlike the other natives, the hornbeam has virtually no folklore associations, but behind Epping Forest's pollarded hornbeams lies a fascinating bit of history. During the 1870s, various wealthy landowners moved to 'improve' parts of Epping Forest, by turning it into more productive farmland. The commoners banded together, proclaiming their forestal rights to lop firewood and to graze stock there. The determination of the commoners contributed to the passage of the Epping Forest Act in 1878, transferring ownership of the forest into the hands of the Corporation of London, which was empowered to preserve the natural appearance of the forest. However, with the best of intentions, the corporation put virtually all forestry activities on hold with the aim of promoting the 'natural aspect of the forest'. Pollards were allowed to become overgrown, shading out much of the ground flora and shrub layer, and most open heathland was lost beneath colonising saplings. The corporation eventually terminated the commoners' forestal rights. Latterly, more enlightened management, incorporating a comprehensive re-pollarding regime, is slowly recouping the lost character of both Epping and Burnham.

OPPOSITE: *Hatfield Forest, in Essex, is one of the very best sites to view ancient hornbeams. While some of the old pollards have been cut once again, after many years' neglect, this one has been allowed to grow into a large multi-stemmed tree.*

LEFT: *The annual ritual of asserting the commoners' forestry rights by lopping firewood in Epping Forest at midnight each 9 November* Illustrated London News, *1876*

ABOVE: *Vivid green catkins of hornbeam appear in March.*

LEFT: *Solitary hornbeam in open pasture in early spring.*

LEFT: *Pollarded hornbeams at Bayfordbury, Hertfordshire, from a plate made by Mr. Elsden, at the turn of the century, for Elwes & Henry. For comparison, an overgrown pollard in Hatfield Forest (*OPPOSITE*) presents the typical upright fanning branches and the same distinctive fluted bole.*

Hazel

One tree which forms a significant membership of many different types of woodland, most commonly in association with oak, is the hazel. Rarely growing higher than 25–30 feet, the hazel is usually seen in its coppiced form, a small, bushy, multi-stemmed tree often forming quite a dense understorey.

For thousands of years, hazel has been harvested for its particularly flexible poles and sticks, which are generally cut on a rotation of seven to 15 years, depending on the thickness required. Hazel poles are incredibly pliable and present an incalculable array of uses. Recent resurgences of coppicing supply poles with which to weave wattle hurdles, which in earlier times also formed the frame for wattle and daub walls in many buildings; in addition, they are used for thatching spars and various types of walking sticks and crooks. Very thinly split hazel can even be used as a binder or tying material in basket work. In *Woodland Crafts*, H. L. Edlin listed many uses of hazel which have gradually disappeared since he wrote in the 1940s. One exceptional reverse in the trends he recorded is thatching. Edlin observed a decline in the thatching of houses and reckoned the thatcher's skills would be used mostly for hay ricks; but even as bailers, loaders and vast storage barns have seen the disappearance of ricks, there has been a sustained renewal of house thatches.

The scaling down of hazel coppicing over the last century has been brought about by a mixture of convenience, such as an increasing use of wire fences by farmers, and economics: while the thinnest hazel poles once were used for pea and bean sticks, imports of foreign canes now fill the garden centres. The last few years have seen a renewed interest in coppicing because of its perceived sustainability and benefit to the woodland ecology. Sadly, it is seldom lucrative enough to provide a realistic income for a family, as was once the case, but public enthusiasm for revitalising old skills enhances a valuable element of Britain's cultural heritage. It may no longer be a fundamental part of the rural economy, but it brings pleasure and some recompense to thousands of people while also contributing to woodland conservation. After all, what's going to happen when the oil runs out?

One of the most familiar aspects of hazel is its golden yellow catkins, or lamb's-tails, which adorn both woodland and hedgerow, sometimes as early as late January. While the male catkins toss in the wind, in the late winter sunshine, one can also see the tiny, bright red female flowers which will become the cobnuts of autumn.

Hazelnuts ripen in mid-August: one explanation of the alternative name of filbert is that St. Philibert's Day falls on 20 August. The nuts are known to many a country child, although the opportunity to garner

Right: *A typical coppice stool of hazel as part of the woodland understorey. Fast-growing hazel is harvested for its very useful pliable poles.*

this harvest is often forestalled before it ripens, by grey squirrels. The filberts of the Christmas nut bowl are usually larger, cultivated varieties imported from Europe; they have longer bracts, which sometimes enclose the whole nut, and are a source of another supposed derivation of the name, from 'fullbeards'.

Commercial growing of hazelnuts was once an important part of the Kent orchards, where both filberts and cob-nuts (essentially a meatier version of the wild hazelnut) were once grown alongside other fruits. Less expensive imports and high operational costs have resulted in a decline in this side of orcharding, so that the estimated area of trees is now barely 150 acres compared to more than 7,000 acres back in 1913. Many cultivated varieties were developed during the nineteenth century, the most important of which was Lambert's Filbert, latterly known as Kent Cob, which is still the most popularly grown variety today. The national collection at Brogdale contains an astounding 44 varieties.

John Evelyn would have known the tree as *Corylus sylvestris*, but in the eighteenth century Linnaeus renamed hazel as *Corylus avellana*, after the Italian town of Avella, a centre of nut cultivation. Evelyn understood that his own surname was related to the cultivated hazels and noted that hazel charcoal was once considered superior for gunpowder until alder proved to be even better. He also recorded that hazel chips could be used to purify wine and that 'the coals are used by painters to draw with, like those of Sallow'. Lastly, he notes the use of 'divinatory rods' for:

'...*the detecting and finding out of minerals; at least, if that tradition be no imposter. By whatsoever occult virtue the forked-stick, so cut and skilfully held, becomes impregnated with those invisible steams and exhalations, as by its spontaneous bending from an horizontal posture, to discover not only mines and subterraneous treasure, and springs of water, but criminals guilty*

of murder, &c., made out so solemnly, by the attestation of magistrates, and divers other learned and credible persons, who have critically examined matters of fact, is certainly next to a miracle, and requires a strong faith.'

Even today, particularly in rural areas, hazel wands are not uncommonly used to dowse for springs or long-forgotten water wells.

The Hazel Nut Tree

ABOVE: *The tiny red female flower above this cluster of male catkins will later develop into the hazelnut.*

LEFT: *Hazel as delineated in First Hunter edition of Evelyn's* Silva

LEFT: *Hazelnuts: part of the annual hedgerow harvest... if you can find them before the squirrels seize upon them!*

Birch

While hornbeam and hazel are also of the family Betulaceae, its most influential members in the British landscape are the silver birch and its close relative the downy birch. Birch has been the most opportunistic coloniser since the last ice age. It is still one of the first trees to invade open ground, particularly on acidic soils. As with the ash, its success has made it unwelcome by some foresters and conservationists. The former remove it from land more valued for commercial conifers, while the latter are concerned about its potential to dominate habitats: left alone, it will for example eventually turn heathland into woodland, crowding out many heathland species en route. However, the long-term prognosis for birch, in any given location other than the northwest of Scotland, where it forms pure birch woods, is relatively bleak. Its natural lifespan is rarely more than 80 years, and it is typically shaded out by the larger woodland trees that succeed it.

The silver birch is a particularly airy and graceful tree, from its distinctive smooth bark, seen to best advantage in the low sunlight of a winter's day, to its drooping fronds of delicate spade-shaped leaves, soft green in spring and rich yellow gold in autumn. The name *Betula pendula* literally means a tree with drooping branches. Coleridge was so enchanted that he pronounced the birch the 'Lady of the Woods'. The silver is by far the more common of the two natives, and its more pronounced silvery bark is the main characteristic associated with birches. Downy birch, *Betula pubescens*, tends to a much darker stem, and its habitat is typically either wetter or at higher altitude; it has downy twigs, while those of silver birch are rough and warty.

Gertrude Jekyll warmed to the slender birch. 'Has any tree so graceful a way of throwing up its stems as the birch? They seem to leap and spring into the air, often leaning and curving upward from the very root, sometimes in forms that would be almost grotesque were it not for the never-failing rightness of free-swinging poise and perfect balance.'

A considerable legacy of literary accounts and folklore concerning birch have evolved in Scotland, where the tree is queen of much of the domain. The birch is supreme upon the peaty, rocky tracts, either as isolated specimens bursting from some crevice or composing a stupendous contrast with the rugged Scots pines of the old Caledonian forest. The birch is often termed birk in Scotland and appears in many a poem and ballad.

Birk is the commonest place-name prefix in Scotland, indicating the long-term predominance of birch, not just as the tree in the landscape, but also as an all-round provider of wood to many a Scotsman and his household. In the past it was made into furniture, turned for bowls and dishes, used to construct farm carts and implements, and burned as fuel – in particular the bark. Birch has the unusual characteristic of growing mature bark which is even tougher than the timber within. In Scandinavia, the bark was sometimes used for roofing houses or even making rudimentary footwear. Many of the old uses have fallen away, but some still survive, or are reappearing.

> '*Even afflictive birch,*
> *Curs'd by unlettr'd idle youth, distils*
> *A limpid current from her wounded bark,*
> *Profuse of nursing sap.*'
> Phillips

all haste for the one short period when they may harvest. This appears to do no harm to the trees as long as they are not ring barked.

Birchwood besoms, ideal for sweeping up leaves, are still favoured by many gardeners. Craftsmen are once again making furniture and turnery from birch, and even the Forestry Commission is beginning to concede that it may have a role to play in economic forestry. It is ideal for pulping, the production of high-quality plywoods and wood-block flooring.

The Birch Project, a joint initiative recently set up in South Wales by the Brecon Beacons National Park, aims to establish a productive yet landscape-enhancing and ecologically sound wood industry. The project will offer practical advice on cultivation, grant assistance and marketing prospects, together with a comprehensive research and development programme. With support from farmers and foresters and the necessary product outlets, things could be looking up for the birch.

Wine, for example, has been made from birch sap for hundreds of years in Scotland, but until recently only by a few home brewers using traditional family recipes. Now some of the large wineries have begun to produce serious quantities of the delicate and aromatic birch sap wine. The sap may only be gathered for about two weeks in early March, so the wine makers must cut the bark and fill their containers with

BIRCHES

When I see birches bend to left and right
Across the lines of straighter darker trees,
I like to think some boy's been swinging them.
But swinging doesn't bend them down to stay
As ice-storms do. Often you must have seen them
Loaded with ice a sunny winter morning
After a rain. They click upon themselves
As the breeze rises, and turn many-colored
As the stir cracks and crazes their enamel.
Soon the sun's warmth makes them shed crystal shells
Shattering and avalanching on the snow-crust —
Such heaps of broken glass to sweep away
You'd think the inner dome of heaven had fallen.
They are dragged to the withered bracken by the load,
And they seem not to break; though once they are bowed
So low for long, they never right themselves:
You may see their trunks arching in the woods
Years afterwards, trailing their leaves on the ground
Like girls on hands and knees that throw their hair
Before them over their heads to dry in the sun.
It's when I'm weary of considerations,
And life is too much like a pathless wood
Where your face burns and tickles with the cobwebs
Broken across it, and one eye is weeping
From a twig's having lashed across it open.
I'd like to get away from earth a while
And then come back to it and begin over.
May no fate willfully misunderstand me
And half grant what I wish and snatch me away
Not to return. Earth's the right place for love:
I don't know where it's likely to go better.
I'd like to go by climbing a birch tree,
And climb black branches up a snow-white trunk
Toward heaven, till the tree could bear no more,
But dipped its top and set me down again.
That would be good both going and coming back.
One could do worse than be a swinger of birches.

Robert Frost was inspired by the extreme flexibility of the birch tree to allegorise upon the trials of life.

CHAPTER FOUR

Enclosure: Trees of the Hedgerow

Hedgerows have been a feature of the landscape for thousands of years. *The Oxford English Dictionary* defines a hedge as 'a fence or boundary formed by closely growing bushes or shrubs' or 'a protection against possible loss or diminution [of stock]'. One of the prime purposes of hedges has been to confine livestock, while they have also delineated property boundaries and rights of way. They shelter the livestock in their lee and wild birds and beasts within and, they provide rich resources of wood, fruits and herbs.

In the last 50 years humans have had the technology to obliterate in a few days, growth that might have evolved over many centuries, and hedges have held an increasingly high profile in the media, as public awareness of them has increased. A widespread network of tree wardens and local conservation groups increasingly clamours, from the grass roots level, for the restoration and careful management of these often fragile habitats.

THE HEDGEROW IN BRITAIN

> *If cattle or coney may enter to crop,*
> *Young oak is in danger of losing his top.*
>
> Thomas Turner, sixteenth century

During the Bronze Age the strange linear boundaries known as Reaves were laid out across Dartmoor. They can still be detected from certain critical viewpoints and aerial survey today. While they evidently divided arable and pasture land, it is not known whether they consisted of living hedgerows.

BELOW: *A view eastwards across the Usk valley from the crags of Craig-y-cilau, above Llangattock in Powys. A fine viewpoint to observe old field patterns containing good grazing with a marked delineation from the rough pastures of the open moorland. The predominance of hawthorn in the hedgerows stands out particularly well in mid-May.*

Aerial surveys reveal that there have been divisions of land in Britain since prehistoric times, seldom better illustrated than in some of the wilder parts of Cornwall and Devon. Although more recent hedges and walls may also appear there, the contours of the land bear evidence of ancient raised banks. It is not known whether these were surmounted by living vegetation or dead wood or were simply great earthworks shored up with massive boulders, as they still are today in many parts of Cornwall. Whether or not vegetation is present, these structures are still known as hedges amongst the Cornish people, and the Penwith peninsula is a particularly good place to view the extant Iron Age field patterns.

Various authorities quote the accounts of Julius Caesar encountering hedges made from plashed (woven) stems of young trees mixed with thorns and brambles during his advances through Belgium. The earliest reference to a hedge in Britain, according to Pollard, Hooper and Moore in *Hedges* (1974), was of one 'erected by Ida of Northumbria around his new settlement, Bamburgh, in AD 547 as recorded in the *Anglo-Saxon Chronicle*.' No indication is made here of a living hedgerow, so it might have been more of a fence, even a raised bank, than a hedge.

The word hedge is derived from the Anglo-Saxon *gehaeg*. Both hay and haw are extractions which, in the past, have denoted a hedge, and hawthorn, literally hedge-thorn, is the tree most often associated with British hedgerows. An alternative name for the hawthorn was quickthorn, signifying live or lively thorn, as opposed to a hedging material that was of dead or cut wood.

Hedgerows have originated in different ways. Though most have been planted at some point during the last 1,000 years, some are remnant marginal strips of long-gone woodlands, which almost certainly explains the appearance of such atypical trees as small-

leaved lime or wild service. In the past, hedges may also have evolved from the neglected borders of fields, where seeds fell on uncultivated land and were simply permitted to grow into a hedge.

Oliver Rackham has made detailed studies of many Anglo-Saxon charters but has not come across any specific references to live hedges. He did, however, find the phrase 'the hedge row that Aelfric made' in a charter from Kington Langley in Wiltshire, dating from 940. The words were accompanied by numerous references to hazel-rows and thorn-rows and led him to believe that these signified the planting of a hedge.

The sketchy history of Britain's hedges gains texture from medieval records, which reveal that hedgerows had economic importance. Mention was made of some of the fine timber trees within them and the purposes for which they were felled; theft of firewood was also an issue. The development of strip cultivation left a physical legacy upon the landscape which is still readable today. Much of the practice, which is as near as Britain has ever come to wholesale co-operative farming, persisted for hundreds of years, right up to the eighteenth century in some areas. Whereas much of it was composed of open-field systems, some fields or strips were set apart by hedges. Many vistas still reveal these long-abandoned patterns of ridge-and-furrow in their distinctive reversed-S formations, formed to accommodate the turning of the oxen-drawn ploughs.

In *Silva*, John Evelyn ardently promoted sound hedgerows, to protect his beloved oaks:

'Are five hundred sheep worthy of the care of a shepherd? And are not five thousand Oaks worth the fencing, and the inspection of a hayward?
Let us therefore shut up what we have thus laboriously planted with some good Quick-set hedge.'

A hayward was an official who oversaw the construction and maintenance of hedges. 'Quick-set' means hawthorn saplings: Evelyn urged upon his readers the commercial wisdom of setting up tree nurseries to grow them.

'I have been told of a Gentleman who has considerably improved his revenue, by sowing Haws only, and raising nurseries of Quick-sets, which he sells by the hundred far and near: This is a commendable industry.'

From the fifteenth century onwards, hedge planting accelerated apace. The first Enclosure Act was passed by Parliament in 1603 and the last one as recently as 1903, but the period between 1750 and 1850 is most commonly thought of as the main period of Enclosure. The whole business of enclosing the land was a cumbersome undertaking whereby the lord of a manor and his associate landowners petitioned Parliament for an award. The groundwork of enclosure was then undertaken by surveyors, who mapped out the area that a landowner proposed to set aside. Commissioners were appointed to see the Act through legislation; there were usually three of them, one each

to act for the landlord, the church, and the interests of the local freeholders. This august body was usually appointed by the landowners, so there was not always an unbiased committee working for the interests of the whole community. The rich got richer (including the commissioners) while the poor became disgruntled, as many of their common rights were eroded.

There were ample opportunities for corruption, for the landowners who were pressing for awards often

LEFT: *A clump of aged rambling yews stands along a hilltop boundary-cum-ancient trackway. Originally planted as marker trees, they have grown into part of a huge and unchecked hedgerow.*

PREVIOUS PAGE: *East Anglian elms line hedgerows near Ramsey in Cambridgeshire.*

NEXT PAGE: *Hedgerow of damson trees in blossom in the West Midlands, where they have been traditionally grown for the purple/blue dye used in the cotton industry.*

PENNY HEDGE

A unique traditional custom concerning the construction of a dead hedge is still actively upheld in the town of Whitby in North Yorkshire. Known as the Planting of the Penny Hedge, or Horngarth, it harks back to 16 October 1159.

On that day three noblemen were out hunting in the woods

of Eskdale-side, on lands belonging to the Abbot of Whitby, when they came upon a wild boar, which they pursued. Wounded and exhausted, the boar blundered into the chapel of a small hermitage in the woods. The hermit took pity on it and proceeded to bar the door against the hunters. This

enraged them, and they set about the poor monk so severely that he died of his injuries. The Abbot, incensed, was intent upon bringing the full force of the law against the noblemen, but the mortally wounded monk persuaded the Abbot to impose a penance of his own instead: the Planting of the Penny (Penance) Hedge.

It was decreed that the three men and the successors to their tenancies, in perpetuity, should go to the woods in the early morn of Ascension Day and cut 'Stout-Stowers', or stakes, and 'Yedders' (thinner rods for inter-weaving), with the aid of a 'knife of a penny price'. The party was then to go to the low tide mark on the Whitby foreshore and build a short hedge, by 9 o'clock in the morning, strong enough to withstand the force of three tides. If the penance was disregarded, or the hedge was not strong enough, the tenants would lose their rights to the land, which would be repossessed by the Abbot or his successors. The hedge need not be built if the tide was right in at the allotted time.

More than eight centuries on, the custom is still religiously observed each year.

> Hedges are dividers, between those who own them and those who do not. Ever since the very first hedges were set, the barrier has often been both material and social.

also sat upon the benches of the House of Lords, where Act after Act was passed as a mere formality. In 1845, a Standing Enclosure Commission was appointed by the Government, to make the commissioners more publicly accountable, but by 1869 it appeared that this was still not the case. Londoners were alarmed to discover that, in the previous 24 years, a mere 0.6 percent of the 320,000 acres of land enclosed in and around the metropolis had been set aside for the use of the common folk.

The legislation was accompanied by prodigious hedge planting. Rackham estimates that 200,000 miles of hedges were started between 1750 and 1850, perhaps equalling the planting of the entire preceding 500 years. Whereas the old hedges had followed ancient patterns of farming practice or physical contours, the Enclosure hedges were straight and can often be defined today by the way they cut across old ridge-and-furrow and pay little heed to any physical features in their path. The counties most affected by the Enclosures were those of the East Midlands, from Oxfordshire up to the East Riding of Yorkshire, where 40–50 percent of the open field land was enclosed. Unfortunately, this has led to a rather predictable modern landscape, with mile after mile of hawthorn stretching into the distance. Those who really benefited from this period were the landowners and the nurseries which were established to satisfy the unprecedented demand for

quick-sets. Previously, wild trees growing in nearby woodland had been used for the purpose.

After 1870, when the whole Enclosure boom began to recede, the pattern and extent of hedgerows changed very little until about 1950. Then began another hedgerow revolution, which would overshadow the dramatic changes of the Great Enclosure, not, this time, through the setting of countless new hedges, but through the relentless march of hedgerow devastation which continued until as recently as 1990. Statistics revealed an average rate of 5,000 miles of hedgerow being grubbed out every year, a rate which has only declined in the last few years thanks to increased pressure from conservation groups and to government legislation. Even so, the accumulated losses amount to approximately 240,000 miles of hedgerow, or the equivalent of about 50 percent of Britain's hedges, eradicated since 1945.

After the last war mechanisation arrived on the agricultural scene with progressively larger and more powerful machines that enabled the farmers to cultivate increasingly large fields. With great tracts of land laid to arable production, networks of small enclosures, previously used to prevent stock from trampling over adjoining crops, became redundant. Mixed farming declined as landowners specialised; wherever the land was suitable and the incentives of the cereal subsidies applied, arable held sway. Hedges were in the way.

BELOW: *Springtime in a Devon lane, and a wealth of wild flowers and ferns adorn the sturdy hedge banks.*

Chainsaws dealt with the hedgerow trees, and massive excavators bulldozed miles of often ancient hedgerows. The now prohibited practice of stubble burning also caused serious damage to many a hedge. For a long time there were few voices concerned or informed enough to oppose this dramatic transition.

Then, in 1974, the Collins New Naturalist series of books published *Hedges,* a seminal work by Doctors Pollard, Hooper and Moore, which covered all aspects of Britain's hedgerows in an authoritative yet accessible manner. With a scientific, historic and socially balanced overview, the authors demonstrated the great significance of British hedgerows as part of the natural and functional heritage of the nation. They explained hedgerow decline in the context of the radically altered face of agriculture, but they avoided a pointless ecological belabouring of farmers who viewed hedges simply as functional structures. The authors envisioned a future for hedges based upon informed dialogue and cooperation between the public, the farmers, conservators and the government, retaining as many hedgerows as possible without causing undue commercial inconvenience to those who had to earn their living from the land.

Probably the best-known single statistic to emerge from *Hedges* came from Dr. Max Hooper's research. After studying some 227 different hedges from around Britain, of which evidence could be authenticated

LEFT: *Planted about 150 years ago as a timber tree, this hedgerow oak now serves as a splendid boundary marker.*

from Saxon Charters, he derived an equation for dating them: each species present within a 30-yard length indicated approximately 100 years of age. While Hooper's Rule is broadly applicable to most hedges, ultimately it is the presence of trees and plants which are noted for their inability to colonise easily that are the best indicators of an ancient hedgerow: field maple, dogwood, hazel, spindle and great old pollards of oak, ash or beech. Other hedges with documented history were also surveyed and found to correspond to this theory, at least for the last 1,100 years or so: earlier hedges hold an increased margin for error, for the older the hedgerow the greater the likelihood of agents such as man or beasts bringing extra species into the equation.

BELOW: *Storm brewing behind a pair of native black poplar pollards in an ancient hedgerow. Sadly, hedgerows are now one of the last bastions of this rare but beautiful tree.*

Hedges planted after the Enclosure Act were typically hawthorn plus perhaps one other species, for example elm, ash or oak, grown as an intermittent timber tree. Through the years the elms have suckered so that, even though large trees may have succumbed to Dutch elm disease, the elm is still a lively presence in the hedge. The great majority of hedges in the planned countryside of the Great Enclosure era, typically those of the eastern half of England, are composed of one, two or three species. In the counties of the south and west, where the Enclosures had less effect on the landscape, a much larger proportion of hedgerows can be dated back to medieval times and frequently earlier. The greater diversity in earlier examples might simply correlate to the known practice of gathering whatever mixtures of woodland saplings were locally available and planting them up into a hedge.

Pollard, Hooper and Moore researched the significance of the presence of non-planted trees, shrubs and flora in relation to hedgerow ages. They located a hedgerow at Monks Wood, known as Judith's hedge (she was the niece of William the Conqueror), which was documented as being planted as a boundary to the wood in the eleventh century. At some point during the intervening years, part of the hedge was realigned, the 'shadow' of the original being plainly visible in neighbouring fields; the first evidence of this change appears on a map of 1835. A survey of the existing hedge provided a dramatic dichotomy: the newer central section was predominantly hawthorn with virtually no evidence of the old hedge indicators such as hazel or field maple, while the two older sections at either end were predominantly blackthorn, with varying mixtures of up to ten other trees and shrubs. The other clearcut finding was the virtual non-existence of any of the woodland herbs, such as dog's mercury, bluebell and wood anemone, in the newer sections. It is startling to observe that even after 150 years these plants had not made any substantial inroads into the newer hedgerow. If ever there was an illustration of the sacrosanct ecology of old hedgerows, this is it!

Elm

ABOVE: *A bold woodcut representation of typical English elm form by C. Dillon McGurk.*

OPPOSITE: *Roadside wych elm in Cumbria.*

BELOW: *Morning sunshine glows behind a large wych elm in Nidderdale, Yorkshire.*

The instantly recognisable shape of English elm has long been considered as one of the archetypal forms of the lowland landscape; that is, until about 30 years ago when they began falling like ninepins to the ravages of Dutch elm disease. In many a sketch, engraving or painting, elms regularly feature whether in the well-defined specimens from the brush of George Stubbs, the elmscapes of both English and narrow-leaved elms so familiar in Constable's works, or the impressionist elms of Sir George Clausen and later those of Paul Nash. The predominance of the English elm is also well portrayed in photographic views up until the 1960s, though seldom as a composition in its own right: like all familiar backdrops it was taken very much for granted, looming over a land of reliable fecundity and seemingly endless halcyon days.

Writers' views of elms have dwelt on both the majesty and the grace of the trees and the gloom and foreboding with which they are often associated. Elm

> ### THE FALLEN ELM
> *Old elm, that murmured in our chimney top*
> *The sweetest anthem autumn ever made*
> *And into mellow whispering calms would drop*
> *When showers fell on thy many coloured shade*
> *And when dark tempests mimic thunder made*
> *While darkness came as it would strangle light*
> *With the black tempest of a winter night*
> *That rocked thee like a cradle in thy root*
> *How did I love to hear the winds upbraid*
> *Thy strength without – while all within was mute.*
> *It seasoned comfort to our hearts' desire,*
> *We felt thy kind protection like a friend*
> *And edged our chairs up closer to the fire,*
> *Enjoying comfort that was never penned.*
> *Old favourite tree, thou'st seen time's changes lower,*
> *Though change till now did never injure thee;*
> *For time beheld thee as her sacred bower*
> *And nature claimed thee her domestic tree.*
> *Such was thy ruin, music-making elm;*
> *The right of freedom was to injure thine:*
> *As thou wert served, so would they overwhelm*
> *In freedom's name the little that is mine.*
> *John Clare*

has the strange trait of suddenly dropping substantial boughs which appear to be perfectly healthy. The day may be completely calm, but the elm may suddenly throw a limb without any warning, causing many to regard the tree with due caution.

> *Ellum she hateth mankind, and waiteth*
> *Till every gust be laid*
> *To drop a limb on the head of him*
> *That anyway trusts her shade.*
>
> A Tree Song, *Rudyard Kipling*

Elms do have their benevolent side, and perhaps few writers have praised them more warmly than John Clare: in *The Fallen Elm*, he bemoans the loss of familiar elm friends and expresses his dismay at their demise through the enclosures of the mid-nineteenth century. The sentiment of his poem still has relevance today.

Gerald Wilkinson wrote his splendid tribute to the elms of Britain, *Epitaph for the Elm*, in which he portrayed their cultural, historic, landscape and botanic significance as it began to dawn on everyone that they might never see another mature English elm in their lifetime. His book is both a lament and a celebration:

'The elms are dying. Gaps appear in familiar lines of trees that we never bothered to think of as elms. Half the tall trees, and many smaller ones, in the roadside hedges seem to have been elms, we notice, now they are so unhappily conspicuous...Great elms, landmarks to nowhere in particular, are shattered from green to grey, and will soon be gone, never to be replaced in our lifetime... Houses and cottages once sequestered among two or three friendly trees are now threatened by their dead shapes... We took them for granted as an inseparable part of our characteristic lowland landscape: loved by the poets and tolerated by the farmers.'

English elm was once one of the commonest of hedgerow trees, but to see plentiful elms, of the smooth-leaved varieties, one must travel to East Anglia today. Elsewhere they slumber in reduced circumstances in many a hedgerow, manifesting only as dense suckering

ABOVE: *Although the hedge has been kept neatly trimmed, these English elms have been left to rise up. At about this size (25 feet high) they usually succumb to Dutch elm disease.*

BELOW: *Wych elm coppice stool in spring.*

colonies and often at the expense of other species. The wych elm does not sucker, and is much more a tree of woods and northern valleys, not so prolific as a hedgerow tree.

There are many other members of the genus *Ulmus*, several of which have fared rather better in recent times, due to varying degrees of resistance to the disease.

The Wych Elm

Among the tangled web of the elm family, the easiest member to identify is the wych elm (*Ulmus glabra*), a large and spreading tree which is found throughout Britain. It has had particular success in the lower-lying regions of Wales, northern England and Scotland, often preferring close proximity to rivers and streams, and even appearing on occasions in quite rugged upland situations. Wych elm is considered by many authorities to be Britain's only native member of the *Ulmus* family. Its most distinctive feature is the large rough leaf, sometimes six inches long, shorter-stemmed than other elms, with a central leading tip and shouldered cusps or horns on either side, a shape which may vary radically between regions. Leaf shape has often caused the tree to be confused, at first glance, with the hazel, and hence its common misnomer of wych hazel. Witch hazel (*Hamamelis virginiana*) is in fact a completely different plant of

eastern American origin, known for the soothing decoction made from its bark. The tree appears to have resisted Dutch elm disease better than other elms, although Wilkinson puts its success down to the remoteness of many of its colonies. Whereas all other elms will reproduce by suckering, wych elm does not; many of the larger trees produce fertile seeds that set new young trees in motion. These wych elms are usually thought of as hedgerow trees but the wych elm has shown itself to be equally at home in the woodland.

The Field Elm

Smooth-leaved, narrow-leaved, small-leaved, field elm: these four names are a small sample of those suggested for *Ulmus minor*. Richens gives the species this broad title and then proceeds to define five principal varieties, on the basis of their geographical distribution: narrow-leaved elm (*Ulmus minor* var. *minor*), effectively the East Anglian elms; English elm (*Ulmus minor* var. *vulgaris*), with its late-lamented tiers of foliage which fan out and billow as it ascends to the distinctive crown; Cornish elm (*Ulmus minor* var. *cornubiensis*), also often classified under *Ulmus stricta*; Lock's elm (*Ulmus minor* var. *lockii*); and Guernsey elm (*Ulmus minor* var. *sarniensis*). Most recent opinion, largely as a result of pollen analysis, considers that the East Anglian elms could well be

native, along with the wych, while all the others have been introduced at various times throughout history. English elm is given the status of a species (*Ulmus procera*) by many older authorities, and *Ulmus minor* var. *vulgaris* is the more commonly accepted classification today.

The English elm, in its pure form, is fairly easy to differentiate from its cousins; for whereas most of them have smooth and quite pointed leaves, the English elm has hairy, shorter and often slightly broader leaves.

Hybrid Elms

The Dutch elm (*Ulmus* x *hollandica*) leads into the realm of hybrids, or at least those which can be defined with some sort of classification. Dutch elms, and there are many slightly different varieties, are hybrids of wych elm and smooth-leaved elm, often displaying the habits of both parents so that a large-leaved elm may be seen to throw out suckers. There has been some confusion about the name of this tree since the familiar appearance of Dutch elm disease in the vernacular, and yet there is no particular connection; the tree is so named because it was introduced from Holland in the seventeenth century. Huntingdon elms (*Ulmus* x *hollandica var. vegeta*) are either a sub-species of Dutch elm or a different hybrid outcome from wych and smooth-leaved elm. The Wheatley elm and the Jersey elm are both varieties of the Cornish elm, and have a distinctive conical shape, which sets them apart from other elms.

Of the elms found in the parts of Britain other than those dominated by English or wych elm, the vast majority will be hybrids of some shape or form, often displaying characteristics of two or more varieties. These will reproduce through their suckering roots to form identical clones. A specific or localised hybrid in a remote location may well indicate the activity of some plantsman in the past, who took a favoured form and brought it on for his own purposes.

Anyone encountering elm trees and puzzling over the identity may look for two distinctive features in the leaves which will indicate some form of elm rather than hazel or hornbeam, with which they are sometimes confused. Elm leaves display bilateral asymmetry, in other words the two halves of the leaf, on either side of the midrib, are of different shapes; this is most obviously seen in the stagger where the leaf blade meets the stem. The teeth around the edge of the leaf are also made distinctive by having teeth themselves, creating a double-toothed effect.

The Value of the Elm

The practical uses of elm have been manifold. For thousands of years elm foliage has been a staple fodder for livestock, and has been recorded as being eaten by human beings. In the past,

ABOVE: *The distinctive, if not slightly spooky, bark pattern of an East Anglian elm at Whaddon, Cambridgeshire. As with many East Anglian elms, it is specific to an extremely localised area.*

LEFT AND ABOVE: *Huntingdon elms in a hedgerow near Oundle in Northamptonshire, displaying a remarkable similarity of form to the Huntingdon elm at Westonbirt arboretum, in the 1920s.*

In Richard Jefferies's fable, Wood Magic, *a squirrel warns of the elm's malicious intent:*

'Now, the reason the Elms are so dangerous is because they will wait so long till somebody passes. Trees can do a great deal, I can tell you: why, I have known a tree, when it could not drop a bough, fall down altogether when there was not a breath of wind nor any lightning, just to kill a cow or a sheep out of sheer bad temper.'

ABOVE: *An uncommon sight in an East Anglian hedgerow is this old elm pollard bursting into life once again.*

BELOW: *An old coppice stool of wych elm clings to steep crags above Upper Swaledale in North Yorkshire.*

elms would have been regularly pollarded to provide both this fodder and a steady supply of poles, yet now such managed trees are a rare sight.

The timber had all manner of uses, many now forgotten or obsolete. The yew is most often associated with the making of bows, but elm was also used in great quantity, although often considered slightly inferior to the best European yew. The great durability of elm and its resistance to splitting made it ideal for the hubs of cartwheels and the boards for the beds of wagons. Boards of a slightly more doom-laden nature were the common material of coffins from at least the Middle Ages onwards, but while elm served the needs of ordinary folk at their demise, the last two or three hundred years have seen the gentry favouring oak, and in recent years a few lumps of deal have supplanted elm. Many a timber-framed house was built using large pieces of elm timber, particularly when oak was in short supply. However, the structure of the wood is seldom as impervious to the furniture beetle as oak, which often appears as badly infested as elm yet remains sound at the centre. Elm was an ideal timber for turning, being particularly suitable for large bowls and platters; the splendid large burrs frequently found along elm trunks form delightful figuring. Some turners specialised in turning nests of elm bowls from one single block of wood, an art that was all but lost until a couple of craftsmen recently revived the skill.

Turning with massive long augers to hollow out water pipes from large elm boughs and trunks was a significant industry during the seventeenth and eighteenth centuries: one of elm's most valuable properties is its ability to resist rotting when in contact with water. Great networks of elm pipes formed many an early municipal water supply and drainage system. Elm also found itself deployed for underwater piles for quays and bridges as well as lock gates and water-wheels of mills, and elm boards were incorporated in the construction of much commercial shipping built in the southern counties. The most common timber for the seats of Windsor-style chairs was elm, also adzed into shape by a worker known as a bottomer.

Elm names survive in most parts of lowland Britain, indicating the tree's once widespread distribution, often in parts where it is now seldom seen. There are numerous variations on the theme, from the rather obviously named village of Elm, near Wisbech in Cambridgeshire, through Elmscott in the western-most reaches of Devon to Elmton in Derbyshire, and others such as Elmley, Elmhurst, Elmstone and Elmstead. From the fourteenth century onwards the records show increasing evidence of elms in Greater London and, most particularly from the seventeenth century they were highly fashionable trees for formal avenues and the adornment of pleasure gardens. One of the best-known tree place names in London is Nine Elms at Battersea, which was first mentioned during the seventeenth century; the elms have long since vanished.

The seventeenth century saw the meteoric rise of amenity planting of elms, most notably in both civic and private gardens by Charles Bridgeman, who was responsible for many of the grandest elm designs, such as Kensington Palace Gardens. An introduction from France during this time was the mall, a long narrow pitch or pathway lined with elms; the ball game of pall mall was traditionally played on this. The first of these designs in London was just north of St. James's Park, still known today as Pall Mall. Although the game was played in this country, most of these malls became promenades, where the fashionable went to walk, talk and be seen. By the mid-eighteenth century such formal and geometric designs had become an anathema, detested for their Frenchness. The new breed of landscape designers, including William Kent, 'Capability' Brown and Humphry Repton, preferred the naturalistic style. Croome Park, in Worcestershire, was one of Brown's most splendid designs using English elm, but, alas, although late nineteenth century photographs of the estate show the trees at their very finest, all have since succumbed.

Dutch Elm Disease

Ever since the most widely known tree malady in Britain was first identified in France as long ago as 1818, it has taken its toll of elms in waves of varying severity. The fungus *Ceratocystis ulmi* is the villain of the piece, smuggled into the tree by two bark beetles, *Scolytus scolytus* and *Scolytus multistriatus*. Beetle larvae tunnel through the sap, infecting the vessels by which the sap carries nutrients throughout the tree. Constrictions build up blocking the flow of sap and, gradually the elm begins to show yellowing clumps of wilting foliage in mid-summer, which is a certain precursor to death.

Insecticide sprays or fungicide injections can foil the disease, but they are expensive, time-consuming and not always effective. It is very difficult to attain total coverage when spraying large, mature elms, and the destruction of invading beetles is largely a matter of luck. The one location that pursued an aggressive policy towards the disease in the early 1970s was East Sussex, because it is virtually isolated from infected trees in the west of England by the south downs. There the local council felled and burnt badly diseased English elms while treating its healthy stock. As a result it is now virtually the only place where one can go to view splendid mature English elms.

All elms have been struck to some degree by the disease, but they will eventually bounce back. Certain elm strains will develop resistance to attacks, the fungus itself may be predated upon by viruses, the bark beetles may decline: such evolution is an infinitely cyclic affair.

Women he liked, did shovel-bearded Bob,
Old Farmer Hayward of the Heath, but he
Loved horses. He himself was like a cob,
And leather-coloured. Also he loved a tree.
For the life in them he loved most living things,
But a tree chiefly. All along the lane
He planted elms where now the stormcock sings
That travellers hear from the slow-climbing train.
Years since, Bob Hayward died, and now
None passes there because the mist and the rain
Out of the elms have turned the lane to slough
And gloom, the name alone survives, Bob's Lane.
Edward Thomas, Women He Liked

Woe Elm, at Bagots Hill.

View at the Entrance into Vaux Hall.

ABOVE: *Patterns left by the tunnelling larvae of bark beetles on the carcass of an elm killed by Dutch elm disease. The eggs are laid in the central channel and, upon hatching, the offspring burrow forth in this typical radial pattern taking the lethal fungus with them.*

TOP LEFT: *Wood engraving of 'The Wych Elm at Bagots Mill' by J. G. Strutt.*

LEFT: *A late-eighteenth-century wood engraving of the Pleasure Gardens at Vaux Hall in London shows a typical promenade of the period lined with elms.*

Elms were once a popular choice for avenues, the common lime proving to be one of the best candidates. Where avenues were laid out in broader tracts of land, spreading trees such as horse chestnuts and beeches were preferred. Oaks were also a most traditional option, but took much longer to grow to great size.

BOTTOM LEFT: *Felling a 380-year-old elm on the Hursley Estate, near Eastleigh in Hampshire, in the 1950s.*

BELOW: *A large wych elm on the woodland floor throws forth new growth.*

Field Maple

ABOVE: *Field maple foliage and fruits issue from the mouth of a late thirteenth-century carving of a green man at Southwell Minster, Nottinghamshire.*

BELOW: *The maple in Boldre churchyard in the New Forest, sketched by Strutt, and considered at the time to be one of the largest field maples in England, at 45 feet high and 7 feet in girth. Today's record specimens virtually double these dimensions. Although William Gilpin remarked that it was 'seldom seen employed in any nobler service, than in filling up its part in a hedge, in company with thorns, and briars, and other ditch trumpery,' he still thought this tree a fine enough specimen to make his last resting place beneath it.*

S mall-leaved, common, or field maple (*Acer campestre*) is one of those trees which may be abundant and yet remain undetected in the landscape. A tree equally at home in woodland or hedgerow, it is seldom seen as a large mature specimen because it is either limited by space and light in woodland or cut and laid by man in hedgerows. Given the right conditions, however, field maples can achieve 80 or 90 feet in height. The new palmate leaves often take on a delightful reddish hue when found in hedgerows, as do the double-winged seed 'keys', but they really come into their own in autumn when they glow with one of the most spectacular of all golden displays of foliage.

As a hedgerow species field maple is generally considered to be a good indicator of significant vintage, since it has seldom been planted in the last 200 years. The inference is that it contributes to hedgerows either as a species from the edge of cleared woodland which has become hedge, or that it was planted in earlier periods, before the advent of nurseries, when wild saplings were extracted from nearby woodland to construct new hedges.

In the past field maple was used in many garden schemes, as it takes very well to regular clipping much in the way of beech or hornbeam. There are even accounts of its being used for topiary.

The old English name for the tree is *Mapuldur* or *Mapulder,* which may be found in place names such as Mapledurham and Mapledurwell in Hampshire, while Maplebeck in Nottinghamshire, Maplestead in Essex and Maplehurst in Sussex also indicate maple connections.

The wood is best known for its turning properties, producing bowls and goblets with some of the most splendid figuring and colouring. Maple was one of the traditional woods used for turning the masers, or mazers, the ceremonial communal drinking bowls associated with rituals such as wassailing. The tree was sometimes known as the maser-tree. The maple roots with their marbled and swirling patterns were popular for making snuff boxes and pipes. The wood was also used for the making of harps as well as the inlays for other musical instruments and furniture. The wood obtained from behind flushes of epicormic shoots is known amongst wood workers as 'bird's eye maple' for the swirling grain peppered with many tiny knots or 'eyes'.

The larger knotty patterns of the veining in the timber were known as 'brusca' and 'mollusca', and wood which fortuitously displayed the semblance of some strange or mythical beast was the most prized for tables. The Romans are reputed to have valued such fine tables more than their weight in gold. As a result of this, the story is told that affluent Romans reproaching their wives for profligate purchases of jewels and finery often risked having them 'turn the tables' upon them for their own expensive tastes in furniture.

Carved images of the maple often decorate church architecture, most notably among the virtual forest of foliage at Southwell Minster in Nottinghamshire and accompanying the sycamore, and several other plants, on the tomb of St. Frideswide in Oxford.

Gilpin had little to say of the field maple's picturesque qualities, but perhaps he gave it his own subdued compliment by choosing to be buried beneath a splendid specimen, in Boldre churchyard, upon his death in 1804.

Maple at Boldre

LEFT: *Field maple in winter in the Cotswolds.*

ABOVE: *The nobbly bole of an old pollard field maple in a Welsh hedgerow.*

LEFT: *A golden cascade of autumnal field maple leaves tumbles from a hedgerow.*

Blackthorn

Youth – the vision of a morn,
That flees the coming day:
It is the blossom of a thorn,
Which rude winds sweep away.
Cunningham

ABOVE: *Blackthorn blossom, which appears in April before the leaves emerge.*

BELOW AND ABOVE RIGHT: *A rare but spectacular sight is this overgrown hedgerow dominated by blackthorns which have been allowed to develop into substantial trees up to 30 feet high. Normally the tree is a shrubby constituent of many a mixed hedgerow.*

In much the same fashion as other small and shrubby trees of the hedgerow, blackthorn (*Prunus spinosa*) is frequently present but seldom noticed. It claims the most attention in late March or early April when it is one of the first trees to swell the hedges with dense clouds of pure white blossom. A widespread tradition associates the flowering of the blackthorn with the cold snap which often occurs at this time, usually after a false start to the spring. Gilbert White recognised the phenomenon, remarking, 'This tree usually blossoms while cold North-East winds blow; so that the harsh rugged weather obtaining at this season is called by the country people, Blackthorn Winter'. Indeed, not only are such times bitter and chill, but the blossom often takes on the appearance of great snowdrifts.

Thorn is one of the most common place names across Britain, though it is seldom clear whether it refers to the blackthorn or the hawthorn, alternatively known as whitethorn. Blackthorn takes its name from the dark colour of its stems, which do appear exceptionally black when contrasted with the white blossom, before any leaf has sprung from the bough. The English poet Geoffrey Grigson quotes a fair selection of local names. Bullen (Shropshire), Bullison (Wiltshire) and Bullister (Cumberland, Scotland and Ireland) are all influenced (or confused) by the bullace tree, another name for the wild plum (*Prunus*

domestica), which bears a slightly larger fruit, more like a damson. The blackthorn is widely considered to be one of the original parents, along with the cherry plum (*Prunus cerasifera*), of the rich and varied selection of domestic plums available today.

At the back end of the year, the blackthorn is known for its small and bitter plum fruits, or sloes, valued by many a countryman for the splendid liqueur sloe gin. The fruits, best gathered after the first frosts of late autumn, are reflected in local names for them such as Hedge-picks (Hampshire) and Winter Picks (Sussex).

Blackthorn seldom grows higher than about 30 feet. Most of its energy is directed into spreading outwards, for it throws up prodigious quantities of suckering shoots from its roots. A hedgerow, or even a solitary tree, left to its own devices can soon become a thicket, which in turn may act as a nurse bed for other tree species to develop and create new woodland. Hence blackthorn's French name of Mère-du-bois (Mother of the Wood). For this reason pure blackthorn hedges are fairly uncommon, but it has been a constituent species of many mixed hedgerows for hundreds of years. In a few areas, typically Shropshire, blackthorn was planted, but more often it has occurred as a particularly successful colonising hedgerow plant, largely, one assumes, as a result of bird and small mammal activity.

Although it is exceptionally hard, the wood of the blackthorn seldom reaches any profitable magnitude save for the making of fine walking sticks and, in Ireland, the traditional shillelagh – a durable cudgel and 'a famous means for strengthening an argument'. This implement, which Robert Graves suggests was more often made of oak, is most often a tourist memento these days.

In times past blackthorn leaves were used to adulterate tea, and the sloes were used either as a dye for wine or to create sham vintage port. A yellow dye may be derived from the bark, and both the fruits and the bark have been employed in the making of writing inks.

The spines of blackthorn are long and sharp. If a thorn penetrates human flesh and lodges there, it is the most painful of all wounds. Thiselton Dyer records the following verse chanted to prevent such pricks from festering:

Christ was of a virgin born,
And he was pricked with a thorn,

And it did neither bell nor swell,
And I trust in Jesus this never will.

The same thorns make the tree an excellent nesting site for many small birds. One quite rare bird, the redbacked shrike, uses the thorns as a larder by impaling surplus prey of reptiles, amphibians and insects upon them.

Right: *Rampant suckering of blackthorn causes it to tumble forth from an unmanaged hedgerow. This trait makes it an unpopular hedge tree with many landowners, for the sharp spikes make it hazardous to livestock and vehicle tyres.*

Left: *Sloe Gin: The sloes are pricked, put in a bottle with a few spoons of sugar, and the bottle filled with gin. The mixture is then left to mature. After several months (years are even better) a wonderful deep purple liqueur is ready for sampling.*

Hawthorn

One of the principal species of British hedgerows, the hawthorn has been dominant in mile upon mile of hedges, particularly in the Midland counties, since the numerous Parliamentary Enclosures, between 1750 and 1850, when it was common practice to plant purely with hawthorn. Rackham has estimated that as much as 200,000 miles of hedges were planted during that period. Long before this, however, the tree had been revered for its magical associations, such as those of May Day. It has been regarded as protector from all matters evil and, as an amulet of love, most especially in England and France from as long ago as the first-century BC.

Hawthorn has, like other trees with a wealth of folk tradition, a variety of colourful local names. Geoffrey Grigson gathered a goodly clutch in his *Englishman's Flora*. The Old English names for the tree were hagathorn or haegthorn, whilst the most commonly used alternative names have been quickthorn, whitethorn or May. Scottish horticultural writer J. C. Loudon explained: 'Booth derives the word Haw from hage, or haeg (oe), a hedge; consequently he makes hawthorn signify hedgethorn. Quick signifies live [see also the rowan – quickbeam]; and was, probably, derived from live hedges made of hawthorn being used instead of fences of dead branches of trees'. Whitethorn is named after the profusion of its white flowers and its thorns, or possibly from its white bark, as compared to the bark of the blackthorn.

Many of Grigson's regional names involve may, hag or haw. One evocative name, from Somerset, is Moon-flower, perhaps referring to the custom of plucking sprays of May blossom on May Eve. The rather quaint name of Bread-and-cheese tree has arisen from country children's habit of nibbling the fresh green leaves of spring. Hay, Haw or Hag in a name usually signifies a hawthorn, as in Hay-on-Wye, Hawarden, Hagley and, most obvious of all, Hawthorn itself, in County Durham

Thorn is the most common tree-name constituent of English place names. There are numerous Thornhills and Thorntons as well as Thornbury, Thorncombe, Thornham and plain old Thorns in Suffolk or Thorne, near Doncaster.

Hawthorn has seldom been noted for its great size or longevity and yet it has the potential to achieve both. Generally the tree has been coppiced or pollarded in woodland, if not trimmed and laid in hedgerows, but in some places, such as Hatfield Forest, old pollards have achieved heights of around 50 feet and boles with diameters of three to four feet. One of the reputedly oldest hawthorns in England is at Hethel in Norfolk: Elwes and Henry included information and a fine photograph of the tree from 1912, in their great seven-volume work (published 1906–1913). J. Edward Milner recalls the popular legend of how the tree was planted in the reign of King John, which would make it more than 700 years old, and, indeed, there is a reference to the tree in a thirteenth-century charter. It is currently 'protected as one of the smallest Sites of Special Scientific Interest in the country'.

Oliver Rackham's diligent research through Anglo-Saxon charters has revealed that the hawthorn was by far the most frequently mentioned named tree in the texts of the perambulations. This might have been because there were many large old trees, or even because there were simply plenty of them, to mark the various points of the boundaries. Another explanation may belong to the centuries-old tradition of 'beating the bounds', typically around such tracts of land as described in the charter perambulations. The custom has been closely linked to Rogationtide, which generally falls upon the third week in May. At this time the May blossom would be in full spate, making the trees stand out as particularly effective landmarks and thus easily defined by the perambulating throng.

ABOVE: *Great tufts of sheep's wool have been combed and caught on these tiny hawthorns along a mountain path in the Black Mountains of Wales. These 'trees' have been so severely cropped by the sheep that, although they are probably quite old, they have barely achieved bonsai proportions.*

OPPOSITE: *A splendid example of a hawthorn in blossom.*

PREVIOUS PAGE: *Wind-blown hawthorn erupts from the limestone pavement above Malham, North Yorkshire.*

BELOW: *The May bride and her bridesmaid slip out of the woodland, for Woodland hawthorns they are; the difference is subtle, and not so sweet of scent as her cousin.*

TOP: *Aged, wind-tossed, but still alive and blooming, a horizontal hawthorn above Dovedale in Derbyshire.*

ABOVE: *Before Christmas each year small sprays of the Glastonbury thorn are cut and sent to the Queen and the Queen Mother, presumably as goodwill tokens, reminiscent of all the snippings taken by pilgrims in times past.*

OPPOSITE: *The spiky winter profile of a Dartmoor hawthorn.*

PREVIOUS PAGE: *Hawthorn in autumn, laden with bright red berries which are highly attractive to birds, on the Stretton Hills in Shropshire.*

RIGHT: *The writhing knotted bole of an ancient hawthorn in a hilltop location where the earth has been blown away from the base of the tree. Hawthorn has great tenacity and will hang on long after other trees have succumbed to the elements.*

The thorns have long been held to represent the crown of thorns that Christ was suffered to wear on the cross, and the berries the drops of blood that he shed, a belief that underlies one of the best-known folk legends concerning the hawthorn.

In an early sixteenth-century poem about the life of Joseph of Arimathea arose the first mention of three thorn trees upon Weary-All Hill, near Glastonbury,

> *Thre hawthornes also, that groweth in Werale,*
> *Do burge and here grene leaves at Christmas*
> *As freshe as other in May.*

The miraculous bloom gave rise to the myth that these trees were born of the same lineage as the crown of thorns. There matters rested for more than a century before the legend was resurrected once more with a written account of a visit made by Joseph of Arimathea to Glastonbury during an evangelical mission between

AD 30 and 63. Tiring from his journey (Weary-All) and pausing, he thrust his staff into the ground, whereupon it took root and flourished as the first Christmas-flowering thorn. The rest, as they say, is history – though, in reality, this tree might have been brought from the Middle East, since it has long been known that certain hawthorns out there do readily bloom in the winter.

Those with a more cynical viewpoint might note how Christian legend deflects attention from the pagan associations of hawthorn celebration of the winter solstice as well as the orgiastic May Day revels to come later. Yet the Christians' own marvellous tree almost presaged its own demise. Pilgrims came from far and wide to see the thorn during the sixteenth and seventeenth centuries, hacking pieces from it as souvenirs, amulets and cuttings to replant. The tree survived this, but not the axes of the Puritan army, who were firmly set against any forms of idolatry. Fortunately, by this time, enough cuttings had been struck to continue the line in various corners of the realm.

Crataegus monogyna 'Biflora' still grows in many parts of Britain, although it seldom blooms in winter in the more northerly counties. The present incumbent thorn in St. John's churchyard, Glastonbury, is still fairly young, but the cycle remains unbroken. The tree flowers or sometimes just throws a few leaves with the vagaries of the British weather, any time from November to March.

There are other members of the *Crataegus* clan worthy of note in the British landscape, but certainly none as abundant as *monogyna*. Midland hawthorn (*Crataegus laevigata*) which was until recently, known to many as *Crataegus oxyacantha* presents a similar tree to the common hawthorn. Closer observation, however, reveals lobed leaves with much shallower indentations, twin stigmas in the flowers and twin seeds in the berry. As its name suggests, it is often to be found as part of the woodland understorey, although the pure form is scarce, as (yet more confusion) it is prone to hybridise. 'Paul's Scarlet' is a splendid cultivar of Midland hawthorn, bearing deep pink double flowers, and often seen as an effective street tree.

One of the American thorns which seems to be increasingly planted in urban schemes is the Cockspur thorn (*Crataegus crus-galli*). The tree is about the same size as the British thorns, but the leaves are a completely different shape and give a spectacular autumn show. The stems bear great spines which may be three inches long, and the clusters of large red berries hang on the tree all through winter. This last trait seems strange, and it must be assumed that the berries are not attractive to birds. British haws, on the other hand, are soon greedily swooped upon in the cold lean days of winter.

OTHER COMMON HEDGEROW SPECIES

Hedgerows play host to a rich array of trees. Naturally enough, the greatest diversity of species will be found in the oldest hedges, in landscapes that have seldom been manipulated by man. By contrast, hedges of planned countryside will have been mainly planted over the last 200 years and will generally consist of hawthorn plus a few other species. Large timber trees such as ashes or oaks are the most obvious, and because of its great longevity, oak has been a boundary marker in many hedgerows. Some timber trees were planted with maturity in view while others are saplings which, through neglect or inaccessibility, have been allowed to rise above the hedgerows and become large trees.

Ash has proved to be a timely replacement for the elm in many regions, so that landscapes where hedgerows abounded with skeletal elms some 30 years ago and subsequently became denuded of substantial trees, now look considerably healthier. Ash is a great coloniser, producing copious quantities of seed, which blow into hedges and germinate freely.

Ashes, as well as the oaks which spread above many a hedgerow, were often managed by pollarding to protect them from the animals of the fields and to generate their useful harvest of poles. Pollarding such trees has almost ceased today, though this management in earlier times has secured many an elder statesman's survival in the hedgerow. Rackham asserts that ancient pollard trees in hedgerows are only found in any significant numbers in eastern England. However, a tour of the Welsh border counties soon reveals plentiful pollards along many a hedgerow, and some of these have come under renewed lopping regimes.

Willows and poplars are common constituents of hedgerows, especially where water courses flow alongside. Crack and white willows, because of their tendency to split and tumble on to fields or thoroughfares, are usually pollarded on a regular basis. Tightly packed rows of black poplar clones are ideal for horticultural windbreaks, while the occasional chance discovery of long lines of Lombardy poplars in hedges or avenues can suggest a French landscape.

One of the favourite haunts of the scarce native black poplar appears to be in old hedgerows. Once most of these trees would have been pollarded, and it is cheering to note that regimes have been renewed upon a few specimens; but in recent times many old outgrown pollards have split and broken down, since few people had a use for the wood.

Of the larger non-native trees which appear in hedgerows, the sycamore must be the most predominant, for it finds the protective haven to its liking. In a world which frowns upon this hugely successful interloper when it invades woodland or gardens, the hedgerow is one habitat where it may thrive with impunity. Even if its keys fall in the adjoining fields the seedlings will soon be cropped by livestock, and verge trimming and hedge cutting will also control

BELOW: *Although most of the surrounding hedgerow has disappeared, this old pollard ash, once planted for its useful poles and firewood, is still in good heart in the midst of the Black Mountains.*

BOTTOM: *Typical of most Devon hedges, the beeches grow on an earth bank and the horizontal branches bear witness to laying a long time ago.*

ABOVE: *The fruits of the horse chestnut, conkers, have been avidly collected by children for generations. The game may hark back to a similar one using hazelnuts during the seventeenth century.*

RIGHT: *Unruly Dog rose and honeysuckle scramble across an ancient hedgerow, their honey-sweet scent an intoxicating pleasure on a still June evening.*

OPPOSITE: *The red horse chestnut, a dark pink-flowered version of the tree, seldom achieves the massive size of this specimen.* Aesculus x carnea *is in fact a cross between a horse chestnut and the red buckeye* (Aesculus pavia) *from North America.*

PREVIOUS PAGE: *An ancient hornbeam hedgerow survives on the remnants of a medieval woodbank.*

BELOW: *The sticky buds of the horse chestnut, containing downy soft emerald leaves, are a children's delight and among the first to burst into leaf in spring.*

RIGHT: *A massive old hazel coppice stool on the site of a long-redundant hedgerow.*
FAR RIGHT: *A huge and unusually heavily blossom-clad wild service tree in a hedgerow, where it has most probably survived as a remnant of an ancient woodland.*

its upstart saplings. It is interesting to study a single hedgerow with numerous neighbouring sycamores and observe the radical differences between the colours of the foliage.

Horse chestnut (*Aesculus hippocastanum*), since its introduction from Turkey in the late sixteenth century, has been essentially a tree of plantsmen. It is fair to say that it is now naturalised and frequently insinuates itself into any suitable plot of earth, including hedgerows. The timber is of little use and the tree is seldom held in high regard other than for its amenity value, but few trees can better it in this respect.

The origins of the name horse chestnut may lie in its native Turkey, where the chestnuts were frequently fed to horses as food or as a cure for coughs. Alternatively, 'horse' may signify an inferior but similar fruit to the sweet or Spanish chestnut, much in the same manner as horse radish or even cow parsley. Also, when the leaves break away from the twigs in autumn, they leave horseshoe-shaped scars.

An impressive avenue of horse chestnuts borders the mile-long drive through Bushey Park, leading to Hampton Court Palace. Planted in 1699 by Sir Christopher Wren, many of the veteran trees have survived not only a whirlwind in 1908, but also the horrendous gales of 1987. From the mid-nineteenth century – Chestnut Sunday, usually the second Sunday in May, when the flowering trees are at their zenith – has seen a congenial gathering of folk promenading and picnicking beneath the chestnuts in celebration of this natural wonder.

Several trees end up as hedgerow species by default. Most commonly these are woodland trees which have survived as ghosts of woods grubbed up long ago so that strangers such as small-leaved lime, wild cherry and wild service tree will appear in an apparently alien situation. All manner of fruit trees and walnuts may

turn up either as edge remnants of old orchards, indications of some long-forgotten homestead, or as the sites where bird, beast or man once enjoyed a wholesome snack before tossing the fertile seed carelessly into the depths of the hedge.

Many of the smaller shrubby trees of hedgerows have colonised by means of fruits attractive to birds, so it is little wonder that elder, holly, spindle, dogwood, guelder rose, wayfaring tree and buckthorn abound in many old hedgerows. All of these, apart from elder, are species which need the space and light in order to fare well, so they do not find many opportunities in the dense hawthorn rows of planned countryside.

Holly has often been a planted species in many older hedgerows of the Welsh borders, where runaway trees which have overtopped their hedge are a common sight, wonderful berry-clad beacons in a wintry landscape. With their dense structure, other evergreens such as yew and box are ideal for hedge making, but inevitably these are used in the domestic setting. Ancient individual yews sometimes occur in old hedges or along old wood banks where they were undoubtedly boundary markers. Scots pines are only found in any significant quantity in the old Breckland hedges of the nineteenth century.

CONSERVATION AND THE FUTURE OF HEDGES

ABOVE: *Both dogwood, seen here in flower, and spindle, noted for its startling bright pink fruits, are small shrubby trees most commonly found in ancient hedgerows.*

BELOW: *The grandest of all lime avenues is the three-mile-long double avenue at Clumber Park in Nottinghamshire.*

Widespread concern over loss of hedgerows has grown dramatically over the last 30 years, often resulting in acrimonious encounters between landowners, farmers and the public. The two most contentious hedgerow issues are: destruction vs. conservation; and, if the latter process wins out, which management regime is most appropriate to maintain both an effective agricultural barrier and the best possible wildlife habitat, within a variety of specialised landscapes.

An all too frequent perception is that 'the public' is a well-meaning but ill-advised body with next to no understanding of rural ways and farming practice, while the landowners are set up as the villains of the piece: the ideologists versus the pragmatists. Only when both parties can overcome these preconceptions of one another can there be calm and constructive dialogue. Those fortunate enough to own hedges might claim to have their destiny in their own hands by right, but it must not be forgotten that today's generation merely holds the land in trust for those who follow and that everyone in a community must live with or, in the tragic aftermath of wilful destruction, without hedges in their landscape.

English Nature and the Council for the Protection of Rural England (CPRE) have long been promoting conservation of hedgerows. English Nature has recently coordinated the establishment of the Veteran Trees Initiative, which will help to conserve the ancient trees found along British hedgerows, one of the most important habitats for such veterans. CPRE has led a campaign to promote the retention of hedgerow saplings and thus increase the potential for future hedgerow trees; otherwise, as there is no longer a policy of hedgerow timber tree planting, the present mature trees will eventually die, leaving a landscape bereft of replacements.

Attention has also focused upon urban hedgerows, which are constantly under threat from developers. These hedges are often homes to rare or endangered species, most notably the Plymouth pear which survives (just) amidst the urban sprawl of Plymouth. Trees and hedges of churchyards have recently undergone a national survey by the Living Churchyard Project, based at Stoneleigh in Warwickshire, in order to gain a better knowledge of these often neglected sites.

As with any other type of environment, conservation may be applied to hedgerows in innumerable shades or levels of management intervention. Any hedge is a likely candidate, and various criteria must be considered before any action is taken.

The Ministry of Agriculture, Fisheries and Food (MAFF), with its new and excellent formulation of the Hedgerows Regulations, passed into legislation in June 1997, has at long last provided a framework upon which all interested parties can work towards the progressive future of British hedgerows. These regulations take into consideration archaeological and sociological relevance, wildlife habitat importance, and inherent tree and shrub composition. The new regulations mean that landowners cannot rip hedges out without first seeking permission. After weighing an individual request against the governing criteria, MAFF will then either permit the removal of the hedge or issue a hedgerow retention notice. (Full details of this are available from MAFF.) While the hope is that most people will see this as a positive and impartial move, there will inevitably be some who will view it as a deprivation of liberty. The bottom line is that it has become illegal to destroy hedges, and perpetrators will be heavily fined if caught and convicted.

To encourage responsible stewardship by landowners, Agriculture and Development Advisory Service and Farming and Wildlife Advisory Groups are available to advise on grants and aspects of management, while some local authorities and groups such as the British Trust for Conservation

Volunteers may often be able to give practical assistance. Although most hedges will require at least some degree of management, both to keep their innate ecology in shape and to provide effective field boundaries, there is also a case for non-intervention when old hedgerows or hedgebanks no longer have a practical purpose and where the evolution of vegetable and animal species has turned the whole hedge into a semi-wild habitat.

There are two principal methods of maintaining hedgerows. The traditional way has always been the laying of hedges, in which saplings are partially cut and then laid along and woven into the hedge. The modern way, implemented increasingly over the last 40 or 50 years, has been to trim with tractor-mounted flails. The automated method works well when carried out on a regular basis and with hedges that have already been well maintained by hand: if they have been previously laid, their gaps filled and larger trees lopped, a general overhaul every five to ten years will maintain

ABOVE: *These beeches on the Blorenge, in southeast Wales, were once a hedge, but after many years of neglect, when their purpose as a barrier became redundant, they have grown into substantial trees which have bent in the teeth of many a gale.*

LEFT: *Walnuts are not common trees of the hedgerow, and when they do appear it is usually in close proximity to habitation, where they were planted for their excellent nuts. This specimen was discovered right out in open countryside, so it must be assumed that a bird was most likely responsible for its existence. From its shape it is clear that, as a young tree, it was laid into the rest of the hedgerow.*

NEXT PAGE: *A fine avenue of horse chestnuts at Bushey Park near Hampton Court in Middlesex.*

a sound hedgerow which will respond well to a flail trim every other year. Where possible it is recommended that hedges should be cut in an 'A' shape as this makes them stronger and provides better wildlife cover. When the ground work has not been done and landowners have let their hedges lose condition, so they become overgrown and woody, the flails simply rip everything to pieces, leaving a battle-scarred devastation which is an all too familiar sight. The hedge will usually recover, but in the meantime all cover has been lost for birds and mammals, the smashed and the splintered trees and shrubs are more vulnerable to disease. When hedges were maintained by hand, 'bad hedge' species such as elder were cut out. Now elder prevails and hedges are no longer stock-proof.

One or two other aspects of progress have also marred the vitality of hedges. Spray drift from herbicides used on arable crops can reduce a hedge to a brown corpse. The increasing quantities of salt spread on roads in winter are also detrimental to many trees and the ground flora associated with hedges. Where the salty water is washed into hedge bottoms by passing traffic it doesn't take very long to build up to toxic levels. Whenever utility companies cut trenches, the root systems, particularly of large hedgerow trees, are in jeopardy; this is also a serious problem for many urban-street trees. Although there are penalties for damaging trees in this way, the policing of this problem is a nightmare: when the aim is to lay services as quickly and cheaply as possible, there is little incentive for operatives to look after the trees in their path.

At the other end of the hedgerow maintenance issue is an altogether brighter perspective. The traditional craft of hedge laying is alive and well and even having something of a revival in recent years. A few of the older hedgers and ditchers still quietly pursue their craft in various corners of Britain; it is often a long-standing family affair, a skill to be perfected and passed down through the generations. All over Britain, there were once competitive hedging matches, as keenly contested as any ploughing match or livestock show. Some have been rekindled in recent times, largely due to the increased interest in the craft. Agricultural colleges and conservation groups have begun to retrain hedge layers, and old experts have begun to employ apprentices once again. None of this revival would have any foundation without a realistic commercial demand for the craft and, fortunately, a fair number of landowners have begun to realise the advantages of expertly laid hedges.

There are various subtle regional differences in the way hedges are laid, but the principal stages of the work have a common thread countrywide. The usual period for hedge laying is between October and March, when the sap is no longer rising and the lack of foliage makes the job easier. First the climbing and straggling plants and weeds are cleared away. The large hedgerow trees are lopped for any decent timber. The younger trees and shrubs are cut back to a few of the stronger, straighter leaders, which will be left at least as high again as the hedge, while some of the slightly stouter stems will be cut to hedge height and left at intervals of 12 to 24 inches. These latter uprights form the natural stakes around which the hedge will be woven: in young hedges stakes are introduced and driven into the ground at regular intervals. The tall stems are then partially cut through, so that they remain part of the living tree yet can still be bent over and woven between the stakes. These are the plashers, and the weaving process is known as plashing, or pleaching. To stop the plashers from springing back up, a braided rail of thin rods, known as ethering or heathering, is woven along the top of the stakes; hazel is most often used. Such hedgerow maintenance is an expensive option for a landowner, but the long-term benefits surely make it a worthwhile investment.

CHAPTER FIVE

Wild Fruit and Orchards

Although there is fragmentary evidence from various prehistoric sites in Britain that the native peoples consumed wild forms of various tree fruits such as sloe, plum, cherry and apple, it seems that they did so as hunter-gatherers. The knowledge of how to hybridise varieties in order to produce more palatable fruits was yet to spread from Central Asia via the Greek and Roman Empires.

The Roman occupation of Britain left in its wake radical developments in agricultural methods. The wild fruit trees of Britain have emphatically influenced the evolution of domestic fruit trees and the orchard. To the wild strains of native apples, pears and cherries, the Romans brought the benefits of their fruit cultivation. After more than three centuries the legacy of their orcharding expertise would form the basis of Britain's own fruit culture. Today our native fruit trees are a significant presence in the countryside woodlands and hedgerows, and across the upland valleys.

Crab Apple

ABOVE: *Crab apples.*

RIGHT: *A roadside crab apple tree of typical form and, luckily, high enough to avoid the unwelcome attentions of the hedge trimmers.*

OPPOSITE: *Wild apple tree alongside a trackway through Salcey Forest.*

BELOW: *Crab apples were often purposely planted in hedgerows, particularly around orchards to act as pollinating agents.*

The native crab apple (*Malus sylvestris*) has, without doubt, been the cornerstone of the remarkably diverse array of several thousand domestic apple varieties. Evidence from pre-Roman sites suggests that crab apples were part of the diet, but exactly how is uncertain. The unpalatability of the small, sour fruit would suggest that they may have been used as a flavouring or perhaps combined with honey and fermented into a wine or mead-like drink. Nowadays, the crab has fallen from grace in the common diet, but for those who take the trouble, a rich reward may be obtained in crab apple wine or crab apple jelly, an excellent preserve.

The name crab probably derives from the shape of the tree, which in old age is often hunched and tangled with a mass of spindly, clawlike boughs arched over towards the ground, giving it a crablike awkwardness and appearance. Grigson offers *Scrab* and *Scrogg* as alternative North Country names, and *Gribble* from Dorset and Somerset.

The crab apple is usually found as a tree of hedgerows and woodland, bearing fruit that ranges from golden yellow-green through to deep red with yellow shading. It is often difficult to identify true crabs, because wildings of domestic varieties may recall their genetic roots and revert to the small crabby-type fruit of their ancestors from perhaps centuries earlier. Thus all 'wild' apple trees with small fruits may not necessarily be crab apples.

The crab's familiarity throughout Britain is borne out by numerous references to the tree as a boundary marker and landscape feature in many early charters. There is also an abundance of place names that incorporate references to the apple: the Celtic word for it was *Abhall* or *Abhal*, in Welsh it was *Avall*, in Armoric *Afall* and *Avall*, in Cornish *Aval* and *Avel*. Ancient Glastonbury was called *Ynys Avallae* and *Ynys Avallon* by the Britons, signifying an apple orchard, and *Avallonia* by the Romans. The name 'crab' occurs in the familiar surname Crabtree, which is also a place name, one being in Sussex. Could the village of Scrabster near Thurso, also be derived from the crab apple?

Crab stocks were often used for grafts, particularly in non-commercial cultivation, and a turn around an old orchard can sometimes produce the sight of a domestic variety surmounting a crab apple on a single tree. Often, through neglect, the crab stock has been permitted to throw shoots from below the graft producing a two-tier tree.

Pear

Botanists refer to two kinds of 'wild' pear growing in British woodland and hedgerows. *Pyrus pyraster* is viewed as the true native, and the countryside historian Dr. Rackham has found references to it among landmark and boundary trees in Anglo-Saxon charters. It is far rarer than the crab apple, but like the crab it is often confused with a great array of wildings which also thrive in woods and hedgerows. True wild pear trees tend to have quite spiny branches, and the fruits are particularly small and much rounder than those of wilding pears.

Other wild pears – Richard Mabey calls them feral pears – are given the name *Pyrus communis*. These are possibly crosses between *Pyrus pyraster* and French introductions brought over by the Normans. These latter varieties are almost certainly the pears known today as perry pears, although it is very likely that the Normans also brought their own favourite perry pear trees with them in the eleventh century. Perry pears, used to make the fine alcoholic beverage of the same name, could be seen as the mid-point in the development of the species from the hard and sour little wild fruit to the sweet and juicy dessert pear.

Place names featuring the pear are manifold: Perry Barr, Perivale, Pershore, Pirton, Pyrton and Pyrford are just a few examples. The surname Perry must also derive from one who planted pears.

One member of the *Pyrus* tribe has recently generated a great deal of interest, due principally to its extreme rarity. Plymouth pear (*Pyrus cordata*) is one of Britain's rarest wild trees, being found in a mere handful of old hedgerows around Plymouth in Devon and Truro in Cornwall. Plymouth naturalist T. R. Archer Briggs first noticed the tree in 1865 and was intrigued by the subtle differences between it and the many other wild pears in his local hedgerows. Its later flowering, more often at the same time as the apples rather than as the other pears, made it conspicuous. The denser and spinier nature of the tree along with its own particular fruits, which were barely half an inch long and more elongated than other wild pears, strengthened its case for individuality.

It is highly unlikely that the Plymouth pear is a native species. The tree is, however, native across most of western Europe, and, as there have traditionally been many trade links between northwest France and the West Country, it is generally assumed that at some point the Plymouth pear was introduced. It has been shown to hybridise freely with other pears.

The very limited natural distribution of Plymouth pear is largely due to its self-incompatibility, which results in exceedingly poor sets of fruit and hence minimal amounts of viable seeds. Fortunately botanists studying the two natural colonies of the tree have discovered small genetic variations between them. This has allowed a controlled cross-pollination between the two which has produced fertile seed. Programmes involving the cultivation from seed,

Left: *A large pear tree displays the distinctive bark pattern. Its great size makes it extremely valuable as a timber tree.*

Below: *The flower of the Plymouth pear, one of the rarest wild trees in Britain. Delicate and pretty it may appear, but its scent has been popularly compared to that of 'rotting scampi'. This led some botanists to assume that the tree was attracting flies rather than bees, and yet both insects frequently visit the tree.*

suckers taken from existing trees and micropropagation are all underway. The tree has achieved celebrity status in the last few years: it was the first woody plant to gain protection under Schedule 8 of the Wildlife and Countryside Act of 1981 and it was selected by English Nature as one of the first six species to receive assistance in the Species Recovery Programme launched in 1991. Conservationists in Plymouth have worked hard to protect the existing trees from both industrial developments and over-enthusiastic amateur plant collectors, as well as engendering a sense of pride in the local communities, who now recognise the tree's rarity and botanic value.

THE WORCESTER BLACK PEAR

The Worcester black pear tree's main claim to fame is that its fruit appears on the City of Worcester coat of arms. This dates back to a visit to the city by Elizabeth I. To impress Her Majesty, a heavily laden tree was positioned near the city gateway, and as she entered, this splendid sight was remarked upon. Elizabeth decreed that these fine-looking pears should become part of the city arms. If the black pear has remained true to type, then its fruit has always been a dull green colour. So perhaps the appellation of 'black' has reverted to the tree from the heraldic depiction.

The Worcester black pear is a large, hard and somewhat useless pear now reduced to a handful of trees growing in and around Worcester. It is impossible to eat as a dessert pear and not even much better when cooked. They were sometimes bottled in the past, so perhaps a prolonged marinating improves their flavour.

Wild Cherry

Two native members of the *Prunus* species are cherries: wild cherry, also known as gean or mazzard (*Prunus avium*) and bird cherry (*Prunus padus*). The former is a common tree of the southern half of Britain, particularly in woodlands, while bird cherry is most often found in Wales, the Welsh borders, northwest England and Scotland. A third native *Prunus*, the sloe or blackthorn, occurs most often in hedgerows (see Chapter Four).

Wild cherry resembles the domestic orchard cherries, and indeed there is a long tradition of hybridising with the wild tree and using wild cherry stocks upon which to graft many varieties. The tree demands plenty of sunlight to fare well and will usually produce the finest specimens either on the edges of woodland or in sheltered clearings. Wild cherry is particularly adept at pushing forth numerous suckers by which it may often form quite dense stands within woods. It generally grows modestly when amongst other trees, but has the potential to reach 100 feet.

The fruits are rather small and sour, but can be eaten, if snatched before the bird population makes serious inroads. However, one of their finest uses is to flavour cherry brandy. Cherries, sugar and brandy are left to steep for several months, in much the same fashion as sloe gin, to make a heart-warming liqueur. The wonderful colours of the autumn foliage bring a late delight to the landscape, but wild cherry's real glory is its mass of white blossom billowing across an April scene still largely devoid of colour. To look across a valley and see blossom-clad hedgerows in otherwise brown woodland is a fine sight. Such views should be treasured and not taken for granted, for they will last but a few days of each year.

The bird cherry is a quite different tree to the wild cherry. Its fruit is black and even sharper in taste than that of the wild cherry. Where bird cherry has room to grow unhindered and in a sheltered site, it will form a conical shape up to 36–39 feet high, but where it is exposed it tends to develop into a large, flattened shrub. It is essentially a tree of northern Britain. Bird cherry goes largely unnoticed in the landscape apart from in May when its magnificent displays of white blossom adorn many an upland valley or remote woodland. Most commonly found in the Derbyshire Dales and northwards, it does occur in a few localised sites further south, notably along the Welsh borders and in the unusual isolated enclave in Norfolk. It may have been an introduced species in southern counties; however, it is also possible that it has actually been eradicated from many parts of southern Britain. The tree was thought to harbour a harmful bacterial disease which can attack wheat, thus making it rather unpopular with farmers who would not tolerate the tree in their hedgerows. However, it is now thought that the problem travels from grasses to wheat; the disease does not spread from the bird cherry to wheat fast enough to cause damage.

FRUIT CULTIVATION AND ORCHARDS

The idea of fruit cultivation was first grasped by the inhabitants of Turkestan (Turkey) and the Caucasus region of northern Iran. In these fertile lands, fruit-bearing trees grew in wild abundance and, as the rural populace cleared and managed the woodland, they discovered natural mutations and hybrids of the wild species that had developed fruit of superior size and palatability. These they nurtured, planting cuttings near their homesteads. Evidence of the agricultural use of such fruit trees in the Middle East can be traced back about 5,000 years. Both the Greek and Roman Empires took to growing fruit trees with a passion that may be gleaned from the writings of their poets and historians. In *The Odyssey,* Homer provided one of the earliest (900–800 BC) and most sumptuous accounts of an orchard:

'Outside the courtyard, but stretching close up to the gates, and with a hedge running down on either side, lies a large orchard of four acres, where trees hang their greenery on high, the pear and the pomegranate, the apple with its glossy burden, the sweet fig and luxuriant olive. Their fruit never fails nor runs short, winter and summer alike. It comes at all seasons of the year and there is never a time when the West Wind's breath is not assisting, here the bud and here the ripening fruit; so that pear after pear, apple after apple, cluster on cluster of grapes, and fig upon fig are always coming to perfection.'

By British standards this sounds almost unbelievable, but no doubt the favourable climate of the Mediterranean made such wonders possible.

After the Roman legions left the shores of Britain in AD 410 there was a steep and general decline in British civilisation; the region slipped into the Dark Ages and the developing practice of fruit cultivation fell by the wayside. This state of affairs was not to be remedied until the advent of the monasteries during the seventh and eighth centuries, as part of the reintroduction of the Christian faith after the arrival of St. Augustine in 597.

Various orders of monks and nuns established communities across the country; they desired to be self-sufficient, and to this end they set up their own agricultural systems, which included the growing of fruit trees. Many of the monasteries had connections with religious orders on the Continent, with which they maintained an interchange of ideas, including experimentation and development of various fruit varieties. The earliest pictorial record showing apple trees in a monastery garden is in a document from Christ Church, Canterbury, dating from about 1165. Both apples and pears were frequently grown in such gardens, for dessert and culinary uses as well as cider making. Increasing expertise and production volumes created a market, and trade ensued, bringing with it social and agricultural prosperity.

During the fourteenth century, however, there was a period of intense turmoil and depression, largely fuelled by the Black Death and an overall deterioration of Britain's climate, which caused a temporary decline in fruit cultivation. Effectively a second Dark Age arrived for the fruit growers.

The accession of Henry VIII brought another era of prosperity. In 1533 he commissioned Richard Harris, the royal fruiterer, to plant large experimental orchards in Kent using grafts brought from France and Holland. This inspirational move by the king was the source of a radical revitalisation of Britain's orchards. As varieties proved to be successful, hardy and disease-free, their scions were widely distributed to other growers. For the period, this was quite a visionary approach and set a trend that persists to the present day; Britain's National Fruit Trials are strongly akin to this Tudor experiment.

It is somewhat ironic that France, Britain's source for so many fruit varieties and so much technical expertise, has become, particularly over the last 200 years, one of Britain's fiercest market rivals. During

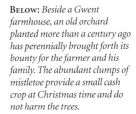

BELOW: *Beside a Gwent farmhouse, an old orchard planted more than a century ago has perennially brought forth its bounty for the farmer and his family. The abundant clumps of mistletoe provide a small cash crop at Christmas time and do not harm the trees.*

the seventeenth century, not only many varieties of apple and pear, but also fine varieties of plums, including the still popular greengage, were brought over from France. In the latter part of the century, the introduction of dwarf and semi-dwarf rootstocks became an enduring influence. For the first time, there was an alternative to large standard trees, one that was convenient for confined garden use. The virtues of the new dwarf stock received the praise of John Worlidge, a resident of Hampshire and a leading writer on horticulture and agriculture in the seventeenth century:

'... land only for the Scythe or Spade, there it is best to plant dwarf on low grafted Trees for several Reasons:

1. You may plant more of them on the like quantity of Land, because the Shadow of the one Tree doth not reach the ground of the other, as that of the tall tree doth.

2. The Low Trees sooner attain to be Fruit-bearing Trees, and grow fairer than the tall.

3. The lower and broad spreading Tree is the greater bearer, by reason the blossoms in the Spring are not so obvious to the bitter blasts, nor the Fruit in the Autumn to the fierce and destructive Winds.

4. Fruits are more easily gathered from a low than a tall Tree, beating or shaking down the Fruit from such Trees, being rejected by all.

5. Any fruit on a low, well spread Tree, is better and fairer than that on a tall Tree by the same reason that the Tree is fairer.'

For better or worse this system is still with us. Admittedly, in small gardens, dwarf trees are a boon, but many of the relentless tracts of them in commercial orchards provide no amenity value and poor wildlife habitat potential, as a result of constant chemical spraying.

During the ten years of the Civil War (1649–59), agriculture was neglected. In a desperate attempt to renew British orcharding, two authorities, Samuel Hartlib (in 1652) and Ralph Austen (in 1653), proposed that the planting of fruit trees be made compulsory. Austen also bemoaned the demise of woodland generally, to the detriment of both shipbuilders and fuel gatherers. He considered that expired orchard trees, together with coppice material from replanted woodland, might meet the need for at least the latter, a theory which was reinforced by John Evelyn in his *Sylva* of 1664. Austen wrote two comprehensive treatises on the subject of orchards, published in 1652 and again in 1657.

Orchard culture thrived through the greater part of the eighteenth century, particularly under the enthusiastic attentions of numerous wealthy gentry who delighted in expansive garden designs which incorporated all manner of fruit growing. Fruit trees were used to decorative advantage, as rambling espaliers or neatly clipped bushy plants for paths and walks.

LEFT: *A fine coloured engraving from the* Pomona Herefordiensis, *of the Cowarne Red, edited by T. A. Knight, and published in 1811. The insect-ravaged leaves, blemished fruit and lichen-snagged branch is much more realistic than the rather sterile illustrations of perfection in many later Pomonas.*

ABOVE: *The magnificent title page of Ralph Austen's* Treatise of Fruit Trees *(2nd edition, 1657) leaves the prospective reader in little doubt as to the comprehensive nature of the work. In the formal orchard the neat rows of fruit trees would appear to be on dwarf stocks; a relatively new fashion at this time. Espalier trees line the walls, whilst solid brickwork and barbed gates defend the garden.*

LEFT: *Typically romantic nineteenth-century images of the apple – with blossoms and in the cottage garden. Chromolitho-graphs from* Familiar Trees *by G. S. Boulger.*

Not only did landowners strive to out-do one another with their grand designs, but many also personally imported, planted, grafted and budded special varieties. Some immortalised themselves by naming their own new varieties.

With an almost cyclic inevitability, the end of the eighteenth century saw another downturn in the fortunes of British orchards. Agricultural literature

HARRIS'S ORCHARD

The benefits accruing to the fruit industry, as a result of Harris's orchard at Teynham, are celebrated in a passage from The Fruiterer's Secrets of 1604:

'One Richard Harris, Fruiterer to King Henry VIII, fetched out of France great store of graftes, especially pippins, before which time there were no right pippins in England. He fetched also out of the Lowe Countrie, cherrie graftes and pear graftes of divers sorts. Then took a peece of ground belonging to the King in the parish of Tenham in Kent, being about the quantitie of seven score acres; whereof he made an orchard, planting therein all those foreign grafts. Which orchard is, and hath been, the chiefe mother of all other orchards for those kindes of

fruit in Kent and divers other places, and afore these said grafts were fetched out of France and the Lowe Countries, although there were some store of fruite in England, yet there wanted both rare fruites and lasting fine fruit. The Dutch and French, finding it to be so scarce, especially in these counties neere London, commonly plyed Billingsgate and divers other places with such kinde of fruite. But now (thanks to God) divers Gentlemen and others taking delight in grafting have planted many orchards, fetching these grafts out of that orchard which Harris planted. And by reason of the

great increase that now is growing in divers parts of this Land of such fine fruit, there is no need of any foreign fruit, but we are able to serve other places.'

ABOVE: *Planting Cox's orange pippin apple trees in the community orchard at Colnbrook, in 1992. The first trees were raised by Richard Cox from two pips which he planted around 1830. The sears, in the shape of a 'C', an 'O' and an 'X' were made by Richard Farrington.*

of the day abounds with diatribes of doom and despair at the state of the nation's fruit trees. A combination of more profitable agricultural products, such as corn and beef, together with increased disease amongst fruit trees, caused largely by indiscriminate importation of blighted rootstocks and scions, contributed to a general decline in orchard quality. By the beginning of the nineteenth century, disorders such as canker and woolly aphid were causing much concern in horticultural circles.

During this period Thomas Andrew Knight, of Downton in Herefordshire, came to the rescue. Realising that disease was being spread throughout the country by the major London nurseries, he advised an innocent public of the risks and also established his own programme of breeding new and superior varieties. His aim, by using rigorously controlled cross-pollination techniques, was to produce better-cropping and superior-quality fruit with more resistance to disease; and, to a large degree, he was successful. Elected as the president of the London Horticultural Society in 1811, he then had the perfect pedestal from which to raise the profile of British fruit growing; and with the publication of his researches and observations most spectacularly realised in his *Pomona Herefordiensis* (*Hereford Pomona*), a beautifully illustrated subscription partwork, he was able to influence a whole generation of fruit growers.

The precarious state of the orchard fruit industry was exacerbated by copious quantities of cheap French imports. During the Napoleonic Wars, imports were suspended, but with the cessation of hostilities in 1815 the problem resumed. The government responded by imposing an import duty of four shillings per bushel (40 pounds) on French apples, thus providing an incentive for growers to invest in new orchards and greatly improving their prospects. The early part of

Queen Victoria's reign saw yet another downturn in orchard fortunes; but, after 1870, the increased prosperity of an industrialised nation with improved transport and communications spawned another orchard boom.

The Victorian age was a time of great expansion, both of commercial orchards and of fruit growing by private individuals. Everyone, from the wealthy landowning classes down to the cottage dweller with his tiny fruit-and-vegetable patch, took an interest in fruit. To satisfy the demand for fruit trees, numerous nurseries offered perhaps the greatest ever catalogues of fruit and developed many new and important varieties. The British Pomological Society was founded in 1854 with the aim of promoting and refining the now numerous apple varieties. The society's secretary, Dr. Robert Hogg, was largely responsible for the splendid *Herefordshire Pomona* (1876–85), for many years an unsurpassed book listing apples and pears with all their relevant merits, characteristics and cultivation information. The society lasted a mere ten years, but the enthusiasm for fruit trees was taken on and promoted by the Royal Horticultural Society (RHS), whose Great Apple Show of 1883 attracted 183 exhibitors. Many self-professed experts offered technical advice to fruit growers, with lectures, pamphlets and correspondence in agricultural society journals.

Early in the twentieth century, the first institutions were set up to approach fruit cultivation scientifically. Long Ashton, near Bristol, originally established in 1903 as the Cider Institute, became the first Fruit Research Station; it was followed ten years later by East Malling, in Kent. In 1904, B. T. P. Barker, a luminary in the realm of orchard fruit, was appointed as the Director of Long Ashton. During his 38 years in that post, he established trial orchards there and at about 50 other test sites in order to conduct comprehensive research into the microbiological and orcharding aspects of cider and perry making.

With the new scientific approach came improved crop yields from a radically rationalised selection of old varieties and specially bred introductions. Scientists delineated the varieties with the best commercial prospects, based on rootstocks, levels of pest and disease resistance, productivity and market appeal. As the twentieth century progressed, fruit became more consistent in quality and crop reliability. Production peaked in the early 1950s, when approximately 273,000 acres were laid to commercial orchards. Since that time, England's substantial orcharding tradition, which has persisted over many centuries, has steadily declined. The statistics for lost orchards are truly staggering: more than 150,000 acres have disappeared since the 1950s.

About half the present national orchard acreage lies in Kent, the 'Garden of England'. In Devon, once the county with the second-largest acreage of orchards, it has been estimated that more than 90 percent have

VINTAGE CIDER

Alongside the Victorian orcharding boom came an explosion of literary discourses on the subject. In the article 'On the Cultivation of Orchards, and the Making of Cider and Perry', published in The Journal of the Royal Agricultural Society of England *in 1843, Frederick Falkner wrote:*

'The advantage of an orchard upon a farm of sufficient size for the supply of cider for the labourers, and the use of the farmer's family, is so generally appreciated in certain districts of the counties of Hereford, Worcester, Glouces- ter, Somerset, and Devon, as to be thought an almost indispensable appendage, and the absence of it a great objection. Through the larger part of these counties, no other liquor, for ordinary use, is thought of; and it would be considered very expensive and troublesome to be brewing malt liquor. An orchard is besides a source of considerable gain, in affording both common and superior cider for sale. In the expressive language of the farmer, it yields a harvest

without a seed-time; and those who have once experienced the benefit and pleasure of its wholesome and luscious supplies, would be sorely annoyed and perplexed by maltsters' bills, and the mysteries of mashing and fermenting, which after all, offer, in their opinion, a very indifferent substitute. An orchard, besides, affords an agreeable variety in the farmer's hopes and pursuits, and no inconsiderable addition to his domestic comforts and enjoyments. It is, indeed, the English-man's proper and natural vineyard, producing him, almost without labour, fruit of rich and various flavour, more beautiful than the grape, and yielding an abundant supply of a scarcely less agreeable and cheering beverage.'

disappeared since the Second World War. The majority of surviving orchards, most of which are given over to apples, are high-density monocultural plantations of a few selected and reliable varieties grown on disease-resistant, dwarf rootstocks. The closely packed, regimented rows of trees, sometimes as many as 300 to an acre, are subjected to regular doses of fertilisers, pesticides and herbicides, to ensure dependable production. The procedures are imperative to maintain economic viability in the face of the amazingly cheap imported fruits. These commercial orchards have a short life span, as within 20 to 30 years most are ripped out. The diversity of varieties is fast disappearing as well: although thousands of apple varieties, for example, have been grown in Britain historically, the commercial orchards are now dominated by nine, mainly Cox's orange pippin and Bramley's seedling. The major cider producers are largely reliant on vast quantities of imported apple concentrate from the Continent, and the quest for home-produced pears, cherries or plums will usually end in disappointment.

The main reason for grubbing up old orchards is pure economics, influenced by generous government subsidies to clear and replant – if a farmer could earn more from a field of cereals, out went the fruit trees. Also, orchards in or around towns and villages have become very attractive to developers, and the price the land will sell for has been irresistible for many farmers. Ultimately orchard demise could even be related to the ease with which a bottle of mass-produced cider can be picked off the shelf and a bag of rather nondescript golden delicious or Granny Smith apples scooped into a plastic bag. An acceptable product at a competitive price diminishes the public's expectations and blunts agitation about the loss of part of natural, agricultural, social and dietetic heritage. If it looks good on the shelf it will sell. Market forces may oblige most fruit farmers to comply with this, but there is another way.

The orchard has been much more than a field in which fruit trees grow. It is the ancient piece of rough pasture behind the old moss-covered farmhouse where sweet and juicy eaters were gathered by the farmer's freckle-faced daughters; it is the clump of sturdy well-pruned trees which proffer their huge green globes to the flour-dusted, sleeves-rolled-up wife who seeks her pie filling. It is the avenue of sharp and bitter-sweet promises shaken, carted, crushed and pressed into bright golden nectar by fellows who can barely wait to sip their prize. It is the greensward where Old Spot pigs scoop up the fallen and fast-fermenting fruit. It is a pale pink, bee-drowsy draught of honey, offering sweet repose to the turtle dove and the swooping woodpecker deep within their rugged homes.

They are the orchards that refresh the soul, if not always the wallet. Old, traditional orchards full of

ABOVE: *The small sharp flavoured fruit of the damson, although rather unfashionable at present, are unbeatable for both wine and jam.*

TOP: *The major plum orchards are in Kent, Cambridgeshire and especially the Vale of Evesham in Worcestershire. The Vale is quite a site to behold in spring, and Blossom Trails, recognised routes for tourists, are signposted every year. In late summer there is feverish activity to gather in the plum crops as they begin to ripen: there may only be a few days when the plums are in absolutely perfect condition for shipping.*

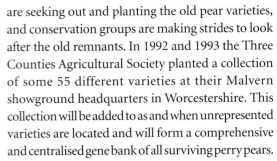

ABOVE: *The cherry pickers in this turn-of-the-century photograph show the scale of this massive cherry tree in a Kent orchard.*

OPPOSITE: *The typical columnal shape with slightly bent stem and arching boughs of the perry pear tree.*

BELOW: *Cherries from an old Worcestershire orchard. The owner no longer bothers to harvest the fruit but the local birdlife has no problem with that!*

BOTTOM: *High above the Teme valley an old cherry orchard remains intact many years after it was nurtured for commercial production.*

gnarled and rambling standard trees are a treasure to be cherished. Although Devon may have lost most of these orchards and Kent is now dominated by commercial plantations, along the Welsh borders in Shropshire, Gloucestershire and particularly Herefordshire there are many of these places still tucked away behind old farmsteads and smallholdings. Usually the bulk of the trees are cider apples and perry pears, with a smattering of eaters and cookers to supply the home.

Fruit trees, both wild and cultivated, offer a varied selection of truly beautiful timber. With the exception of large wild cherries, the trees seldom form broad straight timber of use to such trades as cabinet-making. Cherry wood is splendid when used to make furniture, taking a fine polish and often closely resembling a light mahogany with its wonderful reddish tones. One specialised use of the wood was to make cherry wood pipes for smokers. Plum and damson wood is prized by turners as it often presents rich reds and purples in the heart wood. Pear wood is one of the hardest and most fine-grained of the fruit woods and was sometimes used instead of box by wood engravers, although even pear will not endure as long as box blocks. It has also been favoured by the makers of woodwind instruments and since, like holly, it takes a stain well it has sometimes been dyed black to masquerade as ebony. Crab apple wood was once used to make the heads of golf clubs, but needless to say this has long been eclipsed by modern technology.

Well into the twentieth century, cider and perry were a country currency and had a lucrative market, even for the small-scale producer. A modern resurgence of interest in perry pears has seen smaller cider producers making perry once more, and planting their own orchards as well. Westons of Much Marcle in Herefordshire is typical of this trend. Many enthusiasts

are seeking out and planting the old pear varieties, and conservation groups are making strides to look after the old remnants. In 1992 and 1993 the Three Counties Agricultural Society planted a collection of some 55 different varieties at their Malvern showground headquarters in Worcestershire. This collection will be added to as and when unrepresented varieties are located and will form a comprehensive and centralised gene bank of all surviving perry pears.

The status of apple orchards has changed quite radically over the last 50 years. During the apple holocaust of the 1950s, '60s and '70s, the Ministry of Agriculture, Fisheries and Food (MAFF) was handing out grants to encourage farmers to grub up their orchards. However, since 1988 there has been a return to some semblance of sanity. With concerted efforts from bodies such as Common Ground, English Nature and the Countryside Commission, farmers, community groups and individual enthusiasts have been given grants and information which have helped them regenerate the disappearing orchards. Several commercial nurseries have become involved in rare variety propagation. Restoration work has been carried out in old and dilapidated orchards. New orchards where old varieties are grown as standard trees have been set up as collections and social amenities.

These are developments in all the major apple-growing counties. Kent, where old orchards have been somewhat eclipsed by modern apple farming, has made a great effort to hold on to what few are left, with much support and coordination from Brogdale Horticultural Trust near Sittingbourne. There have also been attempts to keep some of the virtually extinct old cherry orchards, as well as a new initiative to restore a little of the once prolific Kentish cob-nut industry. Gloucester, Hereford and Worcester have all worked hard to regenerate their cider apple and perry pear traditions. Cornwall, Somerset and Devon, which have lost massive proportions of their orchards, have begun the uphill battle of replacing them: in the South Hams district of Devon the local council set up a 'Save our Orchards' campaign to save and catalogue local varieties, to provide advice, and to train tree planters and owners, but above all to make as many people as possible aware of orchards and their significance. Cheshire has a very active population of tree-minded folk. The county's Parish Tree Warden Scheme publishes an excellent newsletter, *The Acorn*, which often concerns itself with orchard issues, while the Cheshire Federation of Women's Institutes has compiled a comprehensive little book on the county orchards. Even less obvious areas such as the northern counties of England, Scotland and Wales all have schemes and activities aimed at maintaining a high profile for old orchards.

Britain's apple heritage is now actively celebrated on Apple Day, which has been held on 21 October every year since 1990.

Apples

SEASONS
The fragrant stones,
the wide projected heaps of apples,
which the lusty handed year,
Innumerous, o'er the blushing orchard shakes;
A various spirit, fresh, delicious, keen,
Dwells in their gelid pores; and active points
The piercing cider for the thirsty tongue.
Thompson

BELOW: *A basket of several apple varieties from a single orchard. The fundamental test for dessert apples is palate appeal. Culinary apples have a greater acidity, which makes them suitable for cooking. Codlings or Codlins (coddling being parboiling) reduce to a soft and frothy texture, and others are favoured for apple dumplings, mincemeat or for drying. Cider apples are yet another group of fruit. 'The best fruit for cider are those which yield the heaviest juice, and these are not, generally speaking, agreeable to the palate.'*
Frederick Falkner, 1845

More than 6,000 varieties of apples have been recorded in Britain. The distinguished nineteenth-century pomologist Robert Hogg divided the fruit into three basic categories related to when they ripen and how long they keep.

In his *Apples of England* (1936), H. V. Taylor differentiated among them on the basis of their evolution from overseas introductions, the wildings, or the process of selection and hybridisation exploited to great effect by horticulturist T. A. Knight at the end of the eighteenth century. Taylor discussed each apple variety in depth, using the Blenheim orange as an example.

Though some authors have cited the sixteenth-century orchards planted in Kent by Richard Harris as the beginnings of apple cultivation in England, specific varieties were being raised and cherished both in monasteries and the gardens of the gentry by the end of the thirteenth century. Taylor denoted two varieties which were known in medieval gardens: the Pearmain and the Costard. The Pearmain was the first variety to be recorded by name, in a deed of 1204 relating to the holding of the lordship of Runham in Norfolk; there a yearly payment of 200 Pearmains was to be paid into the Exchequer on the feast of Michael (29 September in the old calendar). The Pearmain owes its name to the elongated pearlike form of the fruit. It was the principal dessert apple for

hundreds of years; Hogg mentioned it in 1884, and Taylor was aware of the fruit being exhibited in London.

Fruiterers' bills from the court of Edward I, in 1293, contain the first reference to the 'Poma Costard'. Hogg believed the name to have been derived from Costatus (costate or ribbed) because of ribs along the fruit. The traders who sold these apples were traditionally known as Costardmongers, a term which has survived to this day in the form of Costermongers (or Costers), a term for people who sell fruit from a barrow. The Poma Costard is no longer known to exist.

Thomas Andrew Knight, whom Hogg described as that 'zealous horticulturalist', carried out many calculated hybridisation experiments by transferring pollen from one variety to another. In his role as first president of the London Horticultural Society, Knight established gardens where upwards of 1,400 specimens were grown and monitored, including grange apple, downton pippin, red ingestrie, yellow ingestrie, Breinton seedling, Bringewood pippin, Wormesley pippin, yellow Siberian, Siberian pippin, foxley apple, Siberian harvey, Siberian bittersweet. He also developed new pears, cherries, damsons, nectarines, strawberries, potatoes, peas and cabbages. Knight was truly a pioneer in the realm of fruit breeding, and as a result of his well-publicised endeavours numerous plantsmen and aristocratic amateurs took to the task with relish.

The most important variety to be raised by one such enthusiastic amateur was the now world-famous Cox's orange pippin. Mr. Cox of Colnbrook Lawn, near Slough, succeeded in raising this apple in 1830; in 1850, when Charles Truner of the Royal Nurseries in Slough offered it in his catalogue, it became available to a wider audience. A similar story attaches to the Bramley's seedling. In the village of Southwell in Nottinghamshire, at the beginning of the nineteenth century lived the Brailsford family, one of whose daughters, Mary Ann, planted the pip of an apple she had eaten. It flourished in the garden. The Brailsfords departed, and a new owner, Matthew Bramley, bought the cottage in 1846. In 1857 renowned local nurseryman Henry Merryweather (several fruit varieties bear his name, perhaps the best known being a damson) obtained cuttings from Mr. Bramley for commercial propagation, on the condition that the apple was to be given the name Bramley's seedling. (Technically it should have been Brailsford's seedling, but pomological fame has passed over Mary Ann.) The variety is recognised as one of the very best culinary

BLENHEIM PIPPIN

(Syn: Blenheim, Blenheim orange, Woodstock pippin, Kempster's pippin) was first discovered at Woodstock, Oxfordshire, in 1818, and named after Blenheim, the seat of the Duke of Marlborough. An interesting account of the original tree was transcribed in the Gardener's Chronicle:

'In a somewhat dilapidated corner of the decaying borough of ancient Woodstock, within ten yards of the wall of Blenheim Park, stands all that remains of the original stump of that beautiful and justly celebrated apple, the Blenheim orange. It is now entirely dead, and rapidly falling to decay, being a mere shell about ten feet high, loose in the ground, and having a large hole in the centre; till within the last three years, it occasionally sent up long, thin, wiry twigs, but this last sign of vitality has ceased, and what remains will soon be the portion of the woodlouse and the worm. Old Grimmett, the basket-maker, against the corner of whose garden-wall the venerable relict is supported has sat looking on it from his workshop window, and while he wove the pliant osier, has

meditated, for more than fifty successive summers, on the mutability of all sublunary substances, on juice, and core, and vegetable, as well as animal, and flesh, and blood. He can remember the time when, fifty years ago, he was a boy, and the tree a fine, full-bearing stem, full of bud, and blossom, and fruit, and thousands thronged from all parts to gaze on its ruddy, ripening, orange burden; then gardeners came in the spring-tide to select the much coveted scions, and to hear the tale of his horticultural child and sapling, from the lips of the son of the white-haired Kempster. But nearly a century has elapsed since Kempster fell, like a ripened fruit, and was gathered to his fathers.'
Robert Hogg,
British Pomology, 1851

LEFT: *Nancy Harrison in her garden at Southwell, with the 'original' Bramley's seedling.*

LEFT AND BELOW: *Two of the large group of Pippins; the Golden pippin (from Knight's* Pomona Herefordiensis*) and the Ribston pippin (from Hogg's* Herefordshire Pomona*). The name pippin comes from the practice of propagating them from seed.*

apples. The original tree still produces an annual crop of exceptionally large apples in the Southwell cottage garden. In order to preserve the original genotype, botanists have micropropagated tissue from the tree to form new clonal trees, one of which has recently been planted in the same Southwell garden.

The most famous and prolific introducer of new varieties was the nursery firm begun by Thomas Laxton. From the 1890s onwards the firm conducted experimental trials of cross-bred apples, pears and plums. Probably their most famous dessert apple is the Laxton's superb (Wyken pippin x Cox's orange pippin). In the twentieth century the scientists at East Malling and Long Ashton Research Stations and the joint research from the Royal Horticultural Society and Ministry of Agriculture and Fisheries, based at Wisley, have further shaped commercial fruit growing.

LEFT: *'An apple orchard',* a splendid plate, from the Illustrated London News *of 1873. A lively vignette which perpetuated the rural idyll.*

BELOW: *The old driveway to 'Hellens' at Much Marcle, Herefordshire, lined with perry pear trees, much as it would have looked at the end of the nineteenth century.*

CHAPTER SIX

Native Evergreens

This small group of trees represents the select few native evergreens of Britain. Holly and yew will be by far the most familiar of this group. Holly is one of the fundamental ritual elements of any Christmas celebration, and the yew has lately become the focal point for historians and botanists bewitched by the mystic uncertainties of origins in distant millennia. More diminutive and less familiar, at least in its native habitat, box will be known to many only as a hedging plant or topiary medium of formal gardens. Juniper, although it has the greatest world range of any conifer, is Britain's only native member of the cypress family and is found sporadically growing on poor soils from the chalk downs of the southern counties to the rocky glens of Scotland. Scotland is also the home of the Scots pine, which has its native range in the Highlands.

Holly

And as when all the summer trees are seen
So bright and green,
The holly leaves their sober hues display
Less bright than they,
But when the bare and wintry woods we see,
What then so cheerful as the holly tree?
Robert Southey

RIGHT: *As a general rule holly is dioecious, but it is not unknown for both male and female flowers to be borne on the same tree. Engraving from Hunter's edition of Evelyn's* Silva, *1776.*

OPPOSITE: *The large tree, above the shores of Coniston Water in the lake district, is protected from the worst of the elements and has formed the splendid conical shape typical of mature hollies in ideal habitats, despite this one having to cling to a rock face and thrust its root system through a narrow crack.*

BELOW: *In The Thicks, one of the densest pieces of ancient woodland in Staverton Park in Suffolk. These massive holly trees, many of which are more than 150 years old, twist and turn about their decrepit old oak companions.*

I lex aquifolium is by far the commonest native evergreen and it manifests itself in a wide variety of forms in an equally diverse selection of habitats. It dominates many a hedgerow and is ideally suited as a tree of the woodland understorey. It thrives in the low-light conditions acting as a nurse for other broadleaves. In many parts of Britain it is believed that cutting down the holly brings misfortune upon the perpetrator, so many fine trees have been left to flourish above the annual trim of the hedgerow, some of them might even be regarded as boundary trees. Given suitable protection, it will usually form a fine conical tree, often as high as 50 feet, and in a few exceptional cases, as in Staverton Park in Suffolk, it grows to over 70 feet. In more extreme habitats it is found as a straggling, windswept wisp on exposed hillsides, where it sometimes flourishes in the most inhospitable rock crevices; as a small, stunted bush on pebbly beaches such as those at Dungeness on the Kent coast; or even springing vigorously from the aerial forks of oak or ash pollards. It will thrive in all soil types, although it is not encouraged by frequent exposure to hard frosts or particularly water-logged conditions.

Not all of a mature holly's leaves bear the armoury of thorns by which it is so easily recognised and painfully felt. For some reason the tree has managed to evolve a system whereby the crown bears leaves partially or

The Holly Tree.

completely devoid of thorns, while the lower branches bear the typical spiky leaves, capable of rebuffing the unwanted attentions of browsing cattle or, as in the case of the New Forest, the voracious appetites of ponies. Natural selection must have led to this configuration, which encouraged survival. Sometimes odd holly trees may be found which produce barely any spines at all, but this is probably an indication of some random crossing with a spineless cultivar.

The variations of common holly's leaf forms are as nothing when compared with the vast array of cultivar forms of the tree. Many traditionalists and naturalists abhor the gawdy offspring created by the nurserymen and plant collectors, with their sickly yellow berries or spattered and anaemic variegated leaves, sometimes devoid of any spines. All these strangers – and worldwide they number about 400 varieties – have been achieved by crossing back and forth for over 300 years, but with good old *Ilex aquifolium* as the foundation of many, there is a tendency for varietal characteristics such as variegation to 'run out' and revert to the original. Horticulturists with this problem must prune out the green to maintain the variety.

Collections of holly are recorded as far back as the reign of Charles II, yet there is scant detail of specific varieties and how they were obtained. By the late eighteenth century, in the wake of the establishment

of Kew and many other well-organised collections, details of named varieties and their propagation began to emerge. Reference to the notes by Hunter in his edition of Evelyn's *Silva* reveals some delightful varieties, such as Fair Phillis, Chohole, Milkmaid, Chimney-sweeper, Painted Lady, Bench's ninepenny holly and both yellow and silver blotched hedgehog holly (the latter being ferociously clad in spines on all surfaces).

When John Evelyn prepared the first edition of his momentous tome, back in 1664, he was at pains to stress his enthusiasm for the holly, particularly in respect of the mighty hedge at his residence of Say's Court. Although quoted subsequently by several authorities, its sincerity still makes it a worthy inclusion.

'*Is there under heaven a more glorious and refreshing*

RIGHT: *Warm evening sunlight makes the small and insignificant flowers of the holly glow amid the prickles.*

object of the kind, than an impregnable hedge of about four hundred feet in length, nine feet high and five in diameter, which I can shew in my now ruined gardens at Say's Court (thanks to the Czar of Moscovy) at any time of the year, glittering with its armed and varnished leaves? The taller standards at orderly distances, blushing with their natural coral: it mocks the rudest assaults of the weather, beasts, or hedge breakers.'

Apparently while Czar Peter the Great was in Deptford, in order to learn about ship building, he required a nearby residence and thus briefly became Evelyn's tenant. Once installed he is said to have amused himself by encouraging his courtiers to wheel him straight through the hedge while he sat in a wheelbarrow. However, if Evelyn's claims are to be believed, the great holly hedge managed to withstand even this bizarre exhibition of male bravado.

The name holly is generally accepted as a corruption of holy, due to its close association with the celebration of Christmas. However, this seasonal gathering of holly boughs with their bright red berries to adorn homes and churches seems strangely at odds with the other symbolism so often associated with the tree: the berries have long been regarded as representing the drops of Christ's blood, while the spiny leaves were believed to have formed the crown of thorns which Christ was forced to wear before the crucifixion. Surely such gruesome symbolism is at odds with supposedly joyous celebration of Christ's birth. The dangerous practice of hacking down a whole holly tree is believed in many parts of Britain to bode ill luck, yet pruning a few branches for seasonal needs seems to be acceptable.

As with so many other trees, the holly is represented in place names throughout Britain. Even when the hollies are no longer evident, the names signal one-time abundance or presence of some remarkable specimen. Holme or holm, from the Middle English, is one place-name manifestation and Richard Mabey, in his *Flora Britannica*, explores many derived from the holly. One of the more important name associations is hollin, from the Middle English for a group or stand of hollies, which were often pollarded to provide cattle fodder, although there is some dispute as to whether or not this really was part of common farming practice in other parts of Britain. One of the better examples of ancient hollins still in existence is the pure holly woodland on the Stiperstones in Shropshire. Many of these ragged old pollards were once regularly cut and the macerated foliage used as winter feed for the stock. Yet now their main function is to supply the buoyant Christmas market with decorative boughs, many of which are sold through the famous Tenbury Wells market.

The uses of the wood of the holly tree are generally restricted by the small dimensions of the timber, which is a hard, fine-grained wood of great durability. Larger pieces of it were often installed above old fireplaces since, unlike the leaves and green wood, the mature timber is extremely resistant to fire. In accordance with this tradition Mabey remarks upon the existence of a public house in Culmhead, Devon, called the Holman Clavel, which translates literally as 'a holly-beamed open fireplace'. Until the advent of the motor car there was a healthy demand for holly whips, not only because of their tough flexibility, but also because of a widespread superstition that holly had the ability to command power over the wildest of horses. The wood is particularly good for turning purposes. This characteristic, combined with its capacity to take a stain, thereby creating a similar effect to ebony, has caused it to be used for the making of chess pieces. It has also been used by cabinet makers for inlays and there was a vogue in the early nineteenth century for using the knots and burrs from the trunk to make handsome little snuff boxes.

A more sinister use of the holly was in the mixing of a noxious concoction known as birdlime. One of its principal ingredients was the mucilaginous holly bark, which was boiled and bubbled into a glutinous paste and then smeared on tree branches in order to snare small birds.

ABOVE: *The small tree, on the top of limestone crags above Ribblesdale in North Yorkshire, illustrates the amazing tenacity of holly when faced with the double adversity of precarious situation and exposure to the stormy blasts. Holly is a survivor!*

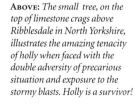

LEFT: *A small selection of some of the hundreds of holly varieties that exist today; clockwise from top left:* Ilex aquifolium *'Bacciflava',* I. x altaclarenis *'Camelliifolia',* I. aquifolium *'Ferox Argentea' and* I. canariensis *'Canary Holly.'*

LEFT: '*The Holly Cart*' *from the* Illustrated London News Supplement, *Christmas 1856. A seasonal vignette of a typically romanticised aspect of a rustic Christmas.*

BOX

RIGHT: *Until part of the cover is cut away, pure box woods are a somewhat gloomy haunt. Here there is evidence of recent coppicing. The previous year's cut stumps are already springing anew in the foreground whilst the tree soaks up a rare splash of sunlight.*

OPPOSITE: *Beneath the dark box canopy, an alien world of twisted green stems stretches far into the distant scrub. Most of the box woods of Boxwell Court are like this – hushed enchanted places through which a questing knight suddenly appears in pursuit of demons, dragons or distressed damsels.*

BELOW: *A wraith-like box tree appears to stagger across a beech-shrouded clearing on top of Box Hill, near Dorking in Surrey.*

To the rover of hill and dale, *Buxus sempervirens* will be anything but familiar, yet box is one of the natives of this land, known far better in its guise as a topiary ornament or diminutive parterre hedgerow of the formal garden.

Confined in the wild to the southern counties of England, and most particularly on calcareous soils, this superficially unremarkable tree is now one of our relatively scarce natives, only existing still in a few isolated enclaves which might be considered as box woodland; many solitary specimens almost certainly derived from garden escapees. Place names in the southern counties include about 25 box related variations, though some of these could correspond to the presence of beech trees (from the Anglo-Saxon *boc*). The great antiquity of many box colonies is particularly well reflected by such settlements as Boxley in Kent, where the name reveals Anglo-Saxon roots (*boc, leah* – a woodland clearing within box), and recent research has uncovered evidence of boxwood charcoal in nearby neolithic camps.

The most familiar extant community is Box Hill, near Dorking in Surrey, where, upon a lease from Sir Matthew Brown to Thomas Constable, dated 25 August 1602, 'the tenant covenants to use his best endeavours for preserving the Yew, Box, and all other trees growing thereupon; as also to deliver half-yearly an account of what hath been sold, to whom, and at what prices'.

A receipt for 'Box trees cut down upon the Sheep Walk on the hill' is 50 pounds, in records for 1608. Prudent management must have been exercised at the site throughout the seventeenth century, for Evelyn was later to celebrate the hill's amenity value: 'whither the Ladies, Gentlemen, and other water-drinkers from the neighbouring Ebesham Spaw, often resort during the heat of summer to walk, collation, and divert themselves in those antilex natural alleys and shady recesses, among the box trees, without taking any such offence at the smell which has of late banished it from our groves and gardens.'

Box Hill is now in the safe-keeping of the National Trust and so the future of its box trees, as well as the numerous yews, whitebeams and the much less publicised large-leaved limes, seems assured.

To the west, the estate of Boxwell Court in Gloucestershire boasts some of the best-kept records of continuous box woodland. In the early years of this century H. J. Elwes visited the Reverend O. Huntley at Boxwell where he was shown the will of an ancestor, Henry Huntley, dated 1556, which reads: 'I will that it shall be lawful for the said Anne my wife, to cut and fell all my boxe, reserving the young store, at any time or times at her pleasure within the space of the said five years'. The coppice produce of this box wood was evidently deemed to be a valuable commercial asset, even 450 years ago. Great quantities of the wood were used in the printing industry, as box is by far the best for wood engraving. The Reverend Huntley recalled that his uncle had once sold a quarter of an acre of the wood for 70 pounds around 1860. By the time Elwes made his visit in about 1912, printing technology had moved on and large-scale commercial wood engraving was a thing of the past. Reverend Huntley bemoaned the fact that the timber was now

ABOVE: *A topiary garden of box at Chastleton House, Oxfordshire, believed to have been originally laid out in the early seventeenth century. Early photographs reveal that there were once peacocks, galleons, teapots and all manner of fanciful beasts, but after neglect the plot was lost and most have reverted to amorphous blobs!*

RIGHT: *Seveneeth-century design for knot garden of clipped box and gravel infill, Moseley Old Hall, Wolverhampton.*

BELOW: *One of the most outstanding testaments to the great durability of box wood is recorded in the biography of the prolific nineteenth-century engraver, Thomas Bewick, who reputedly made one small vignette wood block for a capital letter in the* Newcastle Chronicle *which was used repeatedly over 20 years, at least two million times. Here is one of Thomas Bewick's wood blocks made of box.*

the medium has rediscovered a niche within the critical acceptability of the fine art world.

The other most important boxwood is found on the Chequers estate in Buckinghamshire. This is the most extensive stand of box in Britain and also boasts some of the largest specimens in its three steep chalk coombes. Little is known of its history prior to the last 200 years, but it is a most unusual and evocative place in which to linger. Such dark groves, where sunlight finds little delight in penetration, enhance the serpentine strands of corkscrew-twisted box stems spiralling into the gloomy depths.

Boxwood has had many applications due to its remarkable qualities, 'being so hard, close, and ponderous as to sink like lead in water', according to Evelyn. John Aubrey, writing of the Boxwell trees in 1685, noted that the wood was sold by a Mr. Huntley to 'combe-makers' in London. Evelyn listed box's value to the turner, engraver, carver and makers of pipes, combs and mathematical instruments, as well as cabinet makers:

'Also of box are made wheels or shivers, as our Ship-carpenters call them, and pins for blocks and pulleys; pegs for musical instruments; nut-crackers; weavers' shuttles, hollar sticks, bump sticks, and dressers for the shoemaker, rulers, rolling-pins, pestles, mall-balls, beetles, tops, tables, chess-men, screws, male and female, bobins for bone-lace, spoons, nay the stoutest axle-trees; but above all,

...Box combs bear no small part
In the militia of the female art;
They tie the links which hold our gallants fast,
And spread the nets to which fond lovers haste.'

In the domestic setting, box has found favour as a hedging plant, where the dwarf form has been used to form knot gardens and divisions for formal borders; in its standard form it is not considered to be as suitable for hedges as either holly or yew, but for the art of topiary it is ideal. As far back as Roman times, many garden 'artists' have delighted in the cutting of weird and fanciful designs in the dense foliage.

Like so many art forms, topiary has weathered numerous fluctuations in its popularity according to the whims of prevailing fashion. Although evidence has come to light suggesting that it was once an element of villa gardens during the Roman occupation, it was not until the gardening excesses of Henry VIII at Hampton Court that a prolonged tradition of British topiary was established. Although box was much employed, along with cypress, the preferred tree was generally yew.

worth barely one pound per ton. Elwes noted that the trees generally reach about 25 feet in height and around six to eight inches in girth, yet he did find one or two specimens of about three feet in girth. He also reported that, very much like yew, the dense shade from these trees discourages plant life below.

Little has changed today at Boxwell. The Huntley family still own the estate, and the box trees still thrive in a dense thicket which cloaks the southern slope of the small valley tucked below the Cotswold plateau. Within the wood there is evidence of a renewal of coppicing; the last 15 years have seen a steady revival in the demand for boxwood to supply printing blocks for the numerous artists who have once more taken up wood engraving, as

Yew

No other tree in Britain can be said to attract more fascination and debate than *Taxus baccata*, the yew. Traditionally, this tree of the churchyard has been locked in a sombre role as guardian of the dead; but the yew is also a splendid symbol of renewal and stability in a world that must have changed beyond all comprehension since the oldest survivors first burst above the mould, several thousand years ago. Though many different forms of yew in colours from green to gold now adorn parks and gardens, the original wild and ancient tree still commands respect and instils wonder. In spite of its association with death, one of the most widespread customs allied to yew is the use of the fronds as 'palms', incorporated in the celebration of Palm Sunday often as an alternative or in addition to the male catkins or pussy willows of the goat willow. Writers such as Tennyson and Blair emphasise this theme:

'Old Yew, which graspest at the stones

That name the underlying dead,
Thy fibres net the dreamless head,
Thy roots are wrapt about the bones.

Tennyson, In Memoriam

Well do I know thee by thy trusty Yew,
Cheerless, unsocial plant, that loves to dwell,
Midst skulls and coffins, epitaphs and worms;
Where light-heeled ghosts, and visionary shades,

LEFT: *The greatest yew of all at Fortingall is barely a remnant of a once stupendous tree, yet it is still alive, and measures 52 feet in girth. In 1769 the Hon. Judge Daines Barrington visited the tree and also found the girth to be 52 feet. J. G. Strutt, however, in his Sylva Britannica mentioned a Mr. Pennant who had measured the same tree at 56 feet (perhaps including many of the burrs and buttresses which the yew typically forms). Strutt recalls that it was the custom for funeral processions to pass through the hollowed trunk.*

BELOW: *A small grove of yew trees shelter the tiny village church at Wasdale Head in Cumbria.*

RIGHT: *Viewed across the churchyard the ancient yew at Much Marcle almost blots out the church. The extreme age of the tree means that it stood there long before the church was built in the thirteenth century. The conundrum that surrounds the association of yews with churches still awaits a definitive solution.*

RIGHT: *The deep departed recesses of the ancient yew at Much Marcle, in Herefordshire. Over 30 feet in diameter, the tree is believed to be well over 1,000 years old, perhaps as much as 3,000 years old. Such ancient churchyard giants all lost their heartwood long ago, so accurate dating is impossible. The little seat placed inside the hollowed trunk gives the tree a rather homely appearance.*

DEATH AND YEWS

The relationship between yew and death is widespread, and there are frequent references in early documents and literature to this effect. Richard Folkard, in his book Plant Lore, *asserts that the Egyptians, Greeks and Romans all viewed the yew as a symbol of mourning, suggesting that the Romans brought this belief with them, and thus introduced it into British customs. However, some of the oldest churchyard yews predate the occupation. Wordsworth was moved to write with almost awestruck intonation upon a mighty yew in Lorton, Cumbria: His subject was not located within a churchyard, which is unusual for yews of exceptional age.*

*'Of vast circumference and
gloom profound
This solitary tree! a living
thing
Produced too slowly ever to
decay;
Of form and aspect too
magnificent
To be destroyed.'*

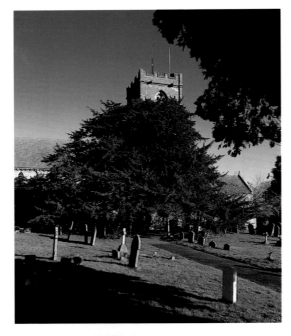

*Beneath the wan cold moon (as fame reports),
Embody'd thick perform their mystic rounds
No other merriment, dull tree is thine.*

Blair, The Grove.

Yew may be found in three principal types of habitat: in mixed woodlands, in the open countryside, and in churchyards. The churchyard, particularly in the southeast and west Midlands of England, and through Wales, tends to support the most venerable yews. Here the trees have for many centuries taken all the time and most of the space that they required to become veritable colossi of the natural world. Deliberations as to the exact age of the oldest yews, along with a rationalisation for their establishment within churchyards, has long confounded academics and naturalists alike.

Since virtually all of the oldest yews have long since become hollow, as the original heartwood rots away, it is well nigh impossible to ascertain definitive dating

through annual ring data. Often the tree has also thrown up new stems which surround and weld with the remnants of the original and further confuse the issue.

As far back as the seventeenth century, observers, already intrigued by the great girth and inestimable vintage of many churchyard yews, began to record and measure them and other trees. More recently, the *Tree Register of the British Isles* has taken on this task, and their records now form a source of fascinating comparative information. Many of the oldest churchyard yews have achieved girths of 30–40 feet, and in trying to establish their vintage, the researcher Allen Meredith has studied archaeological findings in religious sites near yews and the relatively scant documentary evidence of planting dates. He has approximated their ages, putting specimens of around 33 feet at 3,000 years old and those of around 35 feet at as much as 5,000 years old. Of course, this is all largely speculative as no accurate account can be made for growth rates over such vast periods of time. Individual soil types, climatic changes and the tree's tendency to decelerate growth as it ages, could all contribute to growth rate patterns, and although various projections for yew's growth rate have been proposed, there are limitations to the accuracy of any figures.

The question of which came first, the church or the yew tree, continues to stimulate debate. Some of the oldest yews have been confirmed to have occupied ancient burial sites as far back as the neolithic period. Whether or not the yew then bore a spiritual significance, or due to its known longevity simply stood as a memorial, is a matter of conjecture. However, it is recorded that the early Christians in Britain quietly converted the heathen temples into Christian churches and they may have also incorporated yew trees already associated with sites considered to be sacred. Several sites seem to endorse this possibility: Ashbrittle in Somerset, where a yew grows on a neolithic barrow; Darley Dale in Derbyshire and Llanfoist in Gwent, where yews grow on raised platforms within a circle of stones; Fortingall in Perthshire, perhaps the oldest of all yews – possibly as old as 9,000 years – where the remnants of a once mighty tree stand near a Bronze Age tumulus, historically renowned for bearing the great bonfires built to celebrate the pagan festival of Samhain. Such evidence associated with extant ancient yews is not seen anywhere else in the Western world.

Yew yields remarkably fine-grained, durable timber. It is susceptible to a high polish, often rendering a wide range of colours, from cream through an array of pinks to deep orange and red. This makes it admirably suited for cabinet making, in particular for veneers and inlays. Craftsmen in East Anglia are currently using ancient yew dug from beneath drained fenland to make the most exquisite furniture, and similarly some 90 years ago Sir Charles Strickland told of 'yew wood which is occasionally dug up in the bogs and fens of East Yorkshire . . . the most beautiful English wood'. The incredibly dense nature of yew wood assures perfect preservation, even after centuries of submersion.

A popular misconception holds that yews provided the copious quantities of wood required for making British long bows. In fact, the quality of native yew was rarely good enough for this purpose, although many of the old writers perpetuate this myth. Superior yew for bow staves was generally imported for Europe, usually Spain or Italy. W. Dallimore quotes a delightful passage referring to yew bows, but not specifying their provenance, from Pennant's London. It states that 'in 1397, Richard II, holding a parliament in a temporary building, on account of the wretched state of Westminster Hall, surrounded his hut with 4,000 Cheshire archers armed with tough Yew bows, to ensure freedom of debate'.

It is possible that the ancient churchyard yews were planted with no intention of felling. Some would say that the poisonous nature of yew foliage made it an ideal guardian of hallowed ground by discouraging people from allowing their stock to graze upon it. Animals that have never been exposed to yew and are then allowed to binge upon it, or those given access to desiccating cut boughs, where the potency of the toxins is increased, will succumb. However, livestock were often allowed into churchyards, and small amounts

YEW

of browsing yew fronds seem rarely to have proved fatal. Though the seeds of the yew are toxic, the surrounding fleshy pink cup is quite sweet and palatable. Country children of old and birdlife, particularly blackbirds, thrushes, fieldfares and redwings often taste of this feast. Strangely, Prideaux John Selby, in

Left: *A wonderfully vigorous representation of the yew in this 1920s wood engraving by C. Dillon McGurk. The yew has the capacity to grow relatively fast while young, although records of girth measurements of the most ancient specimens sometimes reveal little to no growth over periods of several decades or even centuries. Could this incredible deceleration of growth be the very mechanism of survival?*

Below: *One of the massive, mysterious yews of Kingley Vale. Charles Crocker of Chichester wrote a poem about Kingley and its mighty yews in 1830, reproduced, in part, in* The Gentleman's Magazine *(vol. 32, 1849).*
A thousand charms now open on the view,
O'er which enchanted roves the wonderer's eye
With ever fresh delight.
In stainless blue
Immensity above extends the sky:–
Below, in richest harmony, each dye
Of varied green is blended to adorn
This solitary vale, that seems to lie
Lovely as Eden on Creation's morn,
Ere nature knew decay –
ere pain and grief were born.

his *History of British Forest Trees* (1842), is at variance with the common view of the seeds' toxicity, stating that 'the kernels of these nuts are not deleterious, as supposed by many, but may be eaten with impunity, and they possess a sweet and agreeable nutty flavour'. During his term as warden of Kingley Vale, Richard Williamson meticulously noted the astounding quantities of berries which the birds could devour. Often they would completely gorge themselves until they were obliged to regurgitate 20 or 30 berries, only to return to the guzzling once more, as Williamson observes: 'Like children confronted by bowls of trifle and jelly, they set to work.' He also discovered that both foxes and badgers occasionally ate the berries, again without any obvious harm; the tough seeds must pass harmlessly through the digestive tracts of these animals.

Outside the churchyard wall, the larger individual yews found scattered about the countryside have most probably been marker trees, to signal a boundary or maybe the site of a long-forgotten religious site. The question then arises as to whether these yews were actually planted. Richard Mabey maintains that prior to the Middle Ages, man seldom planted wild tree species: there are no written records to show whether man planted yew trees, back before the dawn of Christianity. These enigmatic memorials to chance or design have endured to confound and amaze to this day.

Yew may also be found in a wide diversity of woodland habitats, although it prefers soils of chalk or limestone above all others. Its great propensity to colonise much of the Hampshire downland has led to it being dubbed the 'Hampshire weed'. The zenith of yew's success may be experienced in Kingley Vale, generally considered the best example of a natural yew wood in Europe. In 1952 it became one of Britain's first designated National Nature Reserves.

Richard Williamson crystallised its essence in his book *The Great Yew Forest*; but no matter how incisive the account, there is nothing that quite prepares the first-time visitor for this magical place. Although the oldest yews, in Kingley Bottom, are thought only to be about 500 years old, their sheer scale and presence are quite overpowering. Local legend claims that upon this site, around AD 900, the men of Chichester massacred an invasion force of Vikings, who are buried in the barrow on the hilltop. W. H. Hudson was moved by the strange to other-worldly atmosphere to write: 'One has here the sensation of being in a vast cathedral: not like that of Chichester, but older and infinitely vaster, fuller of light and gloom and mystery, and more wonderful in its associations.'

Both William Gilpin and Selby applauded the aesthetic glory of the yew tree, hoping to inspire the planting of more yew trees. In the mid-nineteenth century Selby lamented the lack of large fine-quality yew wood for furniture making, conceding that pieces were usually only of use for inlay and veneer work and 'the smaller wares of the turner'. He made a case for the value of yew as a timber tree, in the same breath as oak, and suggested that intermittent planting amongst shorter-term conifers would yield an extremely valuable crop in due course. The sheltering conifers would provide the additional benefit of encouraging the yew to grow straighter and taller, especially when reduced to one leader, thus forming commercially more valuable timber trees. Gilpin owned that yew's exceedingly durable nature made it the ideal choice for certain outside uses: 'Where your paling is most exposed either to winds, or springs; strengthen it with a post of old yew. That hardy veteran fears neither storms above, nor damps below. It is a common saying amongst the inhabitants of New-forest, that a post of yew will outlast a post of iron.'

Boucher offers another invaluable virtue of yew wood, once most reassuring, but happily now of little relevance, 'that the wooden parts of a bed made of Yew will most certainly not be approached by bugs'.

Currently the yew is not grown commercially, since there is relatively little demand and hence value in its timber. Even so, the tree has had an enduring association with the traditions of formal gardening.

A chance discovery around 1760, in the Cuilceagh Mountains of County Fermanagh, Northern Ireland, may well have proved to be a popularist lifeline for the garden yew. One Mr. Willis, a farmer, came upon

BELOW AND BOTTOM: *John Dovaston's pendulous yew at West Felton in Shropshire. Planted in 1777 and still in fine health, with due care and attention it should still be around at the end of the next millennium.*

two young yew trees displaying a fastigiate character. These he dug up, planting one in his own garden, while the other he presented to his landlord, Lord Enniskillen of Florence Court. The farmer's tree perished in 1865, but the Florence Court yew has persisted to this day. During the early nineteenth century, many cuttings were taken from the yew and despatched far and wide, so that by now, some 240 years later, all fastigiate yews across the world can be said to have been derived from that one tree. The main reason for its universal success was its upright shape which made it eminently more suitable for trimming and shaping. Thomas Pakenham, for his recent Meetings with Remarkable Trees, visited and photographed the Florence Court yew, which he found, as his picture shows, to be in a somewhat weary state, many of the lower boughs having sagged to a rather 'semi-fastigiata' form.

The other most important variety of yew is that which displays the weeping form. Although not the very first of its kind to have been known in Britain, the Dovaston Yew, at West Felton in Shropshire, can now lay claim to be the oldest surviving specimen. In 1777 John Dovaston, who was experiencing problems with collapsing earth around a well which he had recently dug in his garden, was fortunate enough to purchase a yew sapling from a local pedlar. His son, J. F. M. Dovaston, recounted the story some 60 years later in his correspondence with J. C. Loudon.

'... he planted near the well a Yew tree, which he bought of a cobbler for sixpence, rightly judging that the fibrous and matting tendency of the Yew roots would hold up the soil. They did so, and, independently of its utility, the Yew grew into a tree of extraordinary and striking beauty, spreading horizontally all round, with a single aspiring leader to a great height, each branch in every direction dangling in tressy verdure downwards, the lowest ones to the very ground, pendulous and playful as the most graceful Birch or Willow, and visibly obedient to the feeblest breath of air. Though a male tree, it has one branch self-productive, and profuse of berries, from which I have raised several plants in the hope that they may inherit some of the beauty of their parent.'

On a tree that is almost invariably dioecious this was a remarkable stroke of fortune, for when the female flowers were pollinated by the tree's own male flowers the splendid weeping form was maintained. The tree still thrives today in the grounds of what was once John Dovaston's glorious nursery garden. In recent years, this derelict site has been developed into a small housing estate, but great care has been taken to preserve the Dovaston yew.

ABOVE: *The fastigiate form, Irish Yew, is the most popular variety for the topiary found in formal gardens and the neatly clipped trees of churchyards, such as these at Painswick in Gloucestershire.*

FAR LEFT: *The wayward annual rings and internal voids of an old yew butt show why it can be such difficult wood to work. The wonderful coloration, enhanced here by evening sunlight, makes it a bold choice for the craftsman seeking something rather special and different.*

ABOVE LEFT: *The bright, pinkish-red berries of yew are not, as many assume, completely poisonous. The fleshy, red aril is edible, but the small, nutty kernel inside is usually considered to be toxic.*

LEFT: *H. J. Elwes was greatly impressed by these avenues and walks of tall, straight yews at Midhurst in Sussex which, he was assured, had been recorded as a fine, mature plantation during a visit by Elizabeth I, some 300 years previously.*

NEXT PAGE: *Several separate stems form this tree in the churchyard at Ashbrittle, Somerset, which is also rather special as it stands upon a Neolithic burial mound.*

Juniper

ABOVE: *Junipers are strange little trees that may grow in completely different shapes even when they are in exactly the same habitat. This one, quite properly, has grown like a cypress, while another, (OPPOSITE), no more than a few feet away, looks rather more like an extra-terrestrial.*

BELOW RIGHT: *A splendid windswept holly commands the foreground, but beyond, in the valley of Little Langdale in Cumbria, the scatter of small green bushes which look like gorse are in fact juniper.*

Small scrubby trees, junipers are rarely more than 15 feet high and from a distance rather resemble gorse bushes: not a very stimulating description of the world's most widespread member of the cypress family. The rugged little native evergreen juniper may be one of Britain's most unappreciated trees.

Juniperus communis has long been abundant in Britain, though concentrations in specific areas may well explain its unfamiliarity. Records reveal that there were once many areas of Scotland with an abundance of juniper, yet regular clearance of land to promote pasture for sheep and commercial plantations of imported conifers have seen the widespread demise of the tree. Rocky mountains and valleys in both Scotland and the Lake District still support many junipers, sometimes in quite substantial stands of predominantly juniper woodland, such as may be seen at Tynron, near Dumfries.

The other stronghold of juniper is along the chalky downland of southern England. Richard Mabey refers to the largest known population, which grows within the confines of the Ministry of Defence Establishment at Porton Down in Wiltshire, where lack of grazing pressure has probably assured their survival. Elwes and Henry refer to several different sites where they found juniper at the turn of the century, not only in Scotland and the southern counties, but also in the Yorkshire Dales and Norfolk; yet even at that time the underlying tone was one of population erosion.

Some 80 years back W. H. Hudson bemoaned the sad decline of juniper on the Sussex Downs. As with much of the Scottish Clearances, he put the eradication of the tree down to increased demand for land for sheep grazing and an increased use of wood for fuel: 'and we may say that in many places the juniper is disappearing from the downs simply because the great landowners have not thought proper to preserve

it'. Hudson concluded his appreciation of the juniper with an appeal to those with the power to waste no time in halting the destruction of the tree. 'If some one of the great landowners of the downs would but create a juniper preserve at some point where the plant grows spontaneously and well, he would deserve the gratitude of all lovers of nature who are accustomed to take their summer rambles in downland.'

The diminutive size of juniper is not in its favour, as a tree with stems seldom growing more than 6–8 inches in girth has no value as timber. However, the wood and the foliage have aromatic qualities; the soft durable wood has been burnt for its incense property indoors, or used to carve spoons, while the berries, which take two years to ripen – green the first year and blue-black when fully ripe – are used as the flavouring for gin. The latter use has long since died out in Britain, and it is doubtful whether there would even be enough berries to be found for the purpose today, as they are now imported from eastern Europe. However, the name gin has derived from geneva, the English construction of the French word genievre, which in turn originally derived from *juniperus*.

The appearance of juniper trees can vary quite radically even within the close confines of a single location. Hudson observed their different shapes: '...the plants are curiously unlike and, viewed at a distance of a couple of hundred yards or so, they have something of the appearance of a grove or wood of miniature trees of different species: alike in colour, in their various forms they look some like isolated clumps of elms, others columnar in shape, others dome-like, resembling evergreen oaks or well-grown yews; and among these and many other forms there are tall straight bushes resembling Lombardy poplars and pointed cypresses.'

The first impression is that these must surely be different subspecies, but this is not apparently the case. The horticulturist Gertrude Jekyll proposed the theory that much of this variation was due to the action of snow. She had observed that junipers which had several years to establish a strong leader without the rigours of bearing great loads of snow seemed to adopt the upright posture, while those which were unlucky enough to be pushing forth through a string of hard winters became broken and stunted in form.

Like Miss Jekyll, Hudson was moved by the beauty of the junipers he found in his Sussex haunts. After citing its common confusion with the furze-bush

NEXT PAGE: *Barren mountainous landscape in Glen Gairn in the Scottish Highlands dominated by sprawling colonies of juniper, one of the few trees tough enough to survive in such extreme conditions.*

BELOW LEFT: *Juniper foliage and berries.*

(gorse), when seen from afar, Hudson continues: '*...the topmost slender sprays, gracefully curved at the tips, are tinged with red; and where the foliage is thick there is a bluish tint on the green that is like a bloom. In some lights, especially in the early mornings, when the level sunbeams strike on the bushes, wet with dew or melting hoar-frost, this blueness gives the plant a rare, delicate, changeful beauty.*'

Although not as widely grown in gardens as it once was, juniper makes an ideal shrub for hedges and parterres. John Evelyn was proud to recall a juniper planted by his late brother, 'having formerly cut out of only one tree an arbour capable of receiving three persons to fit in; it was at my last measuring seven feet square and eleven feet in height, and would certainly have been of a much greater altitude, and farther spreading, had it not continually been kept shorn', and this was a tree barely ten years old.

JEKYLL'S JUNIPER

The wonderful aromatic foliage and berries of juniper. Miss Jekyll's lyrical appreciation of the juniper might appear a trifle gushy, but she clearly valued the tree as an important element both of her garden and the Surrey landscape around her home. In her book Wood and Garden, *first published in 1898, she vividly demonstrates her love of the tree:*

'*Among the many merits of the Juniper, its tenderly mysterious beauty of colouring is by no means the least; a colouring as delicately subtle in its own way as that of cloud or mist, or haze in warm wet woodland. It has very little of positive green; a suspicion of warm colour in the shadowy hollows, and a blue-grey bloom of the tenderest quality imaginable on the outer masses of foliage.*'

Scots Pine

Hugh Johnson's appraisal of the tree in The International Book of Trees is hard to improve upon:

'I find its beauty unrivalled. The richness of its colouring, its wild poise set it apart. No matter how grey the winter's day, the papery bark, flaking in butterfly wings of salmon and green, glows with the warmth of a fire in the sky. And if the bark is richly red, the leaves are no less richly blue. To handle a heavy, resinous spray and see the unripe cones, little jade carvings in the deep sea lights and shadows of the needles, is an intoxication.'

Until the mid-nineteenth century the Scots pine was more often than not referred to as the Scotch fir, which may be confusing to readers of early works, as the tree tended to be grouped together with the numerous other members of the family Pinaceae, while it should be more clearly identified as part of the well-defined *Pinus* species. Scots pine (*Pinus sylvestris*) was for many years virtually the only pine to be found in the greater part of Britain, and it is possible that its numerous cousins were virtually unknown to many early authorities. As introductions of the eighteenth and nineteenth centuries slowly began to make the dozens of other pines more familiar, the name Scots pine began to prevail.

There are now several hundred different conifers which will grow successfully in Britain, and from a distance large stands of such trees – generally of monotone hues all year round, except for the larch – present a rather sombre appearance. Many an upland hillside of Britain has been dominated for the greater part of this century by dark bands and blocks of coniferous plantations, but commercial forestry of this type and on such an enormous scale is a relatively recent occurrence. Virtually all the evergreen softwoods have been introduced by numerous plant collectors, most of whom were active in the late eighteenth and early nineteenth centuries. The majority of introduced evergreens emanated from either the Far East or America. Plants or seeds were sent back to nurseries in Britain where they underwent rigorous trials to ascertain their suitability for cultivation. Although decorative trees were imported from afar it was those species with commercial promise which were of most interest, particularly to the owners of large estates. The reason that such timber was in demand was largely due to an exhausted native supply. So, even though Scots pine was still planted, some of the incomers proved to be even faster-growing and much more adaptable to terrain not suited to pine. Larch, Douglas fir, Corsican pine, as well as Norway and Sitka spruces, were grown aplenty.

The native Scots pine was the only north European pine to survive and recolonise after the last ice age. As the climate slowly warmed, the tree's native range migrated northwards, while the ever expanding range of broadleaf trees took over. Scots pine is still to be found in many isolated parts of England, although these trees would have been planted some time during the last 200 years, or at least be naturally regenerating seedlings from planted trees; and yet there is a small body of opinion that a few native descendants still thrive in some of the more inhospitable realms of northern England, where natural competition might not have been too great.

The Scots pine's native range today is set across the Highlands of Scotland. What remains there is a mere vestige of the once splendid Caledonian Forest. For many centuries, Scots pine was an all-important timber tree for the Scottish construction of buildings and ships, and many other items such as boxes, barrels, fences and gates.

With the influx of marauding Sassenachs from the seventeenth century onwards, the Scots pines began to be depleted in unprecedented quantities. Initially, the wood was converted to charcoal to service the iron foundries; later it was in great demand for shipbuilding during the wars with France. It seems ironic that timber from Scotland, a country closely affiliated to France, should help to build the warships which eventually conquered the French. By the nineteenth and twentieth centuries vast quantities of pine were used for pit props, telegraph poles and railway sleepers. Coinciding with all this relentless timber extraction came the Highland Clearances in the eighteenth and nineteenth centuries. Since much of the countryside was denuded of people there was

Opposite: *A fine mature Scots pine looms above the emerald birches on a misty morning in Glen Affric.*

Right: *A nineteenth-century, stone lithograph of 'Scotch fir' by W. H. Eldridge, 1833.*

Next page: *Red deer beside a Scots pine forest at Braemar on Deeside.*

ABOVE: *Scots pines in the New Forest.*

nobody to manage the regeneration of the pine woods; instead the often absent landowners introduced huge flocks of sheep and herds of deer which soon held sway across the old forests, systematically munching through the emerging seedlings. As old trees were removed wholesale, the source of seed disappeared; and, unlike broadleaf trees, cut pine stumps will not

regenerate any new growth but simply die in the ground. What little management there may have been was totally inadequate. The result was that by 1970 only an estimated 25,000 acres of native pinewood was still standing.

Public perception of the ancient colonies of Scots pine has seen a dramatic turn-around in the last 25 years, and it has spurred a concerted replanting and conservation programme. Deer fences have been erected in many places, and replanting with genetically similar native seedlings is underway. The future looks considerably brighter.

As the nineteenth century opened, various writers were calling attention to the unseemly manner with which great pinewoods were being hacked down. A note of caution was in the air. Was this destruction simply exploitation or was any semblance of sustainable forestry in view? Some pine set its own seed and grew naturally, and there was a measure of organised replanting, yet there were also the new species of spruce and larch which appeared to be potentially much more profitable. Selby, visiting the pine woodland of Glen More to the east of Aviemore during 1824, provided a vivid picture of the scene of desolation:

'*Scattered trees, some of which were in a scathed or dying state, of huge dimensions, picturesque in appearance from their knotty trunks, tortuous branches, and wide-spreading heads were seen in different directions, at*

RIGHT: *Strutt's drawing shows a mighty Scots pine at Dunmore, much in the image of many a forest veteran across the Highlands, with ragged stem and flat top.*

OPPOSITE: *The immensely strong Scots pine is able to bear phenomenal loads of snow through the cold Highland winters without its boughs breaking.*

unequal and frequently at considerable distances from each other; the solitary and mournful-looking relics of the departed glories of this once well-clad woodland scene, and which had only escaped the axe from their previous decay, or the comparative worthlessness of their knotty trunks, while the surface of the ground in almost every direction was littered and bristling with the decaying tops and loppings of the felled trees ...'

Although there are still pine trees in Glen More, many have since been grown in regimented plantation style, so that relatively little of the natural pine forest remains, and what does survive is quite fragmented. What small pockets of the ancient forest of Caledonia do still exist may be found at nearby Rothiemurchus and Abernethy Forests, both also on Speyside; others include the Black Wood of Rannoch in Perthshire, Glen Tanar near Aboyne and Ballochbuie Forest on the Balmoral estate near Braemar, the latter two both on Deeside. In many cases remnants have only survived due to their inaccessibility, where it simply was not practical or viable to extract the timber. This has left a legacy of veritably stunning scenery, perhaps no better observed than in some of the more precipitous parts of Glen Affric, to the southwest of Inverness. Here both pines and birches cling to the sides of the rocky ravine that leads up to Loch Beinn a Mheadhain and eventually Loch Affric. The splendid contrast between the deep blue green of the pine needles and the shimmering translucent green of the birch in spring, or for that matter its burnished gold in autumn, is truly awesome. Even on a misty, rain-clad day the towering pine woods disappearing into the clouds is a spectacular sight.

In many parts of the Highlands, ancient solitary pines may be found as singular remnant trees of old woodland. Many of these are broken and bent, but occasionally handsome specimens may be encountered. Then, the true magnificence of a mature Scots pine can be appreciated; freed from the constraints of plantation emaciation, the tree can fulfil its billowing potential.

All these remarkable woodlands are dominated by the Scots pine and silver birch, with attendant ground cover of bilberry, crowberry, heather and sometimes juniper tumbling hither and yon, resembling some derelict topiary garden. The oldest pines, which have been allowed to carry on growing due to inaccessibility or because they are twisted or stunted and thus unsuitable for timber, have formed the most interesting and unusual trees. Instead of growing ever skyward to develop tall straight stems, these trees have formed great rounded heads with flowing side branches.

Richard Mabey points out that for some elusive reason there is little folklore associated with the pine in its native Scotland and, indeed, it is in the words of English writers that the tree tends to gain more notice. The poet Wordsworth was so enamoured of the Scots pine that he sent to Scotland for some trees

to plant in his Lakeland garden, stating that other than the oak this was his favourite tree, most particularly for 'its beauty in winter, and by moonlight, and in the evening'. Barbellion recognised that; 'It is a noble tree. It has a strength as a giant, and a giant's height, and yet kindly withal, the branches drooping down graciously towards you – like a kind giant extending its hands to a child.'

Although virtually none of the Scots pines in England can claim direct descendancy from native stock, there are still plenty to be found spread widely across the country. In the Breckland of East Anglia, an unusual feature, peculiar to this region, is the abundance of hedges formed out of pine. Most of these were planted during the nineteenth century in the belief that such robust trees would survive well in the poor soils. They were once lopped into shape like any other hedgerow, but when this was abandoned

OPPOSITE: Clump of Scots pine in a parkland setting.

BOTTOM: Mile after mile of great spreading pines are still to be found in Rothiemurchus Forest, part of the ancient Caledonian Forest.

BELOW: A young pine sapling amidst its forebears in the Abernethy Forest.

at the turn of the century, the trees, once bowed and stunted by man, continued to grow, but were in turn moulded by the winds which swept in from the North Sea.

The heathlands of Surrey and Sussex also support many extensive colonies of Scots pines, the most famous of these being Ashdown Forest, known to millions as the haunt of Winnie the Pooh, Piglet, Christopher Robin and their friends. The Enchanted Place at the top of the forest is Gill's Lap and, even though some of the pines have fallen by the wayside since Christopher Milne looked out to the clump from his nursery window, the magic of the place endures.

The regular occurrence of pines as hilltop clumps is a landscape feature that appears right across England and Wales. A long-standing tradition has seen pines planted in prominent positions to serve as waymarks for the travellers of past centuries. When the drovers were moving their stock great distances these clumps were invaluable signposts, particularly in winter when the dark foliage stood out against the snow. It is also believed that the pines signified a place where hospitality and grazing would be available. Alfred Watkins comprehended the significance of the pine clumps and incorporated many of them within his maps of ley lines, linking them with hill-forts, churches and standing stones, and contributing to the theory that they were once a vital element of a nationwide, if not global, network of pathways both physical and metaphysical.

There remains the question of who originally planted the pines. The actual trees in situ today are seldom older than about 100 years, but the prodigious self-seeding capacity of pines indicates that they are the result of continuous renewal down the centuries. Also their locations tend to be exposed and are hence relatively unattractive to rival species. It is possible to infer that some of these hilltop clumps are direct descendants of their native forebears.

The nineteenth-century author and art critic John Ruskin's vision of hilltop pines corroborates the long-abiding power of this tree, not only to guide and to welcome, but also to inspire – shades of the apocalypse which the sunrise stirs in many a breast:

'When the sun rises behind a ridge of Pines, and those Pines are seen from a distance of a mile or two against his light, the whole form of the tree, trunk, branches and all, becomes one frost-work of intensely brilliant silver, which is relieved against the clear sky like a burning fringe, for some distance on either side of the sun.'

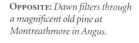

OPPOSITE: Dawn filters through a magnificent old pine at Montreathmore in Angus.

BELOW: C. Dillon McGurk's wood engraving of a Scots pine finely depicts all the key features of a mature tree – the rough bark, clawlike dead boughs, layered clumps of foliage and the slight bend of the windblown top.

ABOVE LEFT: Originally planted to celebrate Queen Victoria's Golden Jubilee in 1887, this great clump of Scots pines stands atop May Hill in Gloucestershire. A distinctive landmark as far as 30 miles away, most of the pines are much too young to be the originals, but have regenerated naturally or by organised replanting. Exactly how long pines have stood on May Hill summit is unknown, for they may well have been there long before the plantings of the nineteenth century. Such a fine landmark would have meshed perfectly into Alfred Watkins's ley system, as would this home rugged clump (LEFT) on a Yorkshire hilltop in Wharfedale.

CHAPTER SEVEN

Trees of Wet Places

All trees need water to succeed in life, and most members of three species in particular demand exceptionally moist domains in order to reach their prime. The larger tree willows, especially the white and the crack and the many varieties and hybrids of the osiers, do best with their roots close to water. The waving cool green fronds of willows lining river and brookside are a definitive landscape feature across lowland Britain. Poplars are also typically found bordering watercourses or in low-lying flood-plain terrain. The five best-known species are aspen, white poplar, grey poplar, black poplar and the towering Lombardy poplar.

The alder completes this trinity of water-loving trees, and is by far the most adaptable to different and difficult habitats. It will do equally well along lowland riverbanks, straddling some rock-strewn stream as it tumbles down a mountain side, or sheltering an upland tarn. As long as the alder has some degree of shelter it can thrive in almost any wet place, where it is easily distinguishable by its dense, dark and glossy green foliage.

Willow

White, Crack and Weeping Willows

Willows can be divided into three different groups. The smaller shrubby osier types often grow in extremely damp conditions. The woodland and hedgerow willows generally known as sallows are small trees with broader leaves, in the main preferring well-drained soils. For many people the archetypal willows are the large trees of the primary grouping, notably the white willow (*Salix alba*), the crack willow (*Salix fragilis*) and the incredibly graceful weeping willow (*Salix babylonica*).

Is it white or crack? That is the question. The form and habitat of both natives tend to be the same, although the white willow has the potential to develop into a slightly taller tree; and from a distance, particularly in the spring, its pale silvery green leaves mark it out from its cousin. Due to willow's rampant capacity to hybridise, and the fact that white and crack often grow in close proximity, the ensuing progeny are difficult to differentiate. The first-generation cross is a variety known as *Salix* x *rubens*, but the infinite variety of crossings and recrossings makes positive identification of many willows a tricky task. The easiest way to establish identity of the crack willow is to put its name to the test: a twig will readily break away with a distinct snap or crack. The lanceolate leaves have the instantly recognisable long, thin and pointed form; crack willow leaves are glabrous (hairless) while white willow's are pubescent (with short soft hairs).

The success of all the willows lies in their ability to regenerate both by seed and vegetatively. The down-clad seeds from the female trees, although they only remain viable for a short time, can often be carried long distances on the wind. When twigs or branches break from trees they may tumble into a watercourse, float downstream and settle on a mud-bank, where, in the twinkling of an eye, another tree is born. Even trees that fall in the wind will find purchase with new roots and send up a host of new stems.

Due to a combination of habitat and growth habit, willows are not noted for great longevity. Many grow on soft damp ground and produce massive top-heavy boughs which eventually cause the tree to split down the middle. Small splits leave it open to the elements and more accessible to wildlife, even creating habitats for other trees and plants to take root. As more birds and insects move in and invasive plants thrust their own roots downwards, the willow trunk splits more and more. Rot and fungal invasion eventually complete the cycle and the tree finally splits asunder. In decay, the tree still participates in the natural chain of life, giving succour to countless other species while also spreading its own seed and tissue upon the land.

Pollarded willows, on the other hand, not only have yielded regular harvests of poles, but, with their bristly compact heads, have also stood for many years longer along rivers and streams. Such management, if regularly maintained, definitely increases the tree's chance of a healthy dotage, though abandoned pollards may fare even less successfully than old maiden trees: the stresses generated by overgrown poles hasten the splitting process and the top-heavy crown provides a settling point for all manner of detritus, rain, snow and alternative wildlife. Because the trees can lay down quite substantial annual growth (sometimes up to an inch thick), the ancient boles can reach remarkable girths relatively quickly. A 70- or 80-year-old willow may have similar proportions to a 300-year-old oak. These somewhat veteran trees have many admirers these days, though in the early nineteenth century J. G. Strutt expressed some dismay about them: '*as they begin to be polled in the third year of their growth, and their decapitated trunks then present an unsightly spectacle, not much improved when they again sprout*

forth. *This is particularly the case in Huntingdonshire, and parts of the adjoining counties, where the uniformity of the low, flat, and often inundated meadows, is only broken by formal rows of Pollard Willows, standing disconsolately by the sides of ditches, over which they have no branches left to bend.'*

Almost 200 years later, due to the relaxation of this type of management in many places, distinctive pollards are now far from commonplace, and have become appreciated as the landscape features of many a lowland riverbank. Though some rivers and ditches have changed their courses over the centuries, it is quite common to see evidence of old waterways mapped out by lines of ancient pollard willows still surviving in old water meadows.

The abundance of willows in lowland pastures has always provided both shade and food for cattle (although pollards are beyond their reach), and the wood of these fast-growing trees has supplied large quantities of sturdy poles on a regular basis. Its tough but extremely lightweight timber found favour in the past with house builders, with some boat builders, as planking for the linings and bottoms of carts, and as cleft post and rail fencing, while thinner rods and shavings were often made into tough baskets and mangers for agricultural purposes. The antiquarian writings of a Mr. Fuller sum up the commercial value of the willow: 'It groweth incredibly fast, it being a byeword in this country, that the profit by willows will buy the owner a horse before that by other trees will pay for his saddle.'

John Evelyn, in typical form, also presented a formidable inventory of the uses of willow in the seventeenth century:

'for boxes, such as apothecaries and goldsmiths use; for cart saddle-trees, yea gun-stocks and half pikes; harrows, shoemakers lasts, heels, clogs for pattens; forks, rakes,

especially the teeth, which should be wedged with Oak; but let them not be cut for this when the sap is stirring, because they will shrink; for pearches, rafters for hovels, portable and light ladders, hop-poles, ricing of kidney-beans, and supporters to Vines, when our English vineyards come more in request: Also for hurdles, sieves, lattices; for the turner, kyele-pins, great town-tops; for platters, little casks and vessels, especially to preserve verjuices in, the best of any. Pales are made of cleft Willow; also dorsers, fruit-baskets, cans, hives for bees, trenchers, trays, &c. and for polishing and whetting table-knives the butler will find it above any wood or whetstone; it makes coals, bavin, and excellent firing, not forgetting the fresh boughs, which, of all the trees in nature, yield the most chaste and coolest shade in the hottest season of the day; and this umbrage so wholesome, that physicians prescribe it to feverish persons, permitting them to be placed even about their beds, as a safe and comfortable refrigerium.'

He concludes by recommending even the rotted heartwood of willow as an efficacious compost for raising garden flowers.

The weaving of cleft willow strips into ropes, the use of bark for tanning and the extraction, from the

ABOVE: *Springtime fronds of weeping willow with tiny leaves unfolding and the emerging catkins.*

BELOW: *C. Dillon McGurk's instantly recognisable woodcut of a pollard willow*

ABOVE LEFT: *Pollards may look somewhat brutalised immediately after lopping. The casual observer may think that they have been killed, but willow regenerates faster than any other tree, and within three or four years will look like this tree in the foreground.*

LEFT: *Any waterside landscape would be the poorer without the presence of willows. Even in winter their sweeping silhouettes make a fine show against turbulent skies.*

OPPOSITE: *Wind-tossed weeping willow on the edge of The Long Water in Kensington Gardens.*

RIGHT: *Bright orange stems of golden willow make a stunning winter display.*

ABOVE: *An Osier bed in Herefordshire which has been planted up as part of a water purification system for a cider company. The term osier is loosely applied commercially for willow species grown for basket weaving.*

ABOVE: *The down-clad seeds of grey sallow on Holme Fen in Cambridgeshire.*

roots, of a purple-red dye are long-forsaken and all but forgotten uses of willow. It is still used as firewood, when well seasoned, and the charcoal is in demand for artists' use; one or two coracle builders also need it. The chemical salicylic acid, which is extracted from the bark, was once the main constituent of aspirin, although it is now synthesised. Geoffrey Grigson suggests two more obscure uses for the white willow wood, namely artificial limbs and polo balls. Recently the tree has also come into fashion for large-scale woven-willow sculptures. Artists around the country are using live willow wands to make living arbours, garden furniture and sculptural installations such as giant human figures or creatures of myth and reality. At first these pieces are simply basketry structures, but they have the potential to grow and evolve as the willow takes root. The construction becomes part of the living landscape, and its future development is then left in the hands of the elements. Such projects are particularly popular with schools and colleges.

The other, and most familiar, large tree willow is the introduced weeping willow. Its Latin signature, *Salix babylonica*, was designated by Linnaeus in the wake of a biblical reference to willows along the rivers of Babylon, which he assumed to be the original source of the tree. It is now believed that the tree originated in China, where a surgeon with the East India company, James Cunningham, was reputed to have collected the first specimen to reach Britain in 1701. By 1730 the tree was available for sale in catalogues published in London, though popular myth traces its provenance to the poet Alexander Pope who, while living at Twickenham in the latter half of the century, received a basket of figs from a lady friend in Turkey. The basket was made of willow, which Pope assumed might have been cut from the Babylonian riverbanks. Noting some buds on one of the willow wands, he decided to plant it, whereupon it grew into a successful tree in the grounds of his villa. After his death this willow

'became the object of so much curiosity that the possessor of his villa cut it down, to avoid being annoyed by persons who came to see it' (Reverend. C. A. Johns, *The Forest Trees of Britain*, 1875). The true species has proved not to be hardy enough to thrive in many situations in Britain, but its hybrid with white willow, introduced around 1800 and known as the golden weeping willow (*Salix* x *sepulcralis nv. Chrysoloma*), tends to be the most commonly grown weeping willow.

Weeping willow has become a particularly popular garden tree, as its vivid green, slender fronds come into leaf earlier than most other trees in spring and stay on the tree until late into the year, and when planted in conjunction with water it is unsurpassed in its graceful beauty. Set against lakes or ponds it provides magnificent cascading mirrors of waving green foliage, while the weeping canopies draped above ornamental falls and cascades emulate the water beneath.

Willow Hybrids

There are dozens of willow hybrids that have been selected for their form or colour, but two of the most striking of these reveal their splendour in the warm low-level sunlight of a winter's day. Scarlet willow (*Salix alba* 'Chermesina') and golden willow (*Salix alba* 'Vitellina') bear the eponymous highly coloured branches which may glow like a beacon in a brown winter landscape. Perhaps the most valuable white willow hybrid is the cricket-bat willow (*Salix alba* 'Caerulea'), now grown almost exclusively in controlled conditions in East Anglia. The tree is usually distinguished by the bluish tinge to its foliage and the more upward thrust of its boughs than in its cousins. Although it has the potential to develop into a large tree, it is rarely left much more than 20 years before being harvested, while the timber is still free of rot and shakes. Its dual qualities of extreme lightness combined with durability have made it the finest willow for cricket bats.

Osier Willows

Still in the realm of water-loving willows are the osier species. Most of these lesser willows have the same long, spear-shaped leaves of the larger trees, but even when allowed to grow to maturity the adult tree rarely exceeds 16 to 20 feet in height. *Salix viminalis* is the true osier, but other willows such as almond willow (*Salix triandra*) and purple willow (*Salix purpurea*) are often grouped in the same type, as they have long been used, along with many of their hybrids, within the osier/basketry industry. When basket wares were in great demand there were osier beds in every lowland part of Britain. Today this industry has become centralised in the Somerset Levels, which produces the vast majority of 'withies' for most of Britain's basket makers.

The whole band of osier-type willows have been

perpetuated or developed for the specific colours of the stems or pliability of the rods, as well as the incredible speed of their regrowth, which may be as much as 12 feet in one season. If the layman is confused by the different types of tree willows then the osiers will cause utter exasperation.

The Sallows

The third grouping of willows is the sallows, widely known colloquially as 'sallies', because of their botanical name *Salix,* from the Celtic sal (near) and lis (water). The sallows are most readily identified by their broader, elliptical or boat-shaped leaves and marvellous golden display of the male catkins known to all as 'pussy willows' which emerge before the leaves in spring. The 'pussy willows' have traditionally been used as 'palms' at Easter, and have been borne in many a parade and used to decorate churches all over Britain, which is why the trees have often been known locally as palms.

By far the most familiar member of the sallows is the native goat willow (*Salix caprea*). Exactly why it is so called is not addressed by many authorities. One possibility might be the similarity of the emerging silvery catkins to the small fluffy toggles found on the throats of many goats, although John White suggests that the literal translation of *caprea* is a female roe deer.

The goat willow is not, strictly speaking, a willow of particularly wet terrain, but much more a tree of hedgerows, woodland edges and waste ground. As this is usually the first of the willow tribe to flower, its silvery male catkins smothered in golden pollen make a welcome glow across the landscape from February through to early April. If warm days have rousted the bees from their winter slumbers, then this is a vital repository of nourishment. The grey-green catkins of the female tree tend to be less conspicuous than those of the male.

One of the closest counterparts of the goat willow is the grey sallow (*Salix cinerea*). The two look very similar, both forming small trees up to about 33 feet in height, yet the grey sallow prefers much damper situations, being found particularly in marshes and fenland. Its flowering is undoubtedly its glory as, for the rest of the year, the sallow, with its rather dull green leaves, blends into the natural backdrop.

The sallows were often coppiced for trug making, handles for implements, hoops for coopering and cleft staves for fencing purposes. The wood is not particularly durable in wet conditions. Today sallows are not viewed as very useful trees, and some foresters even see them as weeds and do their best to eradicate them from commercial woodland, although the trees may have a useful function as 'nurse trees' sheltering other species.

ABOVE: *The male flowering of goat willow – usually known as pussy willows – one of the first harbingers of spring.*

NEXT PAGE: *A most unusual sight, an ancient pollard goat willow, near Braemar in Scotland. Today there are only a few piles of stones where once stood some lonely croft. Many a year has come and gone since this tree was last cut for poles.*

BELOW: *These Oxfordshire willows were pollarded for many years but after long neglect they have been allowed to develop into huge trees.*

Poplar

'The poplars are felled, farewell to the shade,
And the whispering sound of the cool colonnade;
The winds play no longer and sing in the leaves,
Nor Ouse on his bosom, their image receives.
Twelve years have elaps'd, since I last took a view
Of my favourite field, and the bank where they grew;
And now in the grass behold they are laid,
And the tree is my seat, that once lent me shade.'

From The Poplar Field by William Cowper (1731–1800)

ABOVE: *A rendering of the tree at Bury St. Edmunds, drawn by Strutt and then reversed and engraved by Whimper for the Rev. C. A. Johns.*

BELOW: *Most black poplars are found along valleys and usually close to water, but this mighty hedgerow specimen stands on top of a hill, without an obviously damp siting. Traditionally these trees are associated with flood-plain woodland (although little of that habitat remains due to drainage for agriculture), yet the tree is known to do well in upland situations. Their success here is probably due to the existence of underground springs.*

Black Poplar

Two poplars are native to Britain: the black poplar (*Populus nigra* ssp. *betulifolia*) and the aspen (*Populus tremula*). Two others, the white and the grey poplars, are naturalised species. In much the same way as the willows, the poplars, which are part of the same family, are much given to hybridising. As a result there are many ornamental varieties, as well as a new trend for commercially raised clones, grown in vast plantations for their quickly maturing timber.

Arguably the most splendid poplar of all is the native black poplar (*Populus nigra* ssp. *betulifolia*). Much has been researched and written about the black poplar over the last ten years, principally due to its scarcity and its confusion historically with the imported hybrid black poplar (*Populus x canadensis*), of which the clone 'Serotina' grows into large trees that are often confused with the true native. In 1913 the redoubtable Elwes and Henry marvelled at the continuing confusion between the two trees, considering their quite different physical characteristics and seasonal developments.

Recognising a native black poplar may seem a tall order. In winter, from a distance, it is easily confused with a willow, and one of its alternative names is, in fact, willow poplar. However, once identified, it begins to stand out in the landscape. The tree may be as high as 120 feet, with the larger boughs arching downwards in a distinctive manner. The trunk of a mature black poplar tends to lean to one side and has deeply fissured, dark grey bark, often with an abundance of rough burrs and bosses. Its deltoid leaves usually burst forth in bright green rather than the reddish flush of new leaf in its Italian cousin, the Lombardy poplar. Since the trees are dioecious, the males and females bear their respective catkins in March and April; the deep crimson catkins of the male present a striking display in low sunlight. Though there is nothing about its appearance that is absolutely black, its bark is slightly darker than other poplars, the foliage is also darker, and the mature timber often has a black veining which no other poplars produce.

Black poplars have long been regarded as trees of flood-plain woodlands, although because of agricultural improvement, very little such woodland survives. Where they do grow in the close proximity of a wood the trees do not thrive so well as in their now more usual habitats of river banks and hedgerows. The natural distribution is almost exclusively below a line between the Mersey and the Humber, with particular concentrations in the Welsh Marches, the Severn valley, the Vale of Aylesbury and East Anglia. Historic accounts of the trees are confusing. Strutt mentions that they were 'oftener to be found in Cheshire and Suffolk', but Elwes and Henry knew of none in Cheshire, while recent studies led by Jonathan Guest in association with the Cheshire Wildlife Trust indicate the presence of about 200 specimens there. Such anomalous accounts are almost certainly due to the aforementioned confusion with the hybrid black poplar. During the mid-1970s, the botanist Edgar Milne-Redhead began to look more closely at the native black poplar. His survey originally suggested barely 1,000 specimens nationwide, and yet today, as other naturalists and tree enthusiasts seek them out, it is estimated that there are more than 2,000 of the trees, and Richard Mabey expects the figure to rise as more surveys are logged.

Up to the middle of the last century, black poplars were still being planted, usually with cuttings from male trees as females were less favoured due to the copious amounts of white fluff they produced as they set seed. Hence the predominance of male trees today, many of them well over 100 years old: of more than 2,000 trees identified nationally, only about 15 percent are females. Due to the rare coexistence of the two

sexes in close proximity, regeneration by seed is virtually unknown; this has probably been the state of affairs for a long time. The widespread draining of agricultural land has also deprived the black poplar of the continuously damp mud-banks of streams and ditches most suitable for the success of its seedlings, although some natural regeneration takes place from fallen boughs and detached sticks which fetch up in an ideal damp location. A programme of vegetative reproduction with human assistance is now under way to give the tree a better chance. The last 20 years have seen a reawakened public interest in the native black poplar, which had survived more as a result of good fortune than from concerted conservation. The tree is now being monitored nationally, and a strategy for regeneration and conservation has been adopted. There are, for example, schemes to plant female trees near isolated groups of males and well away from interloping hybrids, with the intention of maximising the tree's opportunity to reproduce true seed.

Most of the existing mature trees are pollards. A recent survey of black poplars across Herefordshire (Watkins, Holland and Thomson) found 68 percent to be pollards and 31 percent standards; the remaining two trees were coppiced. Although most of the pollards were well grown out, it was encouraging to find that a few were still being harvested. A lot of the surviving trees, especially the pollards, are situated close to farms and cottages where they have been conveniently situated for lopping.

In Herefordshire, the timber of poplar was valued for the beds of wagons and for the interiors of hop kilns, where its resistance to fire and the hot steamy conditions while the hops were being dried made it admirably suited. Gradually, various researchers are beginning to discover more and more evidence of black poplar used in the construction of early buildings, particularly from the medieval period. It has long been known that the timber's fire retardant quality made it ideal for floorboards, but now some of the great crucks in barns and houses, which were once assumed to be oak or elm, have been identified as poplar. The graceful sweep of the mature trunk makes the tree an obvious contender for this role. In general, the timber was not one of the most valued, being used in the past for baskets, packing-cases, matches, cotton-reels and clogs. It does not endure well in damp conditions, but when kept dry:

> *'Though heart of oak be e'er so stout,*
> *Keep me dry, and I'll see him out.'*

A few large black poplars have survived as landmark trees on village greens or along parish boundaries, but they are exceptional. Undoubtedly the most famous black poplar in the land was the Arbor Tree at Aston on Clun, in Shropshire, which was decorated with flags on 29 May each year. The ceremony was reputed

ANCIENT ARROWS

A surprising use of poplar was discovered when some arrows recovered from Henry VIII's flagship, the Mary Rose, were found to be made of the wood. It is amazing that they survived beneath the Solent for 400 years.

PREVIOUS PAGE: *This magnificent male black poplar at Longnor Hall in Shropshire is the tallest of its species in Britain.*

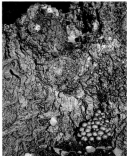

ABOVE: *The rugged bole of an ancient black poplar harbours a clump of honey fungus.*

LEFT: *An old pollard black poplar bears a fine stock of poles. The smaller trees on the right rise from a stream directly below the old tree, and so they may be derived from suckers or may have sprung from broken twigs which managed to lodge in the stream bank.*

OPPOSITE: *A splendid old black poplar makes a stunning landmark on the village green at Blakemere in Herefordshire.*

The White Poplar Tree

to have begun in 1786 when the Lord of the Manor, John Marston, married Mary Carter. The bride was much taken by the decorated tree and left money in trust so that it might be similarly decorated in perpetuity. Sadly, gales in September 1995 proved that the tree was no more immortal than Mary. Down it came: but the local community had seen fit to strike cuttings from the tree over the last few years, a rooted cutting being presented each year to the 'bride' in a re-enactment of the original wedding parade, performed by local children. In December 1995 the best of these young trees was selected and planted by way of a replacement, and thus a genetically identical tree will flourish in the place of its fallen parent.

The hybrid *Populus* x *canadensis* was originally the result of crossing American black poplar with varieties of European black poplars; early English settlers first made the cross in America during the eighteenth century. An alternative and perhaps more accurate name for these poplars introduced from America is *Populus* x *euramericana*. The trees grow remarkably fast, which makes them popular for screen and windbreak functions; the young trees can be harvested for biomass and the older poles cut for uses that require strong, lightweight timber. Some of the newer clones which have been developed in northern Europe have massive leaves as big as dinner plates. Plantations of these 'farmed' poplars are springing up all over the country, and the sight and sound of their fluttering foliage makes a welcome relief from endless stands of conifers. The clonal poplars are capable of phenomenal growth in coppice rotations as short as three years. Government-sponsored studies are exploring the agricultural logistics and industrial potential of their commercial production in order to create an alternative, and environmentally friendlier, fuel source for electricity. Countries such as Sweden and the United States already have wood-fired power stations. Ultimately it will be the economics of the venture which will decide its future: farmers need to know that the income from their land laid to coppice will at least match the returns from more traditional crops and, similarly, electricity producers need to be able to generate competitively priced power.

White Poplar

The white poplar (*Populus alba*) is an introduced tree from southern Europe, although exactly when it arrived is uncertain. Turner, in his Herball of 1568, did not recall seeing any examples in England: nearly 100 years later Samuel Hartlib only had knowledge of 10,000 abeles being imported from Flanders. Here lies the root of a confusion between the white poplar and the grey poplar (*Populus canescens*), a hybrid of white poplar and aspen: these two cousins must surely figure in the same stanza. The name abel or abele comes from the Dutch word *abeel*, which in turn is taken from the Latin *albus* (white). Since, as Hartlib confirms, many of these abeles were imported during the seventeenth century, it might be assumed that this was the era of the white poplar's original introduction. However, in this case, the Dutch name *abeel* refers to the grey poplar.

Oliver Rackham has also found evidence of the word abel appearing in East Anglian medieval records. Today the white poplar is not particularly common in those parts, which is strange since its great capacity for suckering should have assured its proliferation. Here abel might well have referred to the grey poplar, which is often confused with the white poplar. Or was the grey poplar simply brought into Britain before its white parent? Since early authorities had limited botanical knowledge, and dozens of isolated communities across Britain adopted either tree as white poplar, the answer still remains a mystery. It is

LEFT: *This solitary Lombardy poplar certainly makes a mark in the landscape, whether for better or worse, with the vast expanse of the plain of Worcester stretching beyond. A small group of pollard black poplars nestle peacefully beneath this giant invader.*

not lessened by John Miller's delineation of 'The White Poplar Tree' in Hunter's 1796 edition of Evelyn's *Silva:* Miller depicted a leaf which appears to be an artistic hybrid of the white and the grey, although the small detail on the plate shows eight stigmas and is thus actually the grey. The drawing corresponds to what Evelyn describes in his text as, in fact, 'a finer sort of White Poplar, which the Dutch call Abele, and we have of late much of it transported out of Holland'. This latter tree is what is now regarded as white poplar. Confusion seems to reign!

From a distance both trees have a similar profile, and in spring they present the same creamy pale green canopy. The white poplar has palmate leaves more like maples, while the grey poplar has the usual populus deltoid leaves, but with the round-toothed edges of the aspen. Even in the leaves there can be confusion, as leaves on less vigorous shoots of white poplar often closely resemble those of the grey.

Of the two trees the white poplar with its splendid woolly white leaves is the more attractive, particularly when moved by the wind to show the creamy undersides, no better viewed than in low sunlight against a leaden sky. The tree has often been planted in landscape schemes, in much the same way as whitebeams, to contrast with darker foliage trees. Evelyn recommended both poplars for 'walks and avenues about grounds which are situated low, and near the water', while Gilpin confirms that their 'chief use in landscape is to mix as a variety, in contrast with other trees'.

Practical uses of the timber of white and grey poplars, as with other members of the tribe, were for boarding of carts and floors, making boxes and cases, for turning small objects such as bowls and toys, but also, because of the extreme light colour of the wood, for applications where the finished article was to be painted or stained. The trees no longer fill a vital commercial role, but as fine amenity trees they are an invaluable aspect of the lowland landscape.

Lombardy Poplar

Poplar to most people means the Lombardy poplar, that towering spire of a tree that lines many virges and noble drives. The Lombardy is generally assumed to have originated as a sport of a normal black poplar, in the Po valley of northern Italy, around the beginning of the eighteenth century. The tree arrived in England in 1758, brought back by Lord Rochford, who was then ambassador in Turin. He planted his cuttings at St. Osyth's Priory in Essex.

Lombardy poplars grow very fast, often attaining 60–80 feet in about 30 years; larger specimens may reach 120 feet. Though the trees rarely last longer than 60 or 70 years, while they are tall and healthy they provide prominent landmarks discernible from great distances. Its stark contrast with almost every other tree in the British landscape has brought the Lombardy poplar more than its fair share of detractors as well as admirers. Avenues of these great green ninepins have long been planted as massive weather screens or as bold heralds of many a mighty residence; as Hugh Johnson avers, 'There is no tree which creates an architectural incident so quickly and surely.'

Mr. Gilpin had his contribution to make also, being much impressed by the movement of the tree in breezy weather.

'One beauty the Italian poplar possesses, which is almost peculiar to itself; and that is the waving line it forms, when pressed by the wind. Most trees in this circumstance are partially agitated. One side is at rest; while the other is in motion. But the Italian poplar waves in one simple sweep from the top to the bottom, like an ostrich-feather on a lady's head.'

OPPOSITE: *Some of the first Lombardy poplars to be planted in Britain were at Blenheim, in Oxfordshire, c.1760.*

Aspen

Right: *Autumn aspens,
Glen Angus, Scotland.*

The aspen (*Populus tremula*) is the other poplar native to Britain. *Tremula* is a most apt name for this tree. Constantly quivering and rustling in the lightest of breezes, the rounded leaves with their regular wavy indentations and long, flattened stems pitter-patter against one another.

From an early treatise entitled 'The Forester' comes a lyrical description of aspens:

'*There stands a group of fine aspen poplars, about sixty feet high, with their clean grey stems and rugged horizontal branches stationary as the earth upon which they stand, with leaves all in motion like an agitated sky, without a breeze of wind below. Although the evening is so still that the sound of a burn fully a mile off is easily heard, the leaves cannot remain quiet; and now and again, as the air rises into the most gentle breeze, and almost brings with it the sound of the very minnow's flip upon the surface of the water in the far-off pool, the leaves of the aspen vibrate to the sound, and their rustling*

Below: *Seemingly a small group of aspens these are actually all suckers of the same tree.*

falls upon the ear sweeter than any music.'

The tree has several local names which derive from aspen, such as aps, asp and apsen. Other names allude to its quaking and shaking leaves, often likening them to garrulous old women, as in Gerard's description: 'seeing it is the matter whereof womens toongs were made, as the Poets and some others report, which seldome cease wagging'. Aspen is called Old wives' tongues in Roxburghshire, Woman's tongue in Berkshire, and Women's tongues in North Wales (*tafoden merched*) and the Isle of Man (*chengey-ny-mraane*).

Aspens would first have recolonised Britain about 9,000 years ago and, in much the same way as Scots pine, gradually migrated northwards with the warming of the climate. The tree's greatest concentration is now in the north of Britain, where it is perfectly at home growing at altitudes as high as 1,600 feet. While seldom reproducing by seed, the aspen has a phenomenal capacity to sucker, so that in a relatively short space of time a few trees can turn themselves into a dense clump of woodland. One example of this, near Chepstow, has developed more than 1,000 stems.

Aspens tend to prefer porous, well-drained, moist soils, but not waterlogged conditions. They do not tolerate shade well; in Scotland they may sometimes be found colonising felled woodland with, or instead of, birch. In fact aspen can coexist rather well with birch, as both trees are quite light and airy, grow to similar sizes, and have equal life spans of 50 to 80 years.

Aspen timber is similar to many other poplars, being pale, lightweight and strong. The wood is often used for veneers and plywood, as well as pulp for the paper industry.

Alder

Of the same family as birch, alder (*Alnus glutinosa*) may be found wherever there is water in every part of Britain. The tree bears little resemblance to its cousin the birch, yet the trees are both excellent pioneers, producing massive amounts of seed which will germinate with alacrity. From the marshes and fenlands of East Anglia to the rocky mountain streams of Wales and Scotland, alder is equally at home. Where it grows on a site with unhindered space and light, in a mineral-rich soil, a fine specimen tree may reach a height of 100 feet, but this is exceptional. More often alders grow amongst the two other water-loving trees, willows and poplars, and in such habitats they tend to defer, often barely growing to 40 or 50 feet high. On mountainous terrain the alder proves itself a tough contender, unmoved by the wildest of weather conditions, although the tree must sometimes compromise its size in such rugged places, so that where it is particularly exposed, a mature tree may only develop to the size of a shrub. In marshy lowland sites alder often appears as the first line of colonising woodland, sometimes in conjunction with sallow, forming swampy woods known as alder carr. From the outside these might seem like rank and uninviting places, but they are valuable wildlife habitats and give a wonderful sense of an ancient woodland type. Geoffrey Grigson was clearly a fan:

'Once enjoyed, an alder swamp along a Cornish stream, for example, remains perennially and primevally enchanting – the trees alive and dead, moss-bearded and lichen-bearded, the soil and the water like coal slack and blacksmith's water, in between the tussocks of sedge.'

Alder puts on one of its best displays in late winter when the countryside is generally devoid of much colour. A riverbank lined with alders covered in dark red male catkins is quite something to behold. Close

inspection also reveals woody little cones which are the remnants of the previous year's female catkins, the tiny winged seeds having long departed downstream or blown on the wind. The foliage of alder is handsome in spring when the leaves are a rich green with a very glossy, even glutinous (hence the scientific name) surface, but after midsummer they turn to a rather dull and dark green and there is no autumnal show.

As with so many trees the alder has had its advocates and its detractors. Loudon considered that 'it is one of the most melancholy of deciduous trees'. Reverend C. A. Johns thought it a 'stiff, heavy, and even gloomy' tree, but in its favour he noted that 'it has been observed that their shade is much less injurious to vegetation than that of other trees'. Selby claimed that the tree has a great capacity to retain water in the surrounding earth through its powerful root system; he warned foresters that an over-abundance of alders on a tract of land will eventually convert it into swamp.

ABOVE: *The alder festooned in bright red male catkins is a dramatic sight to behold¶.*

LEFT: *Rich green spring leaves bedeck this old alder coppice stool.*

Alder can germinate on many types of terrain, even some that are seemingly almost devoid of decent loam. Whether it be rock, gravel, sand or clay, the tree will thrive and has become an ideal candidate in recent years for landscaping reclaimed industrial sites where soil quality is seldom very good. Alder overcomes such adversity by drawing nitrogen from the air through nitrogen-fixing bacteria contained in nodules on its roots. The relationship is symbiotic, as the bacteria feed on the sugars manufactured by the tree's leaves. At low water on a riverbank these nodules can be seen as small hedgehog-like clumps on the exposed roots.

Many place names derive from alder and reaffirm its longstanding widespread distribution. Most commonly it is known as *eller* or *aller*, with many slightly different forms abounding: Ollerton, Northallerton, Aldershot and Alderley. Grigson quotes some rather more obscure local names, such as dog-tree in Lancashire, wallow-wullow in Shropshire and whistlewood in Northumberland. Due to the reddish-brown colour of the seasoned timber it has also been known as Irish mahogany and Scots mahogany. (Scots mahogany was alder wood buried in peat for some time before use, to deepen and fix the colours.)

Alder timber is strong yet light in weight. It will not split easily and as such is an ideal wood for turners, still being made into broom heads and handles for tools. Two of the most important uses of alderwood are now all but forgotten. It was always chosen as the best wood for making clogs, which were worn by thousands of millworkers, miners and farmworkers across the north of England, up until World War II. The alder was cut when the tree had attained about two feet in girth, and then logged, cleft and shaped into rough blanks by the coppice workers, after which they were stacked to season, before shipment to the cloggers who would trim the blanks to fit the individual requirements of their customers. Tough leather uppers and steel cleats nailed to the soles produced virtually indestructible footwear. One or two of the old cloggers still survive in remote corners of the decaying industrial north, although clogs are no longer a fundamental necessity of working-class life. Alder was also the very finest wood from which to make charcoal for the manufacture of gunpowder, and to this end it was often planted in extensive coppice woods close to gunpowder factories.

Alder timber is remarkably resilient when submerged in water and was thus often used in the construction of underwater pilings and supports for quays, buildings and bridges. Evelyn wrote that 'they used it under that famous bridge at Venice, the Rialto, which passes over the Grand Canal, bearing a vast weight'. Its sturdy root system made it much recommended for planting along riverbanks in order to strengthen them.

The bark is rich in tannin and used to be of great value to tanners. The bark also makes a versatile dyeing medium, once being used to make many shades of yellow, red, brown and, when combined with sulphate of iron, an exceedingly rich and permanent black. Edlin also notes that the fresh wood gives a fawn dye, the female catkins give green and the young shoots produce yellow. The freshly cut stumps of alder can

The Alder Tree

LEFT: *Miller's delineation of Alder from First Hunter edition of Evelyn's* Silva.

NEXT PAGE: *Alders overhang the rushing waters of the Grwyne Fawr above Usk valley in Wales.*

ABOVE: *Nodules are clearly visible on these exposed alder roots above the Walkham river on the edge of Dartmoor.*

BELOW: *Deep in the heart of long-neglected coppice woodland above the Usk valley lies this old alder stool. Such veteran trees develop into great old characters.*

ABOVE: *The golden yellow catkins burst forth to release their pollen. Close by on the twig are the tiny bright red female flowers which will swell, turn green, and by the following winter look like these small brown cones.*

RIGHT: *Typical alder carr woodland at Coed y Cerrig, north of Abergavenny in Gwent. The alders have been coppiced in the past, but after recent years of neglect many large stems have toppled. English Nature have now begun to manage the woodland, and by laying a boardwalk across the site have made it accessible to the public. As a distinctive habitat, alder carr is unique and endangered, as many landowners drain such land.*

OPPOSITE: *Willows, winter and water inspire the spirit.*

BELOW: *Riverbank alders remain unmoved by swirling flood waters in the Welsh Marches.*

look quite striking, as the copious amounts of tannin in the bark turn the cut surfaces a bright red-orange colour. In the past this led to a belief, particularly in Ireland, that the tree was 'bleeding', which, in a way, it was; and it was deemed unlucky to cut it.

The harsh reality for today's alder population is a relatively new, serious threat from the root disease known as Phytophthora. First noticed in 1993, on trees as widespread as Kent, Worcestershire and Gwent, the disease has now spread nationwide. The spores of the Phytophthora fungus are water-borne, so it is the alders on riverbanks which are most at risk. The symptoms include impoverished foliage and strips of dead bark from ground level, often bearing tarry or rusty marks. It will almost certainly kill many thousands of trees, but it must be hoped that at least some alders will develop an immunity to the fungus, for, like Dutch elm disease, it will be almost impossible to control if it spreads with similarly epidemic proportions. Survey and research into the disease is ongoing.

The European alders, Italian (*Alnus cordata*) and grey (*Alnus incana*), along with red alder (*Alnus rubra*) from America, are currently being planted in increasing numbers around Britain. The red alder is valued in America as an important timber tree, but in Britain all three are used principally as amenity trees, particularly in inhospitable sites in urban areas or reclamation schemes, where they fare better than most other species. They may all be distinguished from the common alder as they bear larger and more pointed leaves. There are also various ornamental cultivars available as park and garden trees.

CHAPTER EIGHT

Trees, Folklore and Medicine

Trees have been revered since the beginning of recorded time and probably even before that. Although tree lore and associations have largely subsided into legend and superstition, many people today still give credence to the spiritual importance of trees: and the increasing interest in herbal remedies continues an ancient curative tradition.

When mankind first began to reason, communicate and come to terms with the natural world, the most splendid and powerful elements of that world must have commanded respect, if not reverence or even fear. Plants and animals may have been perceived as either the embodiment of spirits or the channels through which their influence flowed: and though trees were among the most ancient life forms on the planet, their life patterns were strikingly similar to those of human beings. Tender young saplings grew into healthy mature trees, which in turn declined into decrepitude and finally death; they required an abundance of sunlight and water to thrive; they were vulnerable to the harshest of elements and, like man, when broken they bled. Traditions the world over reveal that types of trees or specific ancient individuals have long been worshipped: there were trees to which sacrifices must be made and trees that must never be felled for fear of angering the spirits within.

OPPOSITE: A strange fungal tree being emerges from the smooth grey loins of a towering beech.

LEFT: A twelfth-century Norman carving of the Tree of Life, on the tympanum above the south door of St. Mary's church, Dymock, Gloucestershire.

TREE OF LIFE

The Tree of Life has long been universally recognised as a potent symbol. To most Judeo-Christians it is familiar from the text of Genesis, when Adam and Eve were cast out of the Garden of Eden for partaking of the Tree of Knowledge of Good and Evil. The Tree of Life stood cruciform as the tree of redemption. In China the Tree of Life was the Kien-mou, which grew on the slopes of the earthly paradise of Kuen-luen. Buddha gained enlightenment beneath the Bo or Bhodi tree. The Muslim Lote tree signified the boundary between human understanding and divine mystery. In early Egyptian and Assyrian art, paintings and sculptures of trees represent creation and a source of strength and immortality. Soma, the Tree of Life of the Hindu faith, contained a juice which imparted immortality, as did the sap of the Persian Haoma tree. Ygdrassil, the World Tree of the Norsemen, was a mighty ash, representing the life forces of nature; at its feet sat the three Nornas, or fates – Past, Present and Future – watering the roots from the Sacred Well. Tribes of Africa and North and South America all held trees sacred in some way. In Britain the image continues to have social, if not mythical, importance in the family tree.

Certain trees have traditionally held more significance in folklore than others. The oak and ash are pre-eminent, but many trees of lesser stature also have rich associations. Recurrent themes in tree lore relate to love, fertility, death and protection from evil. The notion has long abided, for example, that plants and trees (or more often bits of them) had the capacity to predict partners in love, marriage and the prospects of children.

Many other superstitions associated with trees are purported to presage death, the onslaught of witches or, in the worst cases, the Devil himself. Some trees have been seen as attracting or harbouring evil spirits or witches, while others might be prized as an amulet against the same evils; the elder is associated with both. The yew has gained a uniquely dour reputation from its frequent presence in graveyards and its dark and poisonous foliage: a misplaced image, since it is a treasure both for its longevity and its beauty. It has always been considered unlucky either to damage it

or to chop it down and it is portrayed most often by writers as the gloomy guardian of the dead. Plutarch was driven to warn of the fatal perils of sleeping beneath a flowering yew which is 'venomous when it is in flower, because the tree is then full of sap'. One of the most colourful, if grim, portrayals of the tree comes from Robert Turner, writing in 1664:

'If the Yew be set in a place subject to poysonous vapours, the very branches will draw and imbibe them, hence it is conceived that the judicious in former times planted it in churchyards on the west side, because those places, being fuller of putrefaction and gross oleaginous vapours exhaled out of the graves by the setting sun, and sometimes drawn by those meteors called ignes fatui, divers have been frightened, supposing some dead bodies to walk, etc.'

It is known that the oldest yews were located in sites of spiritual significance long before Christianity came to the shores of Britain and as symbols of the pre-Christian era could have been viewed as sinister or malevolent. However, the early Christian missionaries absorbed this heathen totem into the new faith; by building the earliest churches near the yews they fostered some system of mutual tolerance, while the symbolic importance of the tree became linked to Christ's resurrection. As Christianity gradually spread and ultimately ousted the old faiths, hundreds, if not thousands, of venerable yew trees stood as memorials around the churches. Slowly but surely their original significance was forgotten and they became accepted as an integral element of the Christian sites, so that as new churches were built, yews were automatically planted without any deep concern for their less than Christian origins. The rowan tree is also often found in or around churchyards, in particular in Wales. It is said that the wood was used to make the Cross of Calvary and that the berries symbolise Christ's blood; another belief is that the trees will prevent the slumbers of the dead from being disturbed. But rowan's most important property is its ability to protect from witchcraft. No other tree may be considered so potent for this purpose.

Before they were replaced or, in many cases, recycled by Christianity, pagan traditions in Britain also placed significance upon the oak. The oak has always taken the mantle of king of the forest, for no other tree stands so great and noble, strong and enduring.

Oaks are revered the world over as one of the most sacred of all trees. Early accounts of the Roman Empire report the high esteem accorded oak. It was customary for citizens who saved someone's life to be crowned with oak wreaths, a tradition which has been perpetuated to this day by the use of oak leaves as part of some military decorations. In many cultures, notably the Greek and Roman, oak was closely associated with thunder and the gods Zeus and Jupiter who held sway in the skies with their thunderbolts and lightning. So, when the Romans poured into Britain they cannot

have been surprised to find that the oak was equally special to the Britons and the Celts. The Druids made homage to the oak, in particular when it was adorned with mistletoe.

With the arrival of Christianity the oak theme was absorbed from the old pagan ways into the New Faith. Depictions of acorns and oak leaves figure prominently in many of the earliest churches. The carvers and stone masons used their long-standing traditional iconography alongside the new Christian themes.

The ultimate display of the 'old ways' were the depictions of the Green Man. In essence this is the tree spirit, taken either as the spirit within the oak or even the source of mankind, for many early cults believed that man was first sprung from the oak. These robust faces, wreathed in oak leaves and often spewing leaves from their mouths, represent renewal and fertility, and are to be found in many a church, but often displayed discreetly, high on roof bosses or tucked beneath choir stalls on misericords, cheek by jowl with all manner of odd mythical beasts, symbols and scenarios which bear no reference to any biblical themes. The Green Man is probably most familiar to the clientele of many a public house all over Britain,

LEFT: *A medieval carving of a Green Man in All Saints church, Sutton Benger, Wiltshire. The leaves are hawthorn and birds can be seen plundering the berries.*

SUPERSTITIONS

While superstitions attending trees have never been rationally explained, it is logical that a poisonous tree would attract negative associations. Witches, for example, would use such herbage to lace their cauldron brews. Conversely, the decoctions of parts of other trees provided miraculous cures. In many modern vaccines, a small dose of the disease agent enables the body to build up its immune system, and the same seems to apply in protective plant lore. Culpeper quoted similar occurrences with herbs, such as hawthorn: 'If cloths or sponges be wet in the distilled water (of the flowers), and applied to any place wherein thorns and splinters, or the like, do abide in the flesh, it will notably draw them forth. And thus you see the thorn gives a medicine for its own pricking, and so doth almost everything else.'

T. F. Thiselton Dyer, in the nineteenth century, mentions the same phenomenon:

'...it is noteworthy that many of the plants which were in repute with witches for working their marvels were reckoned as counter-charms, a fact which is not surprising, as materials used by wizards and others for magical purposes have generally been regarded as equally efficacious if employed against their charms and spells.'

Folkard thought that the tree was an anathema to witches since its very name is a corruption of 'holy' and its thorns and berries symbolise the Passion: charm and counter-charm.

LEFT: *Recent evocation of the spirit of the Green Man created in tree branches in Bexley, Kent. Jeff Higley,* Common Ground.

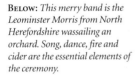

ABOVE: *A Green Man pub sign. This splendid rendition includes the Green Man himself, leaf-clad, sprung from the forest, boldly brandishing his great phallic club, while in the distance the nubile young women dance nimbly about a great maypole.*

since it is a common name. Most likely, some of these inns were connected with a rural group such as foresters or farmers, or maybe they were in close proximity to a ceremonial site such as a maypole. The Green Man, or Jack-in-the-Green as he is sometimes known, is usually depicted as a ribald and lusty-looking fellow.

Trees have always been a fundamental element of the rituals attending the turning of the seasons. Familiar trees such as the oak, holly and hawthorn have significant and symbolic parts to play within the round of traditional annual ceremonies and observances, which combine religious and patriotic holidays with the seasons of the pagan calendar – Samhaine, Imbolc, Beltaine and Lammas – equivalent to winter, spring, summer and autumn. At Samhaine (31 October), the pagan new year begins when the Oak Lord and the Earth Goddess depart for the otherworld, leaving Herne the Hunter, or the Holly Lord, to rule through the winter months. At Imbolc (2 February), the Earth Goddess returns – life springs anew. On Beltaine (May Eve, 30 April), the Holly and Oak Lords compete for the hand of the Maiden. The Oak Lord always wins her and makes her the Mother who will bring forth in the following Imbolc, while the Holly Lord slopes off to the Wild Wood until the next Samhaine. Lammas (1–2 August) is the fourth season, the harvest. The Christian Church has often found ways to integrate the pagan symbols into its own doctrines.

During the chilly dark days of winter, while the trees are still dormant, wassailing shakes and wakes the tree spirits in preparation for the forthcoming spring. It is an ancient custom, at least 1,500 years old; wassail is derived from the Anglo-Saxon *wes hål* meaning 'be whole' or 'good health'. There are two types of wassail. The first was usually confined to the

days leading up to Christmas, when roving merrymakers sang songs of goodwill and gathered round the wassail bowl, often turned from maple or ash and filled with a hot, sweet and spicy brew known as Lamb's Wool. (See Chapter Four.)

Later in the season came the apple wassailing, a tradition which may well stretch back to the days when orchards were first planted and husbandmen prayed for bountiful crops each year. Wassailing is recorded most particularly in the orcharding counties of the west and southwest. The custom has several minor regional variations, but usually follows a similar pattern. On Twelfth Night, the assembled company (sometimes the farmer and his family, or perhaps the whole village) decamped to the orchard. A great deal of noise was made, by beating drums and pans, in order to wake the sleeping tree spirits. Usually one fine tree was selected for attention. Guns were fired through the boughs to drive away the evil spirits. Libations of cider were poured upon the tree roots, while toast soaked in cider was wedged in the forks of the trees to supply sustenance to the tree spirits in the forthcoming year. The company then sang to the tree to encourage the productivity of the whole orchard. The wassail bowl is not usually a feature of this type of wassailing, but it was still customary to drink copious draughts of hot and spiced alcoholic brews, a marvellous stimulant for the carolling of the tree.

The songs appear in many variations, but the sentiment is always the same. Wassailing is essentially pagan in origin, appealing as it does to the spirits of the trees and the forces of nature to bring forth fruitfulness, but it was melded with the Christian calendar, for Twelfth Night is the eve of Epiphany. The intent behind wassailing has long since lapsed, though a few spirited revivalists have done their best to retain or resurrect the custom.

The month of May brings more customs in which trees figure prominently than any other time of the year, featuring most notably the hawthorn and the oak.

May Day has been celebrated with all manner of festivities since time immemorial. These events are all linked to fertility and the rebirth of life, hence the common occurrence of regimes involving garlands of spring flowers, the decoration and perambulation of May dolls, and their real life counterparts the May queens. The maypole was once found on many a village green but is now a fairly rare feature; fair young maids of springtime have traditionally been required to dance about this obviously phallic symbol while the enthroned May queen bears witness.

The maypole was originally a tree hewn freshly from the nearby woodlands; favourites included hawthorn, birch and oak. The pole was typically garlanded with flowers and green boughs, bedecked in streamers and painted in bright colours. Maypoles were often extremely tall. The Church of St. Andrew Undershaft in Leadenhall Street, London, was so

BELOW: *This merry band is the Leominster Morris from North Herefordshire wassailing an orchard. Song, dance, fire and cider are the essential elements of the ceremony.*

named because of the immense maypole (or shaft), taller than the church itself, which was erected each year outside its south door.

In the seventeenth century, the Puritans found May Day rites an abomination; not only were they rife with sexual symbolism, but there was also plenty of licentious activity accompanying the celebrations. Tradition held that on May Eve young lads and lasses repaired to the woods to enjoy each other's favours before returning to the village with garlands and posies on May morning. Maypoles were forbidden in 1644, but after the Restoration of 1660, the King himself ordered the erection of a mighty 134-foot maypole, which was splendidly decorated and set up in the Strand, London.

The blossom of the hawthorn tree is associated with May Day, even though the tree is seldom in flower so early; under the old Julian calendar, which was abandoned in 1752, 12 or 13 May would originally have been the proper day to celebrate May Day, when the blossom is far more likely to have been in full spate. Even though May blossom has long been popular for outdoor garlands and displays, there has still been a long-standing reluctance to bring it into the home, as it is supposed to bode extremely bad luck, a death in the family. Some authorities suggest that the strong smell of the blossom is reminiscent of the stench of London during the Plague: the blossom contains the chemical trimethylamine, which is generated during the process of putrefaction. This smell undoubtedly attracts carrion-seeking flies to perform their pollinating duties.

The British poet Geoffrey Grigson declares that the musky scent of the blossom is also highly suggestive of sex. What better token of fertility rites on May Day? The novelist H.E. Bates sensed the delicious sensuality of the May, noting, 'its flowers are the risen cream of all the milkiness of May-time. Its scent has the exotic heaviness of summer in it, very like the pungent vanilla half-sweetness of meadow-sweet'.

The hawthorn is also the central feature at the annual Bawming of the Thorn at Appleton. Some say it should take place in June, and in recent years the ceremony has gravitated to the Saturday closest to Midsummer's Day (21 June), yet for many years it was the old Midsummer's Day of 5 July which marked the custom of 'bawming', which most authorities agree is a local dialect word for adorning or anointing. The bawming used to involve the whole village, but is now mainly performed by the children. The tree is first decorated with red ribbons, flags and flower garlands before the children dance around the tree singing a bawming song. Nobody seems to know how long this tree dressing has been going on, and history shows several breaks in the tradition. In the nineteenth century it was suppressed because the attending revellers became too rowdy, and during the twentieth there has been fluctuating interest in the custom, which

ABOVE: *Every Christmas since 1947 a massive Norway spruce, decked with lights, has held pride of place in London's Trafalgar Square.*

RIGHT: *If many people today feel that the materialistic, profit-motivated scramble at Christmas has now lost sight of the supposed spiritual significance of the festivity, then it may be interesting to note that even in 1856 others had concerns along the same lines. This allegorical study from the* Illustrated London News *entitled 'The Christmas Tree' depicts an unseemly scramble for the prizes of power and wealth, whilst Jesus, Mary and Joseph watch The Bible shining a light, and yet unheeded by all.*

currently is thriving in the upturn of enthusiasm for folk traditions.

The original Appleton hawthorn was planted by Adam de Dutton to mark his safe return from the Crusades in 1125. The tree was believed to be a cutting taken from the famed Glastonbury Thorn and, like that tree, was known to flower around Christmas time, as well as its normal time of May. Now given the appelation of *Crataegus monogyna 'Biflora'* it is generally considered a sport and currently grows at several different locations around Britain. The present tree in Appleton is of this type, although not a descendant of the original.

Trees figure prominently in traditional customs at the end of the year. In the heart of winter, in an otherwise dead landscape, it has long been the tradition to decorate the house with all manner of evergreens for the winter solstice.

One of the earliest records of such midwinter ritual was the Romans' custom of decorating with laurel, holly and bay to celebrate Saturnalia, from 17th-23rd December. The most popular plants of the British Christmas time are holly, ivy and mistletoe, because they are evergreens and also because they fruit in the midst of winter: holly fitted in with holy symbolism, the thorny leaves reminiscent of Christ's crown of thorns (although this is variously attributed to hawthorn or blackthorn) and the red berries of Christ's blood. Mistletoe never seems to have shed its pagan associations so that, although it finds a place in most homes, it is seldom allowed into churches.

Whatever boughs and sprays of greenery may deck the festive halls, the most widely employed decoration is the Christmas tree, a tradition imported from Germany in the early part of the nineteenth century, though records show that it had already been a typical part of the seasonal celebrations there for several

THE CHRISTMAS TREE

centuries. Although various German immigrants had brought the idea with them, Queen Victoria and Prince Albert made it a fashionable custom from 1841 onwards, as accounts and pictures of the Christmas trees at Windsor were widely published.

Until the end of the last century most trees were decorated with candles, but as electricity made electric fairy lights possible it also spawned the idea of erecting large outdoor Christmas trees. The first ones were in America (reputedly in 1909 at Pasadena, California), and soon the custom spread worldwide. The most famous of those in Britain is the splendid tree erected each year in Trafalgar Square. Every year since 1947 the people of Oslo have sent a mighty Norway spruce to London as a token of appreciation of Britain's help during World War II.

RIGHT: *The annual ceremony of Bawming the Thorn, at Appleton, in Cheshire. This hawthorn was originally reputed to be scion taken from the famous Glastonbury Thorn, but over the centuries many hawthorns have come and gone at the appointed site. At some time in recent history a small error led to this pink flowered specimen being planted as a replacement, when the Glastonbury tree bears white flowers. However, the community does have the 'proper type' planted elsewhere in the village. Even so, the tree acts as an important focal point for the community, and has raised local awareness of folk customs and their historic relationship to tree veneration.*

OPPOSITE: *Winter moon above an ash in the Forest of Bowland in Lancashire.*

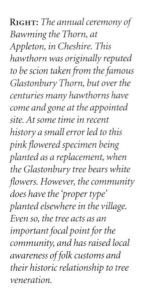

TREE CULTURE

Common Alder: *Alnus glutinosa*

Parts used: Bark and leaves

The common alder, a native of Britain, is a moderately sized tree growing to about 60 feet in damp boggy ground, by streams and rivers. Because it lasts well when immersed in water, it was popular for troughs and sluices, fencing stakes, and props to support buildings as in Venice and Holland. The tannins from the bark were used for dyeing and tanning leather. Until the second half of the nineteenth century alders were grown in coppices for providing charcoal to make gunpowder.

Culpeper said of alder: 'It is a tree under the dominion of Venus and of some watery sign or others, I suppose Pisces, and therefore the decoction, or distilled water of the leaves, is excellent against burnings and inflammations, either with wounds or without... if you cannot get the leaves then make use of the bark.' It is interesting that English farmers used to place alder leaves in their shoes in summer to keep their feet cool. Again Culpeper confirmed this: 'The leaves and bark of the alder tree are cooling, drying and binding... the leaves put under the bare feet galled

LEFT: *The bright red male catkins of the alder.*

OPPOSITE: *The richly coloured autumn leaves of the alder.*

with travelling and a great refreshing to them.'

Alder has been used medicinally since at least the time of the Anglo-Saxons. It was used to stop bleeding of cuts and wounds because of its astringent properties, and to stop gangrene. It was made into poultices for bruises, breast swelling and mastitis.

The bark, made into a decoction, used to be taken to stem diarrhoea, and as a toning remedy for lax muscles. Externally it can be used as a lotion for cuts and grazes, burns and scalds, varicose veins, boils and abscesses, and as a gargle for sore throats. Culpeper also recommended it for getting rid of fleas.

> Alder leaves 'gathered while the morning dew is on them, and brought into a chamber troubled with fleas, will gather them there-unto, which being suddenly cast out, will rid the chamber of those troublesome bedfellows'.
> Culpeper

Orchard Apple: *Malus domestica*

Parts used: Leaves, seeds, fruit

The apple is one of the oldest fruits known to man and was the only fruit available in winter before the days of transportation. It is hardly surprising then that the apple tree has been revered as a holy or sacred tree since antiquity wherever it has been found growing. It features in myths and legends of many different traditions (although in some instances there is controversy as to whether the fruit was apple or quince). In Greek myths, the fruit of the Hesperides which were guarded by the dragon were golden apples (though these were actually said to be quinces). In Arthurian legends, Arthur was taken to a mythical land to heal his wounds; called the Vale of Avalon, or the Apple Vale, it equated with Paradise. In Scandinavian legends apples were the food of the gods which kept them from growing old, and in Celtic tradition apples were to be found growing in Paradise. The fruit eaten by Adam and Eve in the

Garden of Eden was said to be an apple, the fruit of the tree of knowledge, so the apple was the fruit of temptation, the fall and ultimate salvation. This was echoed in the fairy tale of Snow White, where the wicked queen tempted *Snow White* with a poisoned apple which took her into a long sleep, from which she was ultimately saved by the handsome prince. The apple is also a symbol of love, peace and happiness.

In many traditions the apple is an emblem of fruitfulness. In Norse mythology Freya, the queen of heaven and the goddess of childbirth, heard of King Revir's grief due to his childlessness. She took an apple from her store and gave it to Gna, her messenger, to carry to the king. The king duly gave it to his wife to eat and she later bore him a son, Volsung. In Cornwall girls would traditionally sleep with an apple under their pillow to ensure their fertility and a good husband. Apple blossom was often the flower chosen by brides.

The apple was associated with Pomona, the Roman goddess of fruit trees and one of the autumn deities. Her festival is probably what is now called Hallowe'en. Perhaps because apples could be made into intoxicating

BELOW: *In both autumn and winter old apple orchards are magical places.*

drinks, notably cider, the apple represented the unconscious or the underworld, and this is illustrated in the traditional game of apple bobbing at Hallowe'en. This was the time that apples would often be used for divination – they would be tied to string and hurled around by young unmarried people by a fire. The one whose apple fell off first would be the first to marry. The peeling of an apple thrown over the left shoulder would make the initial of the future spouse.

Apples and their juice, as well as cider, have long been used medicinally. The astringent juice of the crab apple, known as 'verjuice', was used to treat diarrhoea and was rubbed onto sprains and bruises. An ointment was made with the juice and 'hard yeast' for burns. Cooked apples used to be applied to soothe sore throats, fevers, inflammatory skin problems, eye problems and rheumatism. The acids in apples help the digestion of heavy, fatty foods, which is why apples were traditionally eaten with rich foods such as pork and goose. Apples were, in fact, regarded by many as a panacea for all ills, which explains the old rhyme 'an apple a day keeps the doctor away'. They have also been used in folk medicine to speed recovery after illness, for fevers, catarrh, coughs, anaemia, anxiety and insomnia. Apple juice has been rubbed on to warts. Apple pulp has been used to soothe sore eyes and, mixed with olive oil, it has been applied to wounds and sores.

Apples are rich in vitamins, minerals and trace elements. They aid digestion, regulate acidity and, by promoting liver function and bowel action, they have a cleansing and detoxifying action in the body. Apples also have a diuretic action, good for people suffering from gout, fluid retention, arthritis, skin problems, headaches and lethargy. Pectin has a particularly detoxifying action as it has the ability to bind toxic metals such as mercury and lead in the body and carry them out via the bowels.

The apple has always had much to do with romance, with the pips and stem often incorporated in games of romantic divination. Pips from a fresh apple were placed on the cheek, each one being given the name of a potential partner. The last pip to fall off was the future spouse. Pips, representing particular lovers, were also sometimes put in the fire, whereupon the following rhyme was chanted, 'If you love me, bounce and fly, if you hate me, lie and die.' The squeezing of a pip between the fingers until the inside shoots forth signifies the direction in which the true love lives. Twisting the stem of an apple once for each letter of the alphabet will reveal the first letter of the true love's first name, while tapping the skin with the broken stem, until the skin punctures, reveals the first letter of the surname. The endearment 'apple of my eye' must be derived from such romantic associations.

Ash: *Fraxinus excelsior*
(Common ash)

Parts used: Leaves, seeds, bark

There is a wealth of myths and folk tales about the magnificent ash tree. In Nordic mythology the entire universe was supported and sheltered by a huge ash tree, called *yggdrasil*, the world tree. In the shade of its branches the gods would hold counsel, presided over by Odin, the supreme deity. Odin and his brothers created the universe and made the first human man and woman from two trees – Ask from the ash and Embla from the elm – from whom the rest of the world's population descended. The Teutons regarded the ash tree as sacred and dedicated it to their god Thor. They used ash wood to make the shafts of their spears, which were so straight, strong and reliable, that they believed it made them invincible. Similarly, the ancient Greeks valued the ash, and legend has it

that Achilles carried spears of ash. In other traditions ash was also seen as the strong protector; if kept in the house or carried about, the leaves or wood would protect from evil and witchcraft and could neutralise spells. Ash sticks were often carried by herdsmen to protect their cattle from witchcraft. In Greek mythology the ash was sacred to Poseidon, the god of the sea, and if it was made into the shape of a cross many believed it would protect those going to sea from drowning.

The ash tree was also used for divination and charms. It was known as 'Venus of the Forest' by some and was frequently associated with love. If the leaves had an even number of divisions on each side, they were a sign of good fortune and if a girl wanted to know whom she was going to marry she would find such an even ash leaf and say:

'Even, even ash
I pluck thee off the tree

The first young man that I do meet
My lover he shall be.'
She would then put the leaf in her left shoe and keep it hidden until she met her man.

Charms made from various parts of the ash tree were used to cure a wide range of ailments, including ague, rickets and whooping cough in children, hernia, warts, snake bites, earache, ringworm, toothache and impotence. This may involve leaving locks of hair or nail clippings attached to the tree's trunk or splitting an ash sapling and passing the sick child or adult through the hole backwards and forwards several times. The hole was then bound tightly so that the split would heal. If it did not then neither would the patient.

Ash leaves were recommended by Greek physicians for curing snake bites and, in the first century AD, Dioscorides said, 'The leaves of this tree are of so sweate virtue against serpents as that they dare not so much as touch the morning and evening shadows of the tree, but shun them afar off.' In medieval Europe St. Hildegarde of Bingen prescribed ash for gout, and Mattioli of Italy recommended it for toothache and deafness. The ash was also a popular remedy for 'dropsy' and to bring down fevers. The seventeenth-century English physician Culpeper said an extract of the leaves would 'abate greatness of those that are too gross or fat'.

Today the leaves (collected in spring or early summer) and the seeds (plucked when young and green) are considered beneficial for their diuretic and laxative effects. They help to relieve fluid retention and sluggish bowels and thus cleanse the body of toxins. They are used specifically for treating gout and rheumatism, which accounts for one of ash's country names, 'the gout tree'. The bark (removed in spring from three- or four-year-old branches) acts as a bitter tonic, stimulating the flow of bile from the liver and aiding digestion. It is also valued for its astringent properties in the treatment of diarrhoea and dysentery and in stemming bleeding. It is a diaphoretic, reduces fevers, and has often been used as a substitute for quinine to treat intermittent fevers and malaria. The father of the French herbalist Maurice Mességué referred to the ash as the 'Quinine of Europe'.

In folk medicine an infusion of the dried leaves taken every four weeks was said to improve general health and prolong life. The Greeks have long said that ash keys boiled in water have aphrodisiac powers. Apparently, eaten in salads, ash keys are delicious!

Aspen: *Populus tremula*
(Quaking aspen, American aspen, abbey tree, the shiver tree)

Parts used: Bark, twigs, leaves, buds

Aspen is the most common poplar in the British Isles and can grow to 60 or 70 feet high. It is happy in most soils, even on high mountain ledges. Its continuously fluttering or trembling leaves explains its botanical name, *Populus tremula*, and that of its American cousin, *P. tremuloides*. According to the Doctrine of Signatures, which expressed the idea that the beneficial or curative properties of trees and plants were suggested by their physical characteristics, both were said to cure fevers and agues that cause the patient to shiver and shake. The sufferer had to pin a lock of hair to the aspen and recite:

'Aspen Tree, Aspen Tree
I prithee to shake and shiver instead of me.'

The aspen's leaves are dark on one side and light on the other and so the tree has come to symbolise night and day, and the passing of time. To the Chinese it represents yin and yang, moon and sun and the duality of life. Another belief was that after the summer solstice, aspen and white poplar (*Populus alba*) turn their leaves to the opposite quarter of the heavens, and that wet weather is heralded by the upturned pale undersides of the leaves.

There are widespread legends that account for the aspen's trembling nature. One says that the tree was used to make Jesus's cross on Calvary, and so it has shuddered endlessly with guilt and horror ever since. Another says that the aspen was the only tree that would not bow down to Jesus as he passed through the wood to Calvary, and that as a punishment the tree is condemned to shudder for eternity.

Twigs and branches of the aspen were used for making shafts of arrows and were held to have divinatory qualities. The buds and leaves were carried to attract money and apparently used by wizards in flying ointments.

The flower buds, leaves and bark all have a long history of use for their medicinal benefits, mainly to stimulate the appetite and enhance digestion and absorption. They can be taken for a range of digestive problems including indigestion, wind, acidity, colic, diarrhoea and liver problems.

One constituent of aspen is salicin, with actions similar to aspirin, making aspen a worthy remedy for headaches, pain and fevers. It was considered a good alternative to Peruvian bark for intermittent fevers or malaria. Its anti-inflammatory and diuretic action helps to relieve urinary problems such as cystitis, irritable bladder and fluid retention. It is often prescribed for painful inflamed joints and because of its astringent tannins, it can tighten weak vaginal and uterine muscles. Externally it tones the skin and makes a good styptic for cuts and grazes, an anti-inflammatory for eczema and ulcers and an astringent for excessive perspiration.

ABOVE: *The autumn foliage takes on a great variety of colours.*

Sweet Bay: *Laurus nobilis*

Parts used: Leaves, berries

The lovely aromatic bay tree, also known as the bay laurel, is a native of the Mediterranean shores, where it can grow as high as 50 feet tall. In Britain's colder climate, it is usually much smaller (up to 20 feet high) and needs to be grown in a sheltered sunny position to survive the harsh English winter.

Throughout history the bay tree has been honoured and revered for its aromatic properties and mythical origins. In Greek mythology, a nymph called Daphne was transformed by the goddess Gaia into a laurel tree to escape the unwanted amorous advances of the god Apollo. Apollo wore a crown of laurel leaves in her memory. Warriors, victorious athletes and poets in ancient Greece and Rome were honoured with laurel crowns, and letters announcing victories were wrapped in laurel leaves. The priestesses at the oracle of Apollo in Delphi were said to have chewed the leaves and inhaled the heady, aromatic smoke of burning laurel leaves, which was responsible for their prophetic states. So the Greeks and Romans believed that bay could stimulate clairvoyance and creative inspiration, and would bring health and happiness. Sprigs used to be hung in houses to protect against bad luck, ghosts, the evil eye and lightning (it was said never to be struck by lightning). They were used to sprinkle cleansing and protective water at purification rites and religious ceremonies.

A bay tree was often planted outside the house to protect the inhabitants from infection, especially the plague, and the leaves were carried about to ward off evil spirits. As far back as the days of Theophrastus (c. 300 BC), people would keep a bay leaf in their mouth to keep away infection as well as intoxication. As an evergreen, the bay was seen by many to represent eternity and immortality. It was carried at Christian funerals as a symbol of the resurrection.

Bay was also used as a love oracle. On Valentine's Day, if a girl placed five bay leaves under her pillow and then repeated the following lines, her future husband would appear to her in a dream:

'Sweet guardian angels, let me have
what I most earnestly do crave –
A Valentine enbued with love,
who will both true and constant prove.'

Since at least the time of the Greek physician Hippocrates (c. 400 BC), bay has been valued for its many therapeutic benefits in treating hysteria, ague, consumption, wind, worms, cramps and fluid retention. Culpeper said of the leaves: 'They wonderfully help all cold and rheumatic distillations from the brain to

LEFT: *Spirals of bright yellow flowers adorn the bay.*

the eyes, lungs or other parts; and being made into an electuary with honey, do help the consumption, old coughs, shortness of breath.'

The leaves have astringent, aromatic and stimulating effects, particularly in the digestive tract. They are traditionally used for flavouring meat, poultry, soups and stews, but have the added benefit of stimulating the appetite and enhancing digestion. Their diuretic effect assists the elimination of excess fluid and toxins from the system and will help to ease rheumatism and arthritis. The leaves also have a warming and stimulating effect in the respiratory system and help to lift the spirits and dispel lethargy. The essential oil diluted in a massage oil is excellent for a warming massage, for poor circulation, aching cold joints, sprains, strains and bruises. It can be used in inhalants for coughs, colds, sore throats and catarrh.

In India the berries are used medicinally, notably for cramp, colic and painful periods, and externally as a linament for sprains.

Beech: *Fagus sylvatica*

Parts used: Leaves, nuts

The beech is a magnificent statuesque tree with a huge outspreading crown. It is one of Britain's largest and most beautiful trees, and is easily recognised by its grey trunk resembling smooth stone. It is well loved by children for its catkins that appear in April and May and its tasty beech nuts in autumn. Because the rust-coloured leaves stay on the tree throughout the winter it is a popular tree when small among gardeners for hedging.

Some say that if you carve a beech stick with the words of your greatest wish, it will come true. Beech is an old symbol of prosperity.

Used medicinally in the past, beech leaves were valued for their cooling effect on the blood and for clearing the skin. Culpeper said they were 'cooling and binding' and recommended them to be applied to hot swellings and skin conditions. The wood from the tree when distilled yields substances called quajacol, cresol and cresoline, known collectively in medicine as creosote. This was used externally for inflammatory and infected skin conditions and as a general antiseptic.

Beech nuts are a good source of protein, minerals and trace elements including potassium, magnesium and calcium. Their nutritious value makes them a good food source, popular for park deer, and for pigs and poultry. At one time beech nuts were frequently eaten by the poor. In World War II, German children were given time off school to pick them to supplement their diets. When roasted, the nuts' slight bitterness disappears and they can be sprinkled on salads or ground up as a coffee substitute. The oil extracted from the nuts can be used for cooking and salad dressings. Interestingly, beech's Latin name, *fagus*, comes from the Greek word meaning 'to eat'.

BELOW: *'The Trysting Tree'. The beech was traditionally regarded as a tree for lovers, because of its smooth grey bark in which letters of devotion, hearts and arrows of desire have long been scribed.*

Illustrated London News, 1875

Birch: *Betula pubescens, Betula pendula*

Parts used: Leaves, buds, bark, sap, oil

This graceful weeping tree with its smooth, shiny, silver and white bark is understandably popular in European and British gardens. A native of Britain, Europe and Asia Minor, it is a deciduous tree bearing catkins in April and May.

Birch branches have long been found decorating the church at Whitsun, stuck into holes at the tops of pews or tied to pillars. They were there as symbols of rebirth, the renewal of life, perhaps because they bear both male and female catkins together on the same tree in spring when nature returns to life after the winter; a wealth of sap rises in the tree at that time. Certainly birch trees have long been associated with fertility and such rites as dancing round maypoles, which were often made from birchwood. Bundles of birch twigs were also given to newly-weds on their wedding night to help ensure their fertility. Another explanation for the presence of birch branches in church at Whitsun is that the sound of their rustling leaves, moved by currents of air in the church, was interpreted as symbolic of the 'rushing mighty wind' as the Holy Spirit descended. So birch was associated with the Holy Spirit and seen to afford protection against the forces of evil. Birch twigs were even used to beat deviants to drive out the evil spirits. Birch wood was popular for making babies' cradles, both to provide protection against malignant forces and also for its association with fertility and new life. Some say a birch tree is never struck by lightning.

In springtime the sap that flows from the tree when

BELOW: *The silvery carcass of an expired birch leans forlornly against her younger sisters in the depths of Holme Fen.*

tapped is copious. This 'birch water' has long been used medicinally. Its antiseptic action and affinity for the kidneys and bladder made it a popular remedy for urinary infections. Apparently, at the time of Napoleon, birch sap was regarded as a universal panacea with a special efficacy for treating skin infections, rheumatism, gout and bladder problems. Mattioli, the Italian physician, called birch 'the kidney tree of Europe' and recommended it for stones in the kidney and gall-bladder. In Germany a tisane made from the birch has long been popular for treating skin problems. The oil from the buds and bark contains methylsalicylate, which explains its use as a liniment for arthritis and rheumatism. It has also been used successfully to treat psoriasis and eczema and as a disinfectant for cuts and wounds.

In the twelfth century, St. Hildegarde of Bingen recommended birch flowers for healing ulcers and wounds. In the sixteenth century, Culpeper advised that the juice of the leaves, the distilled water of them, or the sap of the tree when taken internally would 'break the stone in the kidneys and bladder and is good also to wash sore mouths'.

Indeed, birch leaf tea is not only good for dissolving kidney and bladder stones, but its effective diuretic action reduces fluid retention. It is also antiseptic and can be put to good use when treating urinary infections. By aiding elimination of toxins via the urinary system, birch leaves make a good cleansing remedy, helping to relieve arthritis and rheumatic conditions. Birch leaves also have a beneficial effect on the liver, stimulating the flow of bile and aiding digestion and bowel function. They lower blood cholesterol levels, too. A hot infusion of birch leaves will help allay colds and fevers.

Blackthorn: *Prunus spinosa*
(*Sloe*)

Parts used: Bark, twigs, leaves, buds, fruit

Blackthorn has long been considered a holy tree capable of warding off evil spells of witches and evil spirits, especially when planted in thickets. It was said to bloom at midnight on old Christmas Eve, and that Jesus's crown of thorns was made of blackthorn. There was an old tradition in many parts of the country of making a plaited crown of blackthorn on New Year's morning, scorching it in the fire and then hanging it up with mistletoe to bring luck. It has, however, often been associated with

bad luck; many a country person was and still is afraid of bringing blackthorn in bloom into the house for fear of bringing bad luck, even death, to a member of the family. It was seen to represent the union of life and death as the white flowers bloomed on bare, apparently dead branches, devoid of leaves. Since the Middle Ages a blackthorn stick was the emblem of the Town Mayor and presented to him on his first day of office. It was often made into walking sticks, and small branches were used to sweep chimneys.

The leaves and flowers of the blackthorn were used by country people for their diuretic properties, to clear excess fluid and toxins from the system – a good spring clean. They also have a mild laxative action, augmenting their detoxifying effect.

The fruit is rich in vitamin C and has a high tannin content. It was considered best to pick it after the first frost as then the skins were softer and more permeable for the preparation of drinks, liqueurs, jams and jellies. The syrup, made from covering sloes with sugar and a little brandy and leaving them in a sealed jar from October till Christmas, was considered an excellent tonic to the stomach, as well as a tasty liqueur. Sloes were traditionally made into a variety of astringent, medicinal preparations to stem bleeding, urinary incontinence, diarrhoea and nose bleeds. They have been used as a styptic since the time of Dioscorides (c. the first century AD). A tincture of the wood, bark and leaves has been used homoeopathically for ciliary neuralgia (eye pain) and inflammatory eye problems.

SLOE MEDICINE

All the parts of the Sloe-Bush are binding, cooling, and dry, and all effectual to stay bleeding at the nose and mouth, or any other place; the lask of the belly or stomach, or the bloody flux, the too much abounding of women's courses, and helps to ease the pains of the sides, and bowels, that cometh by overmuch scouring, to drink the decoction of the bark of the roots, or more usually the decoction of the berries, either fresh or dried.

RIGHT: *Startling white blossoms of the blackthorn.*

Cherry: *Prunus* spp.

Parts used: Bark, buds, fruit

In the Christian tradition, cherries were a fruit of Paradise, a symbol of sweetness and goodness and an emblem of Jesus the child. It is said that

Christ gave a cherry at one time to St. Peter, telling him not to despise small things. To the Chinese, the beautiful cherry blossom borne in spring symbolises youth, rebirth, hope, fertility and virility, feminine beauty and yin. In Japan it has long been the image of perfection and is the national flower. The brief but enchanting life of the flower which suddenly but gracefully comes to an end, is a symbol of our brief visit to this earth, and a reminder to live life as fully and joyfully as we can, then willingly surrendering to death when it comes. The Japanese cherry (*Prunus serrulata*) became the symbol of remembrance for all the people who died from the atom bombs dropped on Hiroshima and Nagasaki in 1945. At the Japanese Festival of Dolls in honour of girls, on the third day of the third month, cherry and peach blossoms are lavishly displayed. As a love spell, Japanese girls would tie a strand of hair to a cherry tree in full bloom.

For centuries the fruit and the stalks of the cherry tree have long been valued for their cleansing properties. They are a diuretic, hastening the elimination of toxins and excess fluid from the system. They have been used for rheumatism, arthritis and urinary infections, and cherry stalk tea is famous for lowering uric acid levels and relieving gout. The stalks have an astringent action, helping to protect the lining of the digestive tract from irritation and infection, and to treat diarrhoea, while the fruits have a more laxative effect and make a good remedy for constipation. Culpeper recommended the gum that exudes from the bark of the tree for colds, coughs, hoarseness, kidney stones and gravel and wind, as well as to sharpen eyesight and provoke appetite.

Cherries have a reputation for rejuvenating mind and body and for soothing the nervous system and relieving stress. Certainly cherries are nutritious: they are rich in vitamin C and a good source of beta carotene, calcium, magnesium, iron and phosphorus, potassium and zinc. They contain a wealth of assimilable sugar, laevulose, which is safe for diabetics. They also help the body to combat infection and are easily digested, making a good food for people who are run down or recovering from illness.

The sour or morello cherry has smaller, more bitter fruit and comes from Persia and Kurdistan. The bark of the tree has been used to bring down fevers,

and for coughs and catarrh. The stalks of the fruit are more powerfully diuretic than the wild cherry and again used for bladder problems, fluid retention, gout and arthritis.

The bark of the American wild cherry (*Prunus serotina*) is used today in herbal medicine. It has a sedative effect on the nervous system and particularly on the cough reflex, making it an excellent remedy for harsh, irritating or paroxysmal coughs such as whooping cough and croup. It has also been used for nervous palpitations and digestive problems but should only be used on the advice of a qualified medical herbalist as the bark contains prussic acid.

Culpeper said: 'the cherry is a tree of Venus', and that sour or tart cherries 'are more pleasing to a hot stomach, procure appetite to meat, to help and cut tough phlegm and gross humours; but when these are dried, they are more binding to the belly than when they are fresh, being cooling in hot diseases and welcome to the stomach, and provokes urine'.

ABOVE: *Blossom of the wild cherry.*

BELOW: *A splendid chromolithograph from Boulger's* Familiar

Elder

I n his portrait of the elder in the Tree Alphabet, Robert Graves finds little of merit to say of this tree. He mentions such superstitions as babies in elderwood cradles pining away or 'being pinched black and blue by fairies', the scent of elder causing death and disease, and the popular myths that it was the tree of the Crucifixion and/or Judas's hanging. Ultimately, Graves suggests that because elder has this doom-laden reputation, and because it is the thirteenth letter of the Tree Alphabet (R for Ruis), it has charged the very number thirteen with its reputation for bad luck.

Foresters have long believed that misfortune would strike them if they felled an elder, but if it cannot be

avoided then permission must be sought from the elder dryad. Although the timber is seldom of any commercial value, old elder heartwood is actually one of the hardest of woods, with stumps sometimes standing in as impromptu fence or gate posts for many years. The young boughs are soft and have a centre of spongy pith. Traditionally, this was hollowed out by little boys to make pea-shooters or pop-guns. These hollow stems were also converted into rudimentary whistles or pipes; the scientific name *Sambucus* is derived from a Greek primitive instrument, called a 'sambuca', reputedly similar to a bag pipe. The pith itself has found a use

ABOVE: *A fine solitary elder in flower.*

in the science laboratory as an ideal material for holding small and delicate specimens as they are finely sliced for mounting on to microscope slides.

The elder is most noticeable in the early summer when huge, creamy bundles of blossoms cover the tree, their strong, sweet scent obscuring the somewhat bitter odour of the foliage. At this stage, the prolific quality of the tree is easy to appreciate. Elder prefers rich, nitrogenous soils around rubbish heaps and the little, outside brick privvies in old country gardens. It is equally

at home in many a hedgerow or woodland edge.

Elder has a phenomenal ability to regenerate that should not be underestimated. Hack a large tree back to the ground and, like Medusa, it springs anew with even more stems and added vigour. A great, old elder stump, wrenched from its position, the corner of a renovated house, lay above ground for the summer with barely any soil to protect it. In the early spring, the stump began to throw out young shoots in every direction, and it was replanted, with profuse apologies to the tree spirit. The following year, it produced enough fruit to make two gallons of wine, and the tree continues to expand rapidly promising yet more sumptuous harvests to come.

There appear to be several marginal subspecies of elder, as the flowering cymes are sometimes of slightly varying density and size, later resulting in marginally different berry shapes and cluster characteristics. Foliage may also vary. Some plants have red, if not purple, stems to the clusters of berries, and foliage may sometimes stay green until leaf fall, or in other plants turn into all shades of delicate creamy yellows and pinks. These last traits might be the effect of different soils, although both manifestations may often be found growing close together.

Elder offers a range of different dyes: black from the bark, green from the leaves and a variety of blues and purples from the berries. The tree also provides good things to eat. The berries make a welcome addition

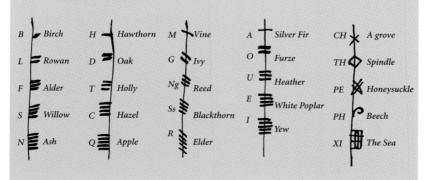

THE TREE ALPHABET

The Tree Alphabet, sometimes known as the Beth-Luis-Nion tree alphabet, or Ogham, harks back to Druidic lore and consists of five vowels and 13 consonants. Each letter takes the name of a tree, so that

Beth-Luis-Nion is literally Birch-Rowan-Ash. Graves discovered it in the writings of one Rodefick O'Haherty, and, as he says: 'I noticed at once that the consonants of this alphabet form a calendar of seasonal tree magic, and that all the trees figure prominently

in European folklore.' He then expounded upon the virtues, vices and associations of each letter tree, in relation to Celtic mythology and the ancient cultures of Scandinavia, Europe and the Middle East. This example is based on Core Hazel's Ogham.

B	Birch	H	Hawthorn	M	Vine	A	Silver Fir	CH	A grove
L	Rowan	D	Oak	G	Ivy	O	Furze	TH	Spindle
F	Alder	T	Holly	Ng	Reed	U	Heather	PE	Honeysuckle
S	Willow	C	Hazel	Ss	Blackthorn	E	White Poplar	PH	Beech
N	Ash	Q	Apple	R	Elder	I	Yew	XI	The Sea

to apple or plum pies and tarts, and have also been used as ingredients in sauces, chutneys and jams. The flowers will delicately perfume sweet dishes such as syllabubs and sorbets, while several writers heartily recommend elderflower fritters, the flower heads being dipped in sweetened batter and then deep-fried. Birds also enjoy the dark berries so that sometimes the human harvesters have to move in quite smartly before the trees are stripped bare.

For a tree which has usually been attended by much derision if not downright antagonism the elder has recently been somewhat rehabilitated. The increasing interest in herbal remedies, naturally derived cosmetics, as well as drinks and comestibles from nature's store have encouraged a few enlightened people to actually plant groves of elders.

Elder: *Sambucus nigra*

Parts used: Flowers, berries, root, bark

Reviled by many as a rank-smelling weed, a scrubby invasive bush of garden and hedgerow, this strange little tree does have a multitude of virtues and folklore associations, and its array of medicinal uses is huge. Indeed, it is one of the most important trees to herbalists. The elder tree is probably more associated with myths, folk-tales and superstitions than any other tree. Bringing its wood into the house has always been thought to be unlucky, even capable of causing a death in the family if burnt on the fire. It is supposed to be the tree that Judas hung himself from and so became the emblem of sorrow and death, and yet the Serbs used to take a stick of elder into wedding ceremonies to bring good luck. In Britain it was said to protect from lightning. In Denmark there is an old belief that if you stood under an elder tree on Midsummer Eve, you would see the king of fairyland ride by, attended by all his retinue. In Scandinavia and Germany the tree was the abode of the elder dryad or mother, whose permission was vital before the tree was cut down or any part used medicinally. If permission was not sought with the words, 'Hyldemoer, Hyldemoer, permit me to cut thy branches', it was feared the dryad would cause harm to home or family.

ELDER
Feigns dead in winter, none lives better.
Chewed by cattle springs up stronger;
an odd Personal smell and unlovable skin;
Straight shoots like organ pipes in cigarette paper.
No nurseryman would sell you an Elder
'not bush, not tree, not bad, not good'.
Judas was surely a fragile man
To hang himself from this 'God's stinking tree'.
In summer it juggles flower-plates in air,
Creamy as cumulus, and berries, each a weasel's eye of light.
Pretends it's unburnable (Who burns it sees the Devil),
cringes, hides a soul of cream plates,
purple fruits in a rattle of bones.
A good example.
P. J. Kavanagh

All parts of the elder tree have been used as medicine for thousands of years and it has had so many different healing properties that it has been called 'the medicine chest of the country people'. John Evelyn writing in 1664 said, 'If the medicinal properties of its leaves, bark and berries were fully known, I cannot tell what our countryman would ail for which he might not fetch a remedy from every hedge, either for sickness or wounds'.

The flowers make a wonderful remedy when taken in hot infusion at the first signs of colds, sore throats, fevers and flu. They stimulate the circulation and cause sweating, thereby cleansing the system through the pores of the skin. At the onset of childrens eruptive diseases like measles and chickenpox they are excellent for bringing out the rash and speeding recovery. They make a good remedy for mucous congestion, chronic catarrh and sinusitis when taken as a hot tea, as they have a decongestant action. For these complaints and also for hay fever, they are traditionally combined with yarrow and peppermint and their relaxant effect helps to relieve bronchospasm at the same time which is very useful when treating asthma.

Elderflowers have a diuretic action and so act further to relieve fluid retention in the body and eliminate toxins from the system via the kidneys. They have long been used to treat rheumatism, gout and arthritis, especially that which is aggravated by cold, damp conditions. Country people used to carry an elder twig in their pockets to protect against rheumatism and arthritis. Elderflowers also release tension, allay anxiety and lift depression. Elderflower tea at night will aid sleep and is excellent for soothing restless or irritable children if they are starting a cold or a fever.

Externally, elderflowers can be used in lotions and creams for cuts and haemorrhoids, wounds, burns, chilblains, sunburn, and skin problems to soothe inflammation and to heal ulcers. An infusion or dilute

ABOVE: *While many artists portrayed the forest giants or veteran oaks and elms, few seemed to have considered the elder as a worthy subject for attention. William Marshall Craig (1820) composed this charming, free-form sketch of elder, with its oft accompanying ivy and nettles.*

tincture makes a good gargle for sore throats, a mouthwash for mouth ulcers and inflamed gums and an eyewash for conjunctivitis and sore eyes. Distilled elder water makes an excellent toning and cleansing lotion for the face and was popular at one time for preventing freckles.

The berries can be added to stewed blackberries and apple, and are particularly valued for their laxative effect. Rich in vitamins A and C, they were traditionally made into syrups and wines for autumn and winter for prevention and treatment of coughs, colds, fevers and sore throats. Hot elderberry wine, often made with spices, was sold on London streets on cold winter days and nights until about 100 years ago, to warm the body and cheer the soul.

The fresh leaves when laid on the temples were an old remedy for nervous headaches. Worn in a hat or rubbed on to the face, the bruised leaves were supposed to keep away insects. An infusion can also be used as a lotion to keep away mosquitoes, as well as to stem bleeding and speed healing of wounds, bruises, burns, sprains and swollen joints. Sometimes the leaves can cause a reaction on sensitive skins.

The root and bark are powerful laxatives and are best avoided by the lay person for internal use. Externally they can be made into ointments for skin problems and a decoction can be used to gargle for sore throats.

It is to be avoided during pregnancy.

The tree has an ancient reputation for protection from all matters evil. As G. S. Boulger notes, 'Elder is a tree of such mingled good and evil report, that its commonness in the neighbourhood of farms and cottages is probably an example of the victory of utilitarianism over superstition.' He spiritedly defends the tree:

'Though the German name of the tree, Holdre, is said to signify hollow, it is also said to be mythologically connected with Hulda, the goddess of love; and, like love, the Elder drives away evil spirits and defeats the arts of the sorcerer, being an antidote to all his machinations.'

ABOVE: *Distinctive profiles of two mature elms at Westonbirt in the 1920s.*

Common or English Elm: *Ulmus minor* var. *Vulgaris (procera)*

Parts used: Dried inner bark, leaves

Elms have been part of our lives for thousands of years. Theophrastus, writing in the third century BC, described the use of elm wood for ship building, and Dioscorides in the first century AD wrote about using the young leaves in sauces and hot dishes. The wood is so pliable that in the past one could bend a sapling into shape to use later, for example as a plough. To the early Christians the elm represented dignity, and its great size and strength symbolised the power of the word to believers of the faith. In Norse mythology, the gods, while out walking along the seashore, found two trees, Ask the ash and Embla, the elm. Odin gave the trees souls, Hoehir gave motion and senses, and Lodur bestowed blood and healthy complexions. Thus were the first man and woman on earth created.

The elm has also long been popular as a medicine, and the subject of many charms and superstitions. To rid themselves of the ague, people would nail strands of hair or pieces of fingernail to the tree in the hope of being cured. The inner bark was often chewed or boiled up in water to relieve sore throats and colds, and they would apply the same brew to scalds and burns. When boiled in milk it was given for jaundice. Culpeper recommended the leaves and bark to be applied to heal wounds, burns and broken bones, and as a poultice for gout. The galls often found on the leaves contain a little water which Culpeper said was 'very effectual to cleanse the skin and make it fair'. If the roots were boiled in water and it was then used as a hair lotion, he said it would restore hair to a bald head!

The bark of the elm is rich in tannins and mucilage which make it a good astringent with soothing properties. It was formerly an official medicine in this country, found in British Pharmacopoeia of 1864 and 1867. It was popular for bowel problems, inflammatory digestive complaints, diarrhoea and infections. It has a diuretic action and was often given as a hot tea to bring down fevers. Externally, the decoction, which is healing and styptic, was used as a lotion for ringworm, sores and ulcers and other skin eruptions to stop bleeding.

An American relative of the English elm, slippery elm, (*Ulmus fulva*), is the one favoured by today's herbalists, who use it widely. It has more mucilage

RIGHT: *Autumn glow of English elm on the banks of the river Lugg*

OPPOSITE: *Boxworth elm, a particular form of East Anglian elm is seen here in the middle of the village of Boxworth, near Huntingdon in Cambridgeshire. English elm seldom reaches this stature, due to Dutch elm disease.*

and has a far more demulcent (soothing) action than its English counterpart. It makes an excellent remedy for all irritated and inflamed mucous membranes – for the respiratory system, harsh irritating coughs, the digestive tract, heartburn, nausea, gastritis,

irritable bowel. In the urinary system, it helps to relieve cystitis and irritable bladder. It makes a very nutritious gruel when the powdered bark is mixed with water; easy to digest, it is excellent food for the elderly and convalescents.

Hawthorn: *Crataegus laevigata (oxyacantha)* and *monogyna*
(May tree, bread and cheese tree, moonflower, tramp's supper, whitethorn)

Parts used: Berries, leaves, flowers

Hawthorn is a handsome perennial shrub or small tree reaching a height of 15 feet. It is a member of the rose family with a wonderful display of fragrant white blossoms decorating our hedgerows in early summer and bright red haws in autumn. Hawthorn is common throughout the British Isles. Almost all parts of hawthorn have been used as medicines since at least the Middle Ages. The leaves, flowers and berries were recommended for heart problems, high blood pressure, vertigo, gout, pleurisy, insomnia and haemorrhages. The bark was made into a decoction to bring down fevers, and the berries made an excellent remedy for diarrhoea.

Hawthorn has traditionally been held as sacred and a protection against evil. Hawthorn twigs were fixed to babies' cradles to protect against harmful influences and illness. When worn or carried it was believed to afford psychic protection and to lift the spirits but was never to be brought indoors. The hawthorn tree was believed to be a meeting place of spirits and fairies, and sitting there on May Day, Midsummer's Day or Hallowe'en, one could apparently be enchanted by fairy folk.

Hawthorn has long been associated with May Day festivities and ancient fertility rites: May Day evokes an ancient spring festival named after the Greek goddess Maia. The girl crowned Queen of the May represented the ancient goddess:

'The Fair maid who the first of May,
Goes to the fields at break of day,
and walks in dew from the Hawthorn tree,
Will ever handsome be.'

The maypole ceremony symbolises rebirth, as spring is a time of renewed life. The phallic maypole, together with the discus at the top from which the ribbons hang, a feminine principle, represent fertility. At this time of year in many places in Europe, hawthorns were placed outside the house of that of a sweetheart and branches decorated the porches. Hawthorn has also been associated with fertility since the days of the ancient Greeks and Romans. At Greek wedding banquets the guests would carry sprigs of hawthorn to symbolise happiness and prosperity for the future bride and groom. The bride was sometimes given a sprig of hawthorn to represent her flowery future, intermingled with thorns. In Roman traditions the bridegroom used to wave a sprig as he led his bride to the nuptial chamber, which was lit with torches of hawthorn wood.

With the coming of Christianity the church rededicated the hawthorn to the Virgin Mary. It was said that hawthorn was used for Christ's crown of thorns and because of this it was believed for centuries

LEFT: *Windswept thorn on a Yorkshire hillside, standing resolutely like some raggedy young lass with her tresses raging behind her.*

OPPOSITE: *The bright, pink blossomed, cultivar of Midland hawthorn can seem incongruous in a British landscape, but the strawberry and vanilla effect can be rather appealing.*

ABOVE: *One of those autumns when the haws hang heavy, and country folk still believe it is an omen of a hard winter to come, as nature sets her store in plenty for the birdlife when snows lie deep.*

that lightning would never strike a hawthorn tree, as lightning was the work of the devil and could not strike a plant that had touched the brow of Christ. The robin is said to have got his red breast from a thorn in Jesus's crown that pierced his breast and left a little blood stain.

As a medicine, hawthorn is considered the best remedy for the heart and circulation. The flowers, leaves and berries all act as a tonic: they have a vasodilatory effect, opening the arteries and thus improving blood supply throughout the body. This makes an excellent remedy for high blood pressure, particularly that associated with hardening of the arteries. It has further benefit in its action on the vagus nerve, which influences the heart, so that an over-fast heart rate is slowed down.

The berries have an astringent effect and make a good remedy for diarrhoea. In Russia they are used for amoebic dysentery. The leaves, flowers and berries all have a beneficial and relaxant effect in the digestive tract. They stimulate the appetite, relieve wind and stagnation of food in the intestines. They have an equally relaxant effect on the nervous system, relieving stress and anxiety, calming agitation, restlessness and nervous palpitations and inducing sleep. They also have a diuretic action, relieving fluid retention and dissolving stones, and can be used during the menopause for night sweats.

Externally, a decoction of the flowers and berries can be used as a lotion for skin problems, notably acne rosacea. A decoction of the berries makes a good gargle for sore throats.

ABOVE: *Holly berries may be mildly poisonous but birds find them highly palatable, and trees that were laden with berries are often plucked bare by mid-December.*

RIGHT: *This Holly has been buffeted by the wind and nibbled by the sheep but still appears to be in good heart.*

*Christmastide comes in like a bride
with Holly and Ivy clad.*

*The Holly and the Ivy
When they are both full grown
Of all the trees that are in the wood
The Holly bears the crown.*

Holly: *Ilex aquifolium*
(Holy tree, Christ's thorns, English holly, mountain holly)

Holly was considered by our ancestors to be a tree of good luck and of protection; if growing near a house or farm it was thought to protect against thunder and lightning. Taken indoors and hung over the door, it was believed to defend the place and its inhabitants from lightning, poison, evil spirits and witchcraft. It was carried by men for good luck but considered unlucky to burn, especially when green, and equally inauspicious to bring inside while it was in flower. Holly water used to be sprinkled on newborn babies for protection.

The leaves of several species of holly are used in different countries for making tea. The best known is *Ilex paraguariensis* from Brazil, known as Yerba-de-maté, which is drunk throughout South America as a revitalising tonic (it is high in caffeine). In the German Black Forest holly leaves were commonly made into tea and drunk as a substitute for ordinary tea and as a tonic.

Holly leaves were formerly used as an official medicine in Britain but are little used today. The astringent properties of the leaves tone the mucous membranes throughout the body and make a good remedy for catarrhal problems – for colds, flu, sinusitis and catarrhal coughs. They also act as an expectorant, useful for persistent coughs, and were used in the past for chronic bronchitis, pleurisy and pneumonia. As a diuretic, holly will help to relieve fluid retention and to hasten the elimination of toxins from the body. It has been used for centuries to treat arthritis, gout and rheumatism for this reason. It can also be taken for urinary infections and to prevent urinary stones.

Taken in a hot infusion, holly leaves have a diaphoretic effect, bringing blood to the surface of the body and causing sweating. This helps to bring

down fevers and to bring out rashes in eruptive infections. In the past, holly was used as an alternative to Peruvian bark to treat intermittent fevers and malaria. The bark of the tree and the leaves were at one time used to heal fractured bone. The juice was considered a cure for jaundice. An old-fashioned remedy for chilblains involved thrashing them with a holly branch 'to let the chilled blood out'. It obviously stimulated the circulation as well! The berries of the holly tree are mildly poisonous, despite the fact that birds like them. They cause vomiting and diarrhoea.

Horse Chestnut: *Aesculus hippocastanum*
(White chestnut, Spanish chestnut, buckeye)

Parts used: Leaves, bark and nuts

The beautiful horse-chestnut tree is greatly admired for its pyramidal spikes of creamy-white flowers and for its bright, shiny, mahogany conkers which are loved by children. The rich colour of the conkers has given its name to a certain shade of dark auburn hair:

> *'Orlando's locks are of a good colour*
> *I' faith your chestnut was ever the only colour'*
> Shakespeare, Rosalind in As You Like It

A native of north Asia and the Balkans, the horse chestnut was brought in 1633 to England, where it grows happily, reaching a magnificent height of over 90 feet. The name horse chestnut may come from the tree's historical use in Turkey: flour of the conkers, it is said, was mixed with oats and fed to broken-winded horses – a practice that is still used today. Also, on the small branches of the tree, after the leaves have fallen in autumn, are leaf scars which resemble minute horseshoes.

The conkers and their capsules contain a substance, aescine, which is poisonous when raw. In World War II the conkers were roasted in parts of Europe and ground to make a rather bitter coffee substitute. The flower buds were used to flavour drinks and as a substitute for hops in brewing beer. The conkers are rich in saponins, which, when mixed with water, make a lather: they were used in the past for making soap, washing clothes and bleaching linen, and also for storing with clothes to prevent mould and moth infestation. Conkers used to be carried in people's pockets, until they were so dried out that they were as hard as stone, to ward off rheumatism, haemorrhoids, giddiness, chills and backache. They were also carried to bring money and success, as they were said to be ruled by Jupiter.

During both World Wars, conkers served a more military purpose. They were collected nationally as their starch content was used to provide acetone necessary for the manufacture of cordite.

The bark of the horse chestnut tree is rich in tannins with an astringent action, making it a good remedy for diarrhoea, gastritis and enteritis. From the early eighteenth century it was popular as a substitute for Peruvian bark for malaria and intermittent fevers. It was often used for the circulation, to relieve congestion in the venous system which causes haemorrhoids and varicose veins. It has also been used to treat enlargement of the prostate gland. The buds of the flowers, the bark and the nuts are all used in ointments, creams and lotions for toning and strengthening veins, and are excellent for treating varicose veins, ulcers and haemorrhoids for which Napoleon Bonaparte is said to have used them. Horse chestnut extracts can also be applied to relieve neuralgia, sports injuries, bruises and sprains, and to prevent and relieve sunburn.

LEFT: *Flowering horse chestnut above a river Skirfare in Littondale.*

NEXT PAGE: *The distinctive conical shape of an unhindered horse chestnut graces a London park.*

Juniper: *Juniperis communis*

Parts used: Branch tops, berries

Juniper trees used to be planted near houses for their aromatic and antiseptic qualities. The leaves and wood were burnt to drive away infections, even the plague. It is documented that on Hippocrates' instruction, fronds were burnt in the streets of Athens, to disinfect the atmosphere and keep the plague at bay. During a smallpox epidemic in Paris in 1870, they were burnt in hospitals as a disinfectant. The twigs and branches used to be strewn over floors for similar purposes in Tudor and Elizabethan times. Apparently Queen Elizabeth I's bedchamber was strewn with juniper to sweeten the air and promote sleep.

Juniper has been seen as a symbol of succour

and protection. It is said that when Mary and the infant Jesus were fleeing from Herod into Egypt, they were hidden by a juniper bush from Herod's assassins. When Elijah was fleeing from Queen Jezebel, he was sheltered by a juniper tree and was visited and looked after by an angel while he was there. The powerful aroma is said to put hounds off the scent (useful for poor foxes and hares when hunted). Sprigs of juniper have a wide reputation for protecting humans as well as animals from demons, thunderbolts, the evil eye, ghosts, hexes and curses when hung over doors and stables. It was considered unlucky to cut down a juniper tree, and it was said that the person that committed such a crime was likely to die within the year.

Juniper berries have a long history of use to procure abortion. In some country areas the tree was even known as 'bastard killer'. A tea made from the berries was drunk 'to prevent conception' once a woman had missed her period, and sprigs were worn in each boot for nine days during the same time.

The fragrant tops and berries of the juniper tree make a powerful antiseptic medicine that is best used only on the advice of a qualified medical herbalist. They have a stimulating and disinfectant action on the kidneys and urinary system and make an excellent remedy for fluid retention and bladder infections (but should be avoided in kidney disease). They have long been used to prevent and treat kidney stones and gravel. Culpeper said, 'It is so powerful a remedy against the dropsy, that the very lye made of the ashes of the herb being drunk, cures the disease. The berries break the stone, procure appetite when it is lost and are

BEWARE THE JUNIPER

Juniper has many protective associations, so that anyone cutting down the tree is doomed either to die or to lose a close relation within 12 months. Dreaming of the tree also indicates impending bad luck and foretells the death of someone who is ill.

LEFT: *Squashing juniper berries releases their fragrant aroma.*

excellently good for all palsies and falling sickness.'

Juniper berries have a warming and stimulating effect in the digestive system, increasing the appetite and enhancing digestion. They make a good remedy for weak, sluggish digestion and for wind and colic as well as stomach and bowel infections. In the respiratory system, their antiseptic and stimulating properties are helpful in coughs, colds, catarrh and bronchial phlegm, especially when used as inhalations. In the reproductive system they stimulate menstruation and have long been used to bring on childbirth (they should obviously be avoided during pregnancy). Their cleansing properties are good for treating arthritis and gout. The oil can be incorporated into massage oils to ease gout, rheumatism, arthritis, aching muscles, sluggish circulation, neuralgia and sciatica.

Larch: *Larix decidua*
(Larix europaea, Pinus larix, European larch, Venice turpentine)

Part used: Inner bark

Larch is a conifer, a native of hilly areas throughout southern and central Europe, where it can be found growing in large forests. Augustus used it for building the forum in Rome, and Tiberius employed it for making bridges. It was introduced into England in the early seventeenth century for its value in housing and shipbuilding. The wood is stronger and more lasting than most other conifers and is almost indestructible by fire. Evelyn described it as 'a goodly tree, which is of so strange a composition, that it will hardly burn'. Not surprisingly, larch was used as a charm to protect against fire as well as illness, sorcery and the evil eye. Larch grows six times faster than oak; some of the larch beams found in houses in Venice are over 100 feet long. Understandably, the tree has been revered in many areas of the world as a symbol of immortality and incorruptibility, and also of boldness and audacity.

The larch is not evergreen like other conifers,

and the reappearance of its bright green leaves in spring after their apparent death in winter symbolised renewal of life, regeneration of energy, death and rebirth. The whole of the tree is scented, including

LEFT: *The edge of an upland stand of larch. The contrasting light and shade bring out the true gracefulness of the tree.*

its resin, known as turpentine, which is collected from trees once they are fully grown. It has been used for centuries in medicine and for making varnish, when it is known as the Venice turpentine as it used to be exported exclusively from Venice. If larch wood is burnt, a gum exudes from it; this was used to keep away snakes. Resin and sap were regarded as the soul of a tree, a source of fire and regeneration. Resin represents immortality and the undying spirit.

Medicinally, larch is mainly used as an expectorant, to clear phlegm from the chest in catarrhal coughs and bronchitis. Its astringent properties can be useful in treating heavy periods, diarrhoea and bleeding. As a diuretic, larch can help relieve fluid retention and cystitis and aid elimination of toxins via the kidneys. This makes it a good remedy for arthritis, rheumatism

and gout. A weak decoction makes a useful eyewash for inflammatory eye problems and a lotion for skin problems such as eczema, psoriasis, haemorrhoids and varicose ulcers. Larch has a reputation for its ability to lift the spirits and dispel melancholy.

Hot inhalations of turpentine diluted in water have been used to help clear catarrh and respiratory infections and ease troublesome coughs. Externally, the turpentine has been used in compounds to apply to sciatica, rheumatic and arthritic joints and gout. The new twigs and bark can be boiled and used as an antiseptic inhalant for catarrh and respiratory infections. Turpentine was used in the past in massage oils, for bronchitis, in hospitals, in cases of paralysis and to stop the onset of gangrene.

Lime: *Tilia europaea*
(Linden – a hybrid between Tilia platyphyllos *and* Tilia cordata*)*

Part used: Flowers

The handsome lime tree has an abundance of sweet-smelling flowers, which in summer are seen in parks and gardens all over Britain and Europe. Planted along roads and avenues in town and countryside alike, the lime tree flourishes even in the smoke-laden atmospheres of cities. It is very resilient and can be cut and trained with little trouble, and used to be planted as a mark tree – the village tree which symbolised the cosmic axis – at geomantically significant sites like crossroads. In German tradition the lime tree was the one under which justice was carried out. It was also planted in turf mazes and

pollarded to symbolise *yggdrasil,* the world tree.

In the Middle Ages, Hildegard of Bingen used the lime tree as a talisman to ward off the plague. She wore a ring with a green stone beneath which were some lime flowers wrapped in a spider's web. Good-luck charms were carved from limewood, which is easy to carve, and it is said that Lithuanian women made sacrifices to lime trees to ensure a good harvest.

In Greek mythology, Cronos was in love with Philyra, the daughter of Oceanus, and on the island of Philyra was caught lovemaking by his wife Rhea. To escape he changed himself into a stallion and galloped off. Philyra gave birth to his son, called Cheiron, the learned centaur, half man and half horse. Philyra was so ashamed of herself that she begged the gods to change her into a tree. She became the lime tree, and Philyra is still the Greek word for this tree.

Lime trees and blossoms have a history as symbols of feminine beauty, happiness, conjugal love, sweetness and peace. The flowers have a sweet, honey-like taste and smell and are loved by bees. They were dedicated to the goddess Venus, while Culpeper said they were ruled by the planet Jupiter, representing expansion, learning and wisdom. For many centuries, lime flowers have been used in healing to soothe tension, relieve depression and irritability, and induce peaceful sleep.

Lime flowers make a delicious, beautifully relaxing tea. Excellent for calming excitable children, the tea can be taken for any condition associated with stress.

Lime flowers have an affinity for matters of the heart. They are often used for treating high blood pressure and arteriosclerosis, as their relaxant effects combined with the beneficial action of the bioflavonoids on the arteries relax and strengthen the arterial walls. Not only do they help to prevent clots but they also heal the heart emotionally, unblocking problems from painful experiences in the past and increasing openness and warmth in relationships. Maurice Messegué, a renowned twentieth-century French herbalist, says lime flowers are one of the ingredients

Right: *Mistletoe grows on a great variety of trees. A common host, and often one of the most impressive, is the common lime. This huge parkland specimen bears many huge clumps, and the mighty stature of the tree means that the only way to harvest the mistletoe is to shoot it off the bough.*

Opposite: *Common lime in high summer in an urban park.*

of his 'tea of happiness' that will bring peaceful nights, joyful awakenings and happy days if taken regularly.

Taken in a hot tea, lime flowers have a diaphoretic action, making a good remedy for bringing down fevers and for clearing catarrh. The tea is especially effective combined with elderflowers. A cool infusion has a diuretic action, clearing fluid and toxins from the system via the urinary system.

Externally, lime-flower tea makes a good soothing lotion for inflammatory skin problems, like boils and abscesses, allergies for scalds and burns and as an eyewash. Lime flowers smell lovely mixed with lavender flowers or lemon balm leaves and are excellent for making into sleep pillows. However, dried lime flowers ferment easily and so are best if not kept longer than a year.

RIGHT: *An engraving from a Christmas edition of the* Illustrated London News *shows a jolly rustic scene of mistletoe gathering. The tree would appear to be an oak, and yet seldom do such copious amounts of mistletoe occur on old oaks. Although there are no Druids in sight, the artist, with a notion of the sacred, has included some white oxen being driven away in the background.*

ABOVE: *Every year just before Christmas, the market at Tenbury Wells, in Shropshire, holds the biggest sale in Britain of holly and mistletoe. Numerous rows of both are laid across the auctioneer's yard, and the actual sale is conducted at a furious pace. Prior to the sale buyers wander through the lots on offer, checking the quality and weight.*

Mistletoe and Ivy

Mistletoe is the sole British representative of the large family Loranthacae, most of whose members are associated with tropical regions. It will grow on a wide variety of host trees: a remarkable total of 62 different host species have so far been recorded, yet it has never been found on any of the conifers (a report of a plant on a cemetery cypress in Stratford-on-Avon perhaps proving the exception). On balance it seems to prefer trees of the rose family.

As a semi-parasite, mistletoe does not harm its host, for that would be rather counter-productive to its own success. Instead the plant lives partly off its own chlorophyll while also taking some sustenance from the host tree. The specialised root system enters the cambium layer of the host tree by means of an organ called a haustorium, which fuses with the host tissue, creating the appearance of a graft. Mistletoe roots may encircle the tree branch and radiate inwards, in a pattern closely resembling the spokes of a bicycle wheel. On some trees, particularly old apple trees in unmanaged farm and cottage orchards, the clumps of mistletoe can become so dense and prolific that they give the impression of the winter tree having burst into leaf.

A vivid folklore has arisen around mistletoe. This fascinating plant in its aerial realm has been considered deeply magical for centuries. It has been credited with powers of curing all maladies, providing protection from witches and evil spirits, as well as

RIGHT: *Only the female mistletoe bears the berries, and this detail of a bunch shows what an unusual little plant it is.*

promoting fertility in apparently infertile women. The latter belief perhaps emanates from the theory of plant signatures, projected in its suggestive berries tucked between the splayed leaves.

Tradition holds that any girl standing beneath the mistletoe at Christmas should expect to be kissed and has no right to refuse. Sometimes part of the custom requires the man to remove a berry for each kiss taken, while in other traditions it is the woman who must remove the berry and throw it over her left shoulder. When the berries have all gone there are no kisses left. Geoffrey Grigson recalls that its name in Somerset has often been 'kiss-and-go'. In regions where mistletoe was seldom found, a kissing bough made of other evergreens garlanded together fulfilled the same role and was still often called 'mistletoe' even though the plant itself was absent.

Along the Welsh borders a strong tradition held that the mistletoe bough should reside in the house all year, to promote good luck, the new bough being cut and hung up at either midnight on New Year's Eve, as was the usual custom prior to the seventeenth century, or, more recently, on Christmas Day, whereupon the old one was taken down and burnt.

There is an old and widespread belief that mistletoe was of great importance to the Druids, especially when the plant grew upon oak trees: a rare occurrence except in Epping Forest, according to one of Richard Mabey's sources. The Roman writer Pliny the Elder first described Druids cutting mistletoe from sacred oaks with golden sickles by the light of the moon. The boughs were cut very carefully and caught in a white cloth, so that they would not lose their potent magical powers by touching the earth. To ensure the magic's potency they might also sacrifice two white bulls!

The future of mistletoe might seem precarious

since many of the old fruit orchards, which are one of its most important habitats, are still being grubbed up. However, the plant manages to do very well on trees in gardens, parkland and to some lesser degree in the hedgerows of open countryside, so it should be fine for a little while yet.

Mistletoe: *Viscum album*
(Misty locum, misseltoe, Lignium crucis, birdlime, all-heal)

Parts used: Leaves, young twigs

Bunches of green mistletoe are a familiar sight as it is commonly used to decorate houses at Christmas. It is also a welcome sight growing among the bare trees in winter, still looking green and vigorous. Mistletoe is an evergreen partial parasite found growing on a variety of deciduous trees – principally on apple, poplar, hawthorn, willow, lime and to a lesser extent on many others – in Europe and temperate Asia. The nourishment and medicinal constituents it draws vary according to the tree on which it lives; when used medicinally, it is usually taken from apple trees. The name mistletoe is said to come from the Anglo-Saxon word *mistilton*, from *mistil* (different) and *tan* (a twig), meaning that it was so unlike the tree it grew on. The name *Viscum* (sticky) probably derives from the glutinous juice contained in the berries. The berries and their sticky seed are enjoyed by thrushes. Should the thrushes chance to soil (missel) while sitting in trees, the sticky seeds they pass stick to their feet and are disseminated by them on their travels, to grow on other trees. For this reason the bird is called the missel thrush.

Mistletoe has long been revered as a sacred plant, one closely allied with magic and medicine. To the Druids mistletoe was a symbol of eternal life, staying green and living in winter while the tree supporting it appears dead. They called it 'all-heal' referring to the respect they had for the plant and its great healing abilities. They particularly valued the mistletoe that grew on oaks for its ability to treat 'falling sickness', now known as epilepsy. The oak was the Druids' sacred tree, and mistletoe that grew on it represented its soul. It was used in their religious ceremonies and shared amongst the priests' people. Sprigs were worn around the neck as talismans or hung over the entrances to their huts as a protection against evil and poor health. Mistletoe was said to save a house from being struck by lightning.

Shakespeare called mistletoe 'the baleful mistletoe', in reference to the Scandinavian legend in which Balder, the god of peace, was slain by an arrow made of mistletoe. His mother, Frigg, the goddess of love, had made him immune to injury by any of the elements – earth, water, fire and air – but Loki, the evil one, was determined to get his revenge on Balder for offending him. However, the gods decided to restore

Balder's life, and thenceforth placed mistletoe under the care of Frigg, on condition that it never touched the earth, which was ruled by Loki. Ever since then it has been the custom to hang mistletoe from the ceiling and it was ordained by the gods that all who passed underneath it should receive a kiss as a reminder that mistletoe is no longer a symbol of death but of love. Certainly this is a custom we are all fond of at Christmas time, when it is used to decorate our houses along with holly. Kissing under the mistletoe could also refer back to a fertility ritual, since mistletoe was a symbol of fertility, the milky white juice from the berries representing semen. In the past women wanting to conceive would always carry some berries with them. In Wales, a good crop of mistletoe was said to herald an abundant harvest. On Midsummer's Night Welsh girls would place a sprig of mistletoe under their pillows so that they could see their future husband in their dreams. In old Germany, the Teutons believed that mistletoe could render their warriors invulnerable, and that it restores fertility to sterile animals. In the language of flowers, mistletoe means 'I surmount difficulties'.

The berries are poisonous. Mistletoe should only be used with the advice of a qualified herbalist.

Mistletoe has been valued for many hundreds of years as a sedative, muscle relaxant and tonic to the nervous system. Culpeper said: 'Both the leaves and berries do heat and dry. Misseltoe is a cephalic and nervine medicine, useful for convulsive fits, palsy and

'It (mistletoe) gives great richness to an old trunk, both by its stem, which often winds round it in thick, hairy, irregular volumes; and by its leaf, which either decks the furrowed bark; or creeps among the branches; or hangs carelessly from them. In all these circumstances it unites with the mosses, and other furniture of the tree, in adorning, and enriching it.'
William Gilpin

NEXT PAGE: *A mighty oak on a misty morning.*

vertigo.' It was a popular remedy for epilepsy, convulsions, hysteria, delirium, spleen and heart problems, and nervous debility. It was also used to stop bleeding, whether as nosebleeds or as heavy periods. Modern herbalists use mistletoe to regulate blood pressure, to remedy both hyper- and hypotension. It dilates the arteries and is of great value where arteriosclerosis has narrowed them and restricted circulation. Mistletoe also has a normalising action on the heart, regulating a slow and unsteady pulse and calming palpitations, especially when they are caused by nervousness. Its relaxant and sedative effect on the nervous system and musculature make it a good remedy for tense aching muscles and cramp, nervous headaches, migraines and a whole range of stress-related disorders.

Recent evidence suggests that mistletoe may be effective against tumours: between 1985 and 1990 approximately 60 articles on this subject were published in medical journals. The 'viscum therapy' of injecting cancer sufferers with mistletoe extracts originated in Switzerland based on the findings of Rudolf Steiner, the founder of anthroposophy. Certainly mistletoe has been shown to have beneficial effects upon the immune system, and specifically on the function of the thymus gland and the spleen, accelerating the production of antibodies. For this reason mistletoe is often used by people whose immunity is compromised, as in those with candida, ME and HIV, and after orthodox cancer treatment, when it has been shown to help patients recover energy, health and appetite.

When dried, mistletoe leaves and twigs should be kept in an airtight container, as they lose their properties on exposure to air.

Ivy: *Hedera helix*

Parts used: Leaves and roots

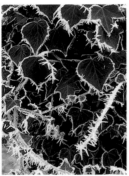

The first year it sleeps; The second year it creeps; The third year it leaps... an old adage about ivy (*Hedera helix*). Anyone who has experienced the rampant colonisation of ivy over wall or house, hedgerow or tree, knows this to be only too true. Its phenomenal capacity to clamber over and cloak almost anything with its glossy green foliage has not made the plant universally welcomed.

Where it grows on buildings there has often been concern that the myriad little suckers with which it clings on will eventually dislodge the mortar. However, in a well-maintained building with sound pointing this need not be a worry. Of more concern to amateur naturalists is the usually erroneous belief that ivy will actively kill trees. It must be stressed that ivy has no parasitic tendencies and is nourished like all other normal plants through its own leaves and root system. However, if left unchecked some large plants can begin to dominate trees. The problems usually occur when the ivy becomes so dense that the host tree's foliage is obscured from the sunlight and the sheer physical weight of the plant makes the tree top-heavy. Sometimes thick stems of ivy can encircle branches and seemingly strangle the tree. The considered opinion of many dendrologists and tree surgeons is that a modicum of ivy management should be exercised, so that rather than stripping a tree bare, it should simply be trimmed to reduce the load.

Ivy's most positive aspects are its contributions to wildlife. Insects take nectar from its autumn flowers. The fruits sustain birds in winter, while the foliage provides cover for nesting in spring.

Ivy has long been employed as part of Christmas decorations, most notably in conjunction with holly. Traditionally the ivy is seen as the lucky plant for women as holly is for men. It also finds a place in many divinations and superstitions. In Scotland, girls were wont to place an ivy leaf on their heart while singing: 'Ivy leaf, Ivy leaf, I love you, In my bosom I put you, The first man who speaks to me, My future husband shall be.'

Various cures have been ascribed to ivy, including treatment of warts, corns, verrucas and sore eyes. The juice of the leaves has been sniffed up the nose to stop running colds, the roots boiled in wine against jaundice, and an ivy wreath has even been worn after illness to prevent hair falling out. One of Culpeper's most intriguing accounts was the long-held belief that ivy was instrumental in preventing the unpleasant effects following overindulgence in alcohol: ivy 'taken before one be set to drink hard, preserves from drunkenness'. He goes on to note 'a very great antipathy

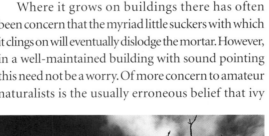

RIGHT: *Springing from a field hedgerow, this great glove puppet of ivy has brought some semblance of life back to the old skeletal oak.*

between wine and ivy; for if one hath got a surfeit by drinking of wine, his speediest cure is to drink a draught of the same wine wherein a handful of ivy leaves, being first bruised, have been boiled'. The stems of ivy sometimes achieve the dimensions of modest timber, and the wood, which is creamy white, is good for turning and carving. Many old accounts tell of ivy goblets having the ability to separate wine from water or, as Richard Mabey recounts, to nullify the effects of bad or poisoned wine.

Medieval drinkers could locate their nearest taverns by the ivy bushes or ivy-clad poles by which they signposted themselves. One might expect that public houses today would have references to the ivy in some of their names, yet it seems not.

It is said that if ivy grows upon a house then it will protect the family within from evil, but if it withers and dies, bad luck will surely follow.

Black and White Mulberry: *Morus nigra/alba*

Parts used: Bark, fruit and leaves

The lovely black mulberry tree is a native of western Asia but has been cultivated in England since at least the mid-sixteenth century. The oldest tree, which was planted in 1548, is still standing at Syon House, Brentford. Some say it was brought here by the Romans. The mulberry grows up to 38 feet high with a short trunk and wide, spreading branches, and bears attractive dark reddish-purple fruit. It is a familiar sight in English parks and gardens, and was grown along with the white mulberry in the sixteenth-century apothecary Gerard's garden. The rich, syrupy fruit was used to impart a pleasing taste and colour to cider.

Mulberries were popular with the Romans who enjoyed them at their banquets – they were supposed to be best if picked just before sunset. In Greek mythology the mulberry is a symbol of misfortune in love. It is said that the fruit turned from white to dark red from the blood of Pyramus and Thisbe, who were killed in the shade of the mulberry tree. Virgil called the tree *Sanguinea morus*. It was dedicated by the Romans to the goddess Minerva, the goddess of wisdom, as according to Pliny: 'Of all the cultivated trees, the mulberry is the last that buds, which it never does until the cold weather is past, and it is therefore called the wisest of trees.'

The mulberry tree was highly valued by our ancestors and it was included in Charlemagne's list of significant therapeutic plants in AD 812 .

Black mulberries have been used medicinally since at least Roman times, when they were popular for soothing sore throats and troublesome coughs. They have laxative properties and were used in syrup form for constipation and as an expectorant for catarrhal coughs. The bark was used for tapeworms, while the bark of the root, according to the sixteenth-century apothecary Gerard: 'purgeth the belly and driveth forth wormes'. It was also recommended to 'open the stoppings of the liver and spleen'.

According to Culpeper, the unripe berries have an astringent action and were popular for stopping bleeding. The decoction of the bark and leaves was considered a good remedy for toothache. In France

LEFT: *Silk moth larvae (*Bombix mori*) gorge themselves on white mulberry leaves.*

MULBERRY TREES
In 1608 James I tried to encourage the cultivation of mulberry trees as he wanted to establish a silk industry. But the venture was unsuccessful because the leaves of the black mulberry were used – silkworms prefer and flourish on the leaves of the white variety, however.

mulberries are still used today as a folk remedy for heartburn. Mulberries are rich in vitamin C and bioflavonoids, and have the ability to strengthen the walls of blood vessels and to help fight off infection. The astringency of the unripe fruits makes them useful as a mouthwash for bleeding gums and mouth ulcers, and as a gargle for sore throats.

Native to China, Japan and the Philippines, the white mulberry grows ten to 15 feet high and bears white or pinkish fruit with a very sweet taste. The white mulberry tree has been grown in China for thousands of years, and the leaves have been used to feed silkworms since at least the time of Si-ling, the Empress of Huangti (2967 BC). The root epidermis, leaves and fruit are official medicines in China.

The root epidermis soothes coughs and acts as an expectorant, and so is useful in treating asthma, bronchitis and other catarrhal or paroxysmal coughs. The fruit (shen, when fully ripe Hsun/t'an) is rich in natural sugars and vitamin C and is used as a tonic for those who are run down and for nervous exhaustion, insomnia and high blood pressure. The juice made from the ripe fruit improves circulation and has a tonic effect on the heart. It has a diuretic effect, which is helpful for fluid retention and cardiac dropsy.

A tea made from the leaves is an effective remedy to bring down fevers. The leaves also act as a sedative to the liver, and are used in the treatment of eye problems, to improve vision and reduce pain and swelling in the eyes; they have a cooling effect, and can be effective for problems caused by 'wind-heat', colds, headaches and coughs.

Oak: *Quercus robur* and *Quercus petraea*
(Jove's nuts, tanners bark)

Part used: Bark

Thousands of years ago, Europe was covered in trees, the most prolific of which was the oak. The woods must have had an enormous influence on the lives and beliefs of the people who lived in those times, in the woods or clearings within them. It is thought that people subsisted on acorns before they started to work the land. In medieval times people in the mountains of Spain lived on acorn bread for two-thirds of the year. Our ancestors also used oakwood for fuel, building, boats and roads; their animals, notably pigs, ate acorns. Understandably, the oak was to them a sacred tree, awe-inspiringly powerful and the abode of gods or goddesses or of sylvan deities. Oaks that were felled for practical purposes were first the subject of religious ceremonies to pacify the tree deity. Oak trees or oak gods were worshipped all over Europe. The Greeks and Romans associated the oak with Zeus-Jupiter, the god of the sky, rain, lightning and thunder. It was Zeus that the Greeks prayed to for rain and the fertility of their crops. The oak was the holiest of trees to the ancient Germans and dedicated to their god of thunder, Donar or Thunar, who in Norse mythology is Thor, the god of thunder, lightning, wind and rain, good weather and the growth of crops.

This ancient worship of the oak survived through many hundreds of years. In Christian times clergy would recite psalms and Gospel truths in the shade of oak trees, notably during a yearly ceremony when the clergyman and his parishioners would visit the boundaries of the parish usually around Rogationtide. Such trees were known as gospel oaks.

> 'Dearest, bury me
> under that Holy oke, or Gospel Tree;
> where, though thou see'st not,
> Though may'st think upon
> me, when yor yearly go'st procession.'
> *Herrick*

King Edward the Confessor (1042–1066) swore to keep and defend the laws of England under a large oak in Highgate, London.

Acorns were traditionally carved on stair bannisters to ward off lightning. They were carried in pockets or little bags to ward off illness and ensure fertility, potency and longevity. If an acorn was planted on a new moon it would mean money was on its way. The acorn is a symbol of hard work and achievement – 'great oaks from little acorns grow' – and of the cosmic egg, the source of all life and of immortality.

The protective qualities of oak are seen in its medicinal benefits. The bark is rich in tannins which protect the lining of the digestive tract from irritation and inflammation. It has long been used as a remedy for diarrhoea and dysentery, gastro-enteritis and colitis.

A decoction of acorns and oak bark mixed with milk was an old country remedy, taken to resist the effect of poisons and infections in the digestive tract and to heal infection and inflammation in the urinary system. The astringent action of oak bark is excellent for relieving catarrhal congestion, sinusitis, and heavy menstrual bleeding and has a toning effect on muscles throughout the body. This is particularly useful for prolapse and to tone blood vessels – it can be used as a lotion to tone varicose veins and haemorrhoids. A decoction of oak bark makes a good gargle for sore throats, a mouthwash for mouth ulcers and inflamed gums and a lotion for cuts and grazes.

On 29 May comes the widespread celebration of Oak Apple Day, also frequently known as Royal Oak Day. The significance of the date is the anniversary of Charles II's triumphal arrival in London upon his restoration in 1660. After the dour times of the Interregnum, the pomp and ceremony of the monarchy was a breath of fresh air, so the British people took this date as their own joyous occasion, making it yet another opportunity for festivities. In many places, Oak Apple Day took on all the customs of May Day, often becoming even more important.

The observation of Oak Apple Day has usually employed the oak in the same manner throughout the country. Oak boughs are cut and used to decorate houses and churches, while smaller oak sprigs are

OPPOSITE: *A classic oak in autumn colours.*

BELOW: *With an incredible girth of 37 feet this ancient sessile oak at Croft Castle is a record tree, and thought to be at least a thousand years old.*

worn by all; branches bearing the oak apples (small round galls), known in some parts as 'shick-shacks', were the most eagerly sought after. Among children, anyone discovered without their oaken sprig was subjected to splashing with water, pelting with rotten eggs or a beating with nettles. The failure to support the monarchy exacted a painful price!

Several Friendly Societies around the country chose 29 May as their own annual day of celebration. Most of these were formed during the nineteenth century to provide mutual assistance within small communities during times of financial, labour or health hardships. A day of marching and merriment was not only good fun, but also helped to raise funds. Oak featured in many of their rituals, as in the village of Fownhope, in Herefordshire, where the Hearts of

Oak Friendly Society still thrives today. Before first light, the chairman of the club will go to the nearby woods and cut a large oak branch, which will then be decorated with red, white and blue ribbons before being paraded at the front of the club and on a walk around the village. Many other members of the society will also carry long staves decorated with garlands of flowers (a prize being given for the best one). A church service, a good lunch and plenty of imbibing at the numerous stops around the village (accompanied by many speeches of welcome and acceptance), finishing with a lively session of dance and song in the evening, complete a rousing day.

At Great Wishford, Wiltshire, an ancient ceremony once linked to Whitsuntide has, since the Restoration, been relocated to 29 May, probably as an expression of loyalty. In 1603 a Court was held in the nearby Grovely Wood to settle a dispute between the villages of Wishford and Barford St. Martin about grazing and wood-gathering rights. These were two of the most basic necessities of any rural community and, several centuries ago, vital for survival: the right to 'lawfully gather and bring away all kinde of deade snappinge woode Boughes and Stickes' for fuel, throughout the year. Once a year the villagers were allowed to take green boughs of oak from the wood, but only those which could be pulled without mechanical aid. In the early hours the young men of

Wishford still march through the village banging drums, blowing bugles and chanting 'Grovely, Grovely, and all Grovely' in order to remind everyone to repair to the woods and bring home the oak boughs. The largest of all which the men haul back is called the Marriage Bough. This is duly dressed with ribbons and hoisted atop the church tower, where it will bring good luck to those who are married in the coming year. Tradition then decrees that four women of the village must go to Salisbury Cathedral, where they dance in the aisle before approaching the altar and once more proclaiming 'Grovely, Grovely, and all Grovely'. Apparently the reason for the latter part of this chant is that the neighbouring village of Barford lost its common rights and now Wishford has them all. This done, and all rights presumed secure for another year, the women and their entourage may return to the village and the day's festivities commence.

Peach: *Prunus persica*

Parts used: Flowers, leaves and fruit

The peach is a small tree which lives for only 20–25 years. It has been grown not only for its delicious fruit but also for its beautiful spring blossoms. It was once believed to come originally from Persia, explaining its scientific name, but in fact it comes from China, where the first records of it date back to 551 BC. It was brought to Europe before the first century BC, probably by Alexander the Great. Much later, in the sixteenth century, the Spanish took the peach to South America; by the seventeenth century it was growing in California, and by the nineteenth century it was to be found in Australia. In Britain we have known the peach since Anglo-Saxon times, and by the Elizabethan era it was being cultivated. Gerard, the sixteenth-century English apothecary, described several varieties growing in his garden, including one double-flowered one.

Throughout its history the peach has had a variety of symbolic uses. In China, peach branches were used to ward off ill health and evil. Children wore necklaces of peach pips to drive away demons. The tree represented youth, spring, marriage, good fortune and long life. The Chinese today still use peach blossoms to decorate their homes at New Year to signify spring. To the Taoists the peach tree was the tree of life, representing immortality, and the peach the food of the immortals. In Japan it is said the deep pink blossoms and tender green foliage in springtime inspire thoughts of immortality, feminine charm and marriage. Branches of peach blossom are used for floral decorations during the Girls' Festival each year in March. The peach is associated with fertility, while the branches are used as divining rods. In Christian symbolism the peach represents the fruit of salvation and virtue. Traditionally those who ate the fruit were said to be given wisdom. In other customs the fruit was thought to inspire love; and Culpeper said it provoked lust.

Various parts of the peach tree have been used medicinally for several centuries. The kernels, which contain prussic acid, as do almonds, should not be eaten in excess, but were recommended in moderation by the first-century Greek physician Dioscorides to benefit the stomach and digestion, and by the second-century Greek physician Galen to stimulate the appetite.

A syrup and infusion of peach flowers was used by apothecaries including Gerard as a mild purgative. The syrup was given as a tonic to weak children, and for jaundice. A tincture of the flowers was used to relieve colic caused by kidney stones and gravel. Culpeper said that the peach tree was ruled by Venus and recommended a syrup of the flowers or leaves 'to purge choler and the jaundice'. He said the leaves laid on the belly killed worms and when powdered and 'strewed on fresh bleeding wounds stayeth their bleeding and closeth them'. He recommended applying the milk of the kernels and the oil expressed from them to the forehead to cure migraine.

Today the medicinal use of the peach tree has fallen from use, but the luxurious fruit is still enjoyed by many and this has a wealth of health benefits. Unpeeled and raw, the peach is a good source of vitamins and minerals, particularly vitamin C, and especially when dried, of iron and potassium. It is easily digested, and is both a mild laxative and a gentle diuretic, aiding elimination of toxins and relieving fluid retention. The juice will soothe an irritated urinary system and help to relieve cystitis. Peaches have a calming influence on the nervous system, and are said to lower blood cholesterol and help to prevent atherosclerosis. As a beauty aid, crushed fresh peaches can be applied to the face for a few minutes each day to keep the skin cool, fresh and young-looking.

STRUCK BY LIGHTENING

The oak is noted as one of the trees most often struck by lightning, possibly due to the trees' great size and height (the lightning hits them before it finds lesser trees), and this has led to many superstitions. "Beware the oak, it draws the stroke." Shards of lightning-singed oak were treasured as amulets and branches of stricken trees were placed in houses to guarantee safety, working on the principle that lightning never strikes twice in the same place.

LEFT: *Peach trees have long been associated with fertility and its branches used as divining rods.*

Pear: *Pyrus communis*

Part used: Fruit

The pear tree with its beautiful spring blossom and enticing golden fruit has been grown since ancient times. There are over 30 different species of pear from different climatic regions; the European wild pear, *Pyrus communis*, is a native of temperate Europe and Asia.

The first-century Roman writer Pliny the Elder wrote of 39 known varieties; by the sixteenth century 232 were available; and today there are over 3,000. It has always been an extremely popular fruit, both raw and cooked. Pears in syrup were good enough for a king: they were served at Henry IV's wedding feast.

Pears were grown in many medieval monasteries. The popular wardon, a cooking pear, was developed in the Cistercian gardens of Wardon in Bedfordshire, where Wardon pies, made with baked pears, originated and whose coat of arms bears three pears. Three pears also appear on the arms of the city of Worcester, to commemorate a visit there by Elizabeth I.

A pear traditionally symbolised good health, fortune and hope. To the Chinese it represents longevity, justice and good judgement. In Christian symbolism it meant the love of Christ for humanity. An old custom in Switzerland called for the planting of a pear tree when a baby girl was born. Apple trees were planted for boys.

Perry is a delicious alcoholic beverage made from small, unpalatable perry pears, akin to cider from apples and, in the seventeenth century, as popular a drink. Perry was enjoyed as far back as Roman times and was considered the best antidote to the effects of poisonous mushrooms. In England, perry has traditionally been drunk while eating mushrooms, for the same reasons. Gerard, the famous sixteenth-century herbalist, wrote that perry 'purgeth those that are not accustomed to drink thereof, especially

when it is new... it comforteth and warmeth the stomache and causeth good digestion'. Culpeper noted that pears, particularly if they were sour, had a binding quality, good to stop bleeding and diarrhoea, and that they cooled the blood.

Pears are a delicious source of fibre and vitamins A, B and C and are rich in minerals and trace elements including potassium, calcium, magnesium, iron and manganese. They are a good source of quick energy as they are high in natural sugars, particularly when dried, and make an excellent food for people prone to food allergies, being one of the least allergenic foods known. They make a good first food for babies when weaning.

Pears have a cooling quality, useful for inflammatory conditions of the digestive tract and a variety of stomach problems, including gastritis, nervous dyspepsia, colitis, irritable bowel syndrome and diverticulitis. Pears' soothing and cooling action will also help to soothe coughs, particularly if cooked with fennel or aniseed with a little honey.

Pears have a mild diuretic action; they aid in excreting uric acid (valuable to gout sufferers) and reducing fluid retention. Their cooling and cleansing properties can help reduce hot swollen joints in arthritis.

MYTHS AND LEGENDS

In the Greek myths the pine tree is dedicated to the goddess Cybele, mother of the gods, to Zeus, Artemis, and Aphrodite; to Poseidon (God of the sea), because the first boats were made of pine; and to Dionysus (God of wine) as pine cones were put into vats to flavour wine. Cybele turns her unfaithful lover into a pine tree to prevent him from killing himself, and then mourns her loss under the branches of the tree until Zeus promises her that the pine would remain forever green, a symbol of immortality and the eternal spirit.

Another Greek myth makes the pine sacred to Pan. Pan courted one of his nymphs, Pitys, who preferred him to Boreas, god of the north wind. Boreas, in a jealous fury, began to attack her and so Gaia (goddess of the whole Earth) transformed Pitys into a pine tree to escape Boreas' wrath. Since then the pine has been a symbol of pity.

Pine: *Pinus sylvestris*
(Scotch pine, Scotch fir, Norwegian pine)

Parts used: Needles, bark, resin, cones, buds, kernels

Pines have been revered for centuries in many parts of the world for their symbolic significance and medicinal value. Scots pine symbolises uprightness, vitality and strength of character, and because it grows so high, it represents the connection between heaven and earth.

The pine tree was often used at funerals and in mourning rites. In China pine trees are planted on graves to strengthen the soul of the deceased and save their bodies from corruption. It represents longevity, courage, faithfulness and constancy in adversity. With the plum and the bamboo it is one of the three friends of winter in Japan, used in New Year celebrations.

Pine has been used as a medicine since ancient times. We know that the Egyptians used it and ate the kernels probably of *Pinus pinea*, as a source of nutrition. Hundreds of years later, Culpeper recommended pine kernels as a restorative after long illness such as consumption (tuberculosis) and prescribed them in a mixture with barley water for urinary problems. The Native Americans traditionally used pine resin for respiratory infections, and extracts of the kernels as a nutritious tonic and as an antiseptic.

The pine tree provides a wonderful medicine for the respiratory system. It has an expectorant action, helping to liquefy and expel the phlegm from the bronchial tubes. It also clears the head of congestion, and its antispasmodic action in the chest helps relieve asthma and harsh tight coughs. Pine has an antiseptic and anti-inflammatory action, excellent for treatment

of urinary and respiratory infections, colds, coughs, flu, sore throats, bronchitis, and even pneumonia and tuberculosis. By aiding the elimination of toxins via the kidneys, it makes a good remedy for arthritis and gout, and due to its anodyne properties it may help to relieve painful conditions such as headaches and toothache.

Pine has a revitalising effect generally and is well worth trying if you are feeling tired and run down. When used in the morning, pine's tonic and invigorating properties will help wake you up, enliven the mind and alert the senses. At the same time pine is calming and refreshing and can be used for exhaustion, debility, anxiety and stress-related problems. It is warming and strengthening, particularly in winter as it stimulates the circulation. It also has a beneficial effect on the digestion.

Externally, the resin and oil have long been used in linaments to massage into painful joints and aching muscles and to increase the circulation. The oil makes a good inhalant for respiratory infections, coughs, colds, catarrh and sinusitis and can be used preventatively to keep infections at bay. The tar is used in lotions for skin problems such as ringworm. Rubbed into the skin, dilute pine oil makes a good insect repellent and deodorant, it is particularly good for excessive perspiration. It can also be rubbed into the hair to clear headlice (but only when diluted in a base oil as it can sometimes cause irritation).

Plum: *Prunus domestica*

Part used: Fruit

ABOVE: *A fine crop of Victoria plums. Gathered slightly under-ripe they are perfect for jam making.*

OPPOSITE: *A plum orchard in glorious bloom.*

Plums have been known at least since the days of ancient Rome for their energy-giving properties and laxative action, particularly when in their dried form as prunes.

Plum blossom is one of the most beautiful and uplifting sights in the springtime. In many cultures, particularly in China and Japan, plum blossom is a symbol of the triumph of spring over winter, of light over darkness and of life over death. With the bamboo and the pine, plum blossom is known as one of the three friends of winter, sho-chiku-bai, symbolising strength, abundance and beauty, and heralding the new year.

> '*Hear the pine: may your prosperity be as constant as the greenness of my mantle, and may your friends stand as I do, steadfast against the adverse winds of the world.*
> *Hear the bamboo; may your lifetime be as long as mine and may you know the joy of living abundantly.*
> *Hear the plum: may your hopes rise fresh and strong like the young shoots that spring from my rugged trunk, and may your life flower with loveliness.*'
> *Alfred Loelm, from a* Gift of Japanese Flowers

Culpeper said that plums were all ruled by Venus 'and are like women – some better, some worse'.

Apparently, plums have long been used as love potions. Culpeper recommended prunes to 'loosen the belly ... to procure appetite... and cool the stomach'. He said the leaves boiled in wine made a good gargle, the gum from the tree was 'good to break the stone' in the urinary tract and the leaves boiled in vinegar would kill ringworm.

Plums are a highly nutritious as well as delicious fruit. They are rich in minerals and trace elements, including iron, calcium, phosphorus, magnesium, sodium and manganese and contain vitamins B, C and E; the latter two have an antioxidant action, helping to protect cells against damage caused from free radicals. This may well help to retard some of the effects of ageing on the body. When dried, as prunes, they provide a concentrated source of energy, fibre and nutrients and make a good remedy for constipation. They have a diuretic action, helping to reduce fluid retention and aiding the elimination of toxins via the urinary system. For this reason plums and prunes have been recommended to detoxify the system, and for gout and arthritis. Both fruits also have a cleansing and tonic effect on the liver which augments their effect as the liver is the major detoxifying organ of the body. Their cooling and cleansing properties help to keep the skin young and clear. The twentieth-century French herbalist Maurice Messegué recommends them for treating urinary stones, liver problems and arteriosclerosis.

Quince: *Cydonia oblonga*

Part used: Fruit

The quince tree with its handsome pear-shaped golden fruit has been grown since ancient times, probably originating from Persia and Greece. Its scientific name is derived from Cydonia, a town in Crete. It was obviously highly prized, for, over the centuries, it has featured in myths and legends, art and statuary. The ancient Greeks called the quince 'chrysomelon', meaning the golden apple, and the mythological golden apples of the Hesperides were said to be quinces, as was the apple that Paris awarded to Aphrodite on Mount Olympus. To the ancient Greeks the quince was a symbol of love, fruitfulness and fertility, as well as temptation and was dedicated to the goddess Aphrodite. Newlyweds in Greece and Rome would share a quince at the wedding banquet

as a symbol of their love and union. Writing many centuries later, Shakespeare has Lady Capulet saying as she orders the wedding feast, 'Call for dates and quinces in the pastry!'

The quince was widely grown in the East since at least biblical times. It is said that the 'Tappuach' of the Scriptures was, in fact, the quince although it was mostly interpreted as an apple. In Canticles it says, 'I sat down under his shadow with great delight and his fruit was sweet to my taste.'

The quince is a small, attractive tree, growing to about 15–20 feet, with pretty pink flowers in spring. Despite the fruit's tempting appearance, when raw it is hard and acid and quite unpalatable. It is mainly used for flavouring cooked apple and pear dishes and for making jams and jellies. Jam made from quinces was first known as 'marmelo', later marmalade, the name that now includes other conserves. Gerard said of marmalade, 'The boiled seeds from the fruit exude a soft mucilaginous substance that is very soothing to hot inflamed surfaces.' Culpeper recommended it for sore mouths, throats and breasts.

The quince is still recommended as a nutritional and medicinally beneficial fruit. It is an excellent cooling, anti-inflammatory remedy for the digestive tract. It is rich in tannins with an astringent action, good for curbing diarrhoea and dysentery, healing gastritis and ulcers, relieving acidity and heartburn, and inflammatory bowel problems such as colitis and diverticulitis. It stimulates the appetite, aids liver function and enhances digestion. Its astringent and anti-inflammatory action is also felt in the respiratory

LEFT: *The Quince, with its solitary flowers and large felted leaves is frequently grown as an ornamental shrub rather than a purposeful fruit tree.*

BELOW: *Large golden quinces hang from the autumn bough.*

system where it can be useful for colds, catarrh, coughs and sore throats. A broth made from the fruit has been used traditionally for vaginal infections and rectal inflammation. A handful of mashed seeds soaked in half a glass of lukewarm water is recommended by Valnet for soothing haemorrhoids, irritated breasts, chapped or cracked skin and burns. A decoction of the seeds has long been used to soothe the digestive tract, for diarrhoea and dysentery, and as an eye lotion and an adjunct to skin lotions and creams.

Rowan

The rowan (*Sorbus aucuparia*), or mountain ash, must be considered one of the most potent antidotes to all works of evil and darkness. Planted near the home it will act as guardian. Incorporated inside a building, such as over the fireplace or in a door frame, it will stop witches entering. Rowan was deemed particularly useful for protecting livestock from 'fascinations', sometimes known as being 'overlooked', of witches. If a stable or byre did not have the tree growing nearby, sprigs would be placed over the door, or even around the horns of cattle. Switches and riding crops were often cut of rowan, and a rowan churn-staff would ensure that the butter would thicken well. The acute concern with the wellbeing of the farm animals harks back to a time when a family's prosperity could be measured in such commodities and, indeed, their very sustenance and survival depended upon them. The making of small rowan crosses to protect the household was common in many parts of Britain, either at Easter time or around May Day. Richard Mabey suggests that this is still much the case in the Isle of Man, although it has generally died out in mainland Britain.

Evelyn noted that rowan was particularly 'familiar in Wales, where this tree is reputed so sacred, that there is not a churchyard without one of them planted in it (as among us the Yew)'. He marked the custom of making rowan crosses: 'On a certain day in the year, every body religiously wears a cross made of the wood; and the tree is by some authors called *Fraxinus cambro-*

OPPOSITE: *high on a Cumbrian hillside a sentinel rowan in autumn.*

BELOW: *Two large rowans in flowering splendour on the edge of a Dyfed woodland.*

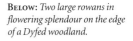

britannica (Welsh ash), reputed to be a preservative against fascinations and evil spirits; whence, perhaps, we call it Witchen, the boughs being stuck about the house, or the wood used for walking-staves.'

Though others, like Evelyn, have correlated the name with 'witch', it is doubtful that the name Witchen is derived from pagan connections; far more likely it has come from wice or wic, which is Old English for a tree of pliant branches or timber. However, Geoffrey Grigson records that the Welsh name for the tree, *cas gangythraul*, translates as 'Devil's hate'.

The Druids are believed to have incorporated rowan in rituals designed to extract answers to difficult questions from demons. Robert Graves, in *The White*

Goddess, explains that in ancient Ireland 'fires of rowan were kindled by the Druids of opposing armies and incantations spoken over them, summoning spirits to take part in the fight'. Another somewhat lurid superstition which Graves records from Ireland is that of hammering a stake of rowan wood through a corpse to immobilise its ghost.

There is no other tree so instantly familiar with its stunning display of scarlet berries than the rowan, for wherever it flourishes it is a little tree that stages a grand show; and it will grow almost anywhere. Its alternative name is mountain ash, because its leaves bear a striking resemblance to those of the ash, although they are not related. The tree has a great variety of regional names. Quicken tree or Quickbeam literally means 'lively tree' – particularly because of its ability to grow well in almost any habitat and the springy, flexible nature of its slender boughs, which allow the leaves to dance in the lightest of breezes.

Another, almost forgotten name for rowan is Fowler's service tree. The Latin name *aucuparia* is derived from auceps, a fowler. An old custom, more so in Europe than in Britain, involved the snaring of birds such as fieldfares, redwings and thrushes, with horse-hair nooses baited with rowan berries. Marcus Woodward observed of the rowan berries: 'The Welsh make of it a kind of perry, and some Scottish Highlanders may still convert it into a sort of cider, or extract a fiery spirit.' According to John Evelyn, the berries made 'an incomparable drink, familiar in Wales'.

Rowan is remarkable for its versatility, being equally at home at over 3,000 feet on a windswept mountainside, where it tends to prefer the slight protection afforded by small hollows and stream gullies, or lining an urban street. In spring it provides a fine display of creamy clusters of tiny flowers, to be followed in late summer with vivid vermilion berries and a show of golden foliage.

While rowan looks splendid in a wild and rugged landscape it must not be overlooked as a contributory species of much woodland. It may thrive in young woodland as an excellent nurse tree for other species, but as the canopy develops, and light and space are reduced, it often bows out in the competition. However, on woodland margins and wherever there are clearings it will continue to flourish. Exceptionally tall and spindly rowans seem to be doing well in some ancient, and relatively dense, woodland such as The Thicks at Staverton Park. In some of the clearings in this wood massive coppiced rowans, which could well be over 100 years old, are also evident.

Since the tree seldom achieves great proportions, its timber is mainly valued for the coppice poles. Perhaps another reason why the tree became known as mountain ash was the fact that its fine-grained hard wood was used for similar purposes to the ash, such as handles for tools and implements.

The rowan's potent protective capacity has made it a popular tree for planting close to the homestead, and to buildings where animals were housed. It must be supposed that this memory lingers on, for rowan and its many cultivars are still frequently chosen for garden ornamentation.

Rowan: *Sorbus aucuparia*
(Mountain ash)

Parts used: Bark, fruit

The handsome rowan tree is a native to Britain, Europe, north Africa and Asia Minor, and can grow to a height of about 50 feet. The wood is relatively hard and has been used for making tool handles and spinning wheels, stakes and pegs. It is a member of the rose family.

The tree has brilliant orange-red berries which are loved by birds and are rich in vitamin C. The berries can be made into delicious jelly which makes a good accompaniment to roast meats such as game and lamb. It was considered a good bitter tonic and antiseptic – useful in the days before fridges. The berries were given to animals to ease the birth of their young.

The name 'rowan tree' comes from the Norse *'runa'* meaning a charm, as it was said to protect against evil. In some areas the tree was known as the witch-tree or wicken tree. 'Runes' or esoteric secrets used to be carved in trunks of rowan trees in Scandinavia and Britain. In Scotland twigs from the tree were crossed and tied with red thread to hang over doorways, explaining the old Scottish saying:

'Rowan tree and red threid
Gar the witches tyne their speed.'

The red berries used to be made into necklaces to protect the wearer from the evil eye and fairies. Rowan Tree Day was celebrated in some areas of the country on the feast of the Invention of the Holy Cross, 3 May, and branches of rowan were cut and hung over the houses and outbuildings to protect them through the coming year. If a tree was found growing near a house it was seen to be lucky and was not to be cut down.

All parts of the tree are rich in tannins that have an astringent action in the body. The bark has been used to protect and heal inflamed or irritated mucous membranes throughout the digestive tract and has been given as a remedy for gastritis, gastro-enteritis, colitis and diarrhoea. Externally, a decoction makes a good lotion to stem bleeding of cuts and to speed healing of sores and ulcers, as well as to treat inflammatory skin problems. As a mouthwash or gargle, rowan bark can be used for bleeding gums, mouth ulcers and sore throats.

The berries, particularly when they are unripe, are also astringent as well as antiseptic. They too can be made into gargles for sore throats and tonsillitis, and lotions for skin problems and haemorrhoids. Taken internally, they make a good remedy for diarrhoea, stomach and bowel infections and inflammatory bowel problems.

BELOW: *Autumnal leaf of* Sorbus aucuparia.

BOTTOM: *Mountain ash in Glen Shee, Tayside.*

SERVICE TREE

Service is thought by some to be a corruption of Sorbus, *the tree's botanical name, while other authorities state that service is derived from* cervisia, *meaning beer. Recent thinking has moved away from the beery theory, as Richard Mabey reports in* Flora Britannica *that Patrick Roper, who has researched the subject, now believes that service derives from the Old English* syfre, *and that the drink made from the service berries was more of a liqueur than a beer.*

RIGHT: *Flowers of the wild service tree deep in a woodland glade.*

Service Tree: *Sorbus torminalis/domestica*

Part used: Fruit

The Wild Service Tree, *Sorbus torminalis*, is a native of Britain. The name service (or servise) tree comes from *cervisia*, meaning fermented beer, as the fruits used to be popular for making an intoxicating drink. Its small brown fruits are very acid but edible when over-ripe. In the past they were widely sold in the south of England, where they were called chequers. Apparently they were very popular among children and were said to taste a bit like dried apricots or tamarinds.

The fruits of the service tree were considered medicinally beneficial due to their high tannin content, which gives them an astringent action, useful when

treating diarrhoea, dysentery and infections in the digestive tract. Culpeper was familiar with their medicinal virtues and said: 'Services, when they are mellow, are fit to be taken to stay fluxes... If they be dried before they be mellow, and kept all the year, they... are profitably used... to stay the bleeding of wounds, and of the mouth or nose.'

A native of north Africa, western Asia and Europe, the True Service Tree, *Sorbus domestica*, has attractive cream-coloured flowers in May and yellowish or red-green fruit which look a bit like apples or pears. Until recently, the tree was widely assumed to have been introduced, but the discovery of several isolated and ancient colonies in south Glamorgan have laid its claim for native status. These can be eaten, preferably when a bit over-ripe, or frosted, otherwise they also taste very sour and astringent. Making a form of ale from the ripe berries was obviously popular at the time of the roman poet Virgil (c. first century BC) who said: 'with acid juices from the service Ash | And humming ale, they make their lemon squash.'

Eight hundred years later, Evelyn said: 'Ale and beer brewed from the berries when ripe, of the true Service tree is an incomparable drink.'

In English folklore the fruit was valued for its ability to protect against witchcraft. The hard fruit was hung up in or outside houses to keep witches away, which accounts for the tree's country name 'witten pear tree'. Witten is the old English word meaning 'to know', indicating that the service tree was the 'wise' tree.

Sweet Chestnut: *Castanea sativa*
(*Spanish chestnut, stover nut, meat nut, Jupiter's nut, Sardian nut*)

Parts used: Nuts, leaves

The sweet chestnut tree flourishes luxuriantly all over Europe but is said to come from Sardis in Asia Minor, from where the fruit gets its name 'Sardian nut'. It was brought to Britain probably by the Romans and can reach the magnificent height of 100 feet and girth of 40 feet.

Sweet chestnuts were known to the ancient Greeks as *dios balanos* (Zeus's acorns) and dedicated to Zeus. The name *castanea* comes from a town called Castanis on Thessaly, an area where chestnut trees were grown in abundance. In Christianity the sweet chestnut tree is a symbol of goodness, chastity and triumph over temptation because of the prickly case that encloses the nut.

Chestnuts are high in complex carbohydrates and contain more starch and less fat than other nuts; as a result, they have less than half the calories of other nuts. They are easy to digest and make a good flour. Roasted chestnuts and chestnut stuffing are highly nutritious old winter favourites, rich in vitamins B

and C and minerals including calcium, magnesium, phosphorus and iron. Cooking reduces their slight bitterness. Marons are a large sort of sweet chestnut grown in France, Italy and Switzerland, famous for the delicious sweet marons glacés.

Evelyn loved chestnut trees and was said to have organised the planting of many stately avenues of them. He described the chestnuts as 'delicacies for princes and a lusty and masculine food for rusticks, and able to make a woman well-complexioned'. Chestnuts were once popular as love spells.

The leaves of the sweet chestnut were traditionally used for relaxing spasm, particularly in the chest where it causes paroxysmal coughing as in whooping cough. They have long been used for bringing down fevers and to treat 'ague' (malaria). The leaves have a soothing and astringent action on the mucous membranes throughout the body; in Culpeper's day chestnut leaves were used to stop bleeding, particularly to alleviate heavy menstrual bleeding. The nuts are warming and strengthening and act as a tonic to the nervous system. The French used to give chocolate made with chestnuts as a restorative to people after a long illness.

The nuts when powdered were a country remedy for haemorrhoids, and gypsies used to wear nuts in a bag around their neck to prevent haemorrhoids.

Tamarisk: *Tamarix gallica*

Parts used: Roots, leaves, young branches

An elegant deciduous shrub that grows to about 10 feet high, with feathery greeny-blue leaves and pink flowers, tamarisk is a native of north China and Japan. It can be found growing in coastal areas in England where it will happily tolerate windy and damp areas. Tamarisk has long been attributed protective powers, able to protect against demons, witchcraft and evil. It was used in purification ceremonies, and the twigs were frequently burnt, for the smoke was said to keep away snakes. Apparently, the wood from the tamarisk tree used to be a popular material for crab-pot bases in the Channel Islands because of its common occurrence there and its ability to withstand the long-term effects of salt-water. In Guernsey, it was used like the hazel twig for water divining.

The tamarisk has been used in medicine at least since the time of the ancient Egyptians. Ineni, a builder to King Tuthmosis I (1528-1510 BC), made a list of the king's orchard trees 200 years prior to Tutankhamun, and ten tamarisk trees were included in this. In Egypt the tamarisk and the date palm were sacred.

According to legend, both were planted by a king in the courtyard of his palace, and a banquet was said to have taken place under the shade of the tamarisk. In a famous Egyptian myth, when Osiris was killed by Seth, his coffin sailed out to sea and across to Phoenicia, where it landed under a tamarisk tree.

Culpeper said the tamarisk tree was ruled by Saturn, and recommended that a brew containing the roots, leaves, young branches and bark, boiled in urine, should be drunk to remedy varicose veins, heavy menstrual bleeding, spitting of blood, melancholy, jaundice, colic, bites of venomous beasts, even leprosy. When applied externally it could ease toothache and ear and eye pain. A decoction mixed with honey, he said, should be applied to gangrene and ulcers and used as a wash for nits and lice, and he recommended the ashes to heal blisters caused by burns.

The branchlets and leaves are official medicines in China, and contain salicin, with an action like the aspirin – effective for treating fevers and headaches, arthritic pain and swelling and as an analgesic to relieve pain generally. Tamarisk has a diuretic action, useful for dealing with bladder problems and fluid retention, and clearing toxins from the system.

ABOVE: *The flower of tamarisk, a tree able to thrive in either salt-laden air or with its roots planted firmly in the brackish shoreline.*

Walnut: *Juglans regia*
(*Jove's nuts, Jupiter's nuts*)

Parts used: Leaves, nuts, bark

The majestic walnut tree, with its huge spreading crown, is probably a native of China, introduced in ancient times to the Balkans and grows happily throughout Europe. It was probably introduced to Britain, where it is at the northern limit of its climatic tolerance, by the Romans. Walnut trees can reach a height of 100 feet and have been known to live for up to 1,000 years. Many of the large trees seen in Britain and Europe were planted by monks in medieval monastery gardens for their nutritious nuts and the medicinal benefits of their leaves and husks. Walnut

timber is highly prized for making furniture, and the nuts are harvested for pickling or eating raw.

Walnut's name means 'royal nut of Jupiter'. It comes from *glans* meaning acorn and *Jovis* or Jupiter, and probably meant that walnuts were food fit for the gods. Its common name, walnut, derives from the German *wallnuss* or *welsche nuss*, which means strange or foreign nut.

In pagan symbolism nuts represented life and fertility, so in ancient Greece the walnut was dedicated to the moon goddess Artemis, and associated with love, marriage and childbirth, as well as wisdom and longevity. Nuts were used in love spells, were given to newly-weds at Greek and Roman weddings and strewn on the floor of the banqueting walls. Later, in Christian Europe, walnuts were used like confetti to shower the bride and groom after the wedding ceremony as a symbol of fertility and happiness. A gift of a bag of walnuts is said to make your dreams come true.

Walnut husks boiled in water used to be popular as a hair dye to cover grey hair, and said to thicken thinning hair. The hair dye makes a good wool dye as it does not require a fixative.

The walnut tree has been used for medicine since at least the time of Mithridates, King of Pontus. Culpeper recommended the juice of the boiled green husks and honey as an excellent gargle 'for sore mouths or the heat and inflammation of the throat and stomach', and the distilled water of the green husks to 'cool the heat of agues... also to resist the infection of the plague'.

Today, the walnut tree is just as valuable. The leaves (best picked before mid-July) are bitter and astringent and are an excellent medicine for the digestion. They protect the gut and stomach lining from irritation and inflammation and can be used for indigestion, gastritis, peptic ulcers, gastro-enteritis, nausea and diarrhoea, as well as worms. In the respiratory system the astringent action helps check catarrh and resolve catarrhal coughs. Externally walnut leaves can be made into a tea which makes a good lotion for cold sores, chilblains, and excessive perspiration from the hands or feet; it can be used for haemorrhoids, for varicose veins and ulcers, and as a douche for vaginal infections such as thrush. As a gargle the tea can be used for bleeding gums, mouth ulcers and sore throats.

The bark of the walnut tree has a laxative effect. It has long been used as a vermifuge for expelling worms. The oil extracted from the nuts is an excellent source of Omega 3 essential fatty acids linolenic acid, which enhances the immune system and has a beneficial effect on the heart and circulation. The oil is delicious on salads. Walnut oil can also be applied to the skin for inflammatory skin problems such as eczema and psoriasis.

The leaves have a diuretic effect; they can be used to clear skin problems such as acne and for swollen glands and lymphatic congestion. They are said to lower blood sugar.

> **KING'S RECIPE**
>
> *Culpeper describes how in King Pontus's treasury, was found a 'scroll of his own handwriting, containing a medicine against any poison or infection; which is this: Take two dry walnuts and as many good figs, and twenty leaves of rue, bruised and beaten together with two or three corns of salt and twenty juniper berries, which taken every morning fasting, preserves from danger of poison, and infection the day it is taken.'*

OPPOSITE: *The walnut tree, long treasured for its many medicinal virtues as well as the highly nutritious and delicious walnuts.*

> **LOVE'S LABOURS**
>
> *An obscure superstition encouraged walking three times around a walnut tree at midnight on Hallowe'en, whereupon one's true love's face would appear among the branches. Sleeping beneath the tree would bring dreams of a future lover.*

White Willow: *Salix alba*

Part used: Bark

There are around 250 species of willow tree, of which the white willow and the black willow are the most commonly used in healing. The black willow is an American tree seldom seen in Britain. The white willow, a handsome tree with silvery white leaves, grows by streams, rivers and marshes and in damp places throughout Europe, north Africa and central Asia. It can reach a height of 80 feet. The botanical name, *Salix,* comes from saline, meaning to leap, for willows are said to grow so fast they almost seem to leap. For this reason the white willow has come to symbolise rebirth and fertility.

In Japan mistletoe growing on a willow tree was traditionally given to childless women who wanted to conceive, and in other places the leaves have been carried in the belief that they attract love. To the Chinese the willow represents spring, femininity, grace, artistic ability, yin and the moon. In ancient Greece the willow was dedicated to Artemis, the moon goddess, whose domain was fertility, pregnancy and childbirth, as well as the growing of crops and the running of rivers and streams. The image of Artemis bound with withies, refers to the branches of willow. In English tradition, Culpeper said the white willow was ruled by the moon, and in pagan traditions willow branches were used to make wands for lunar magic. Branches were traditionally used in the ceremony of 'beating the bounds' (an annual walk of the villagers around the boundaries of the parish) which was a springtime ritual of purification.

To Taoists, willow with its tough, bendy and resilient branches represents strength in weakness, unlike the pine and oak, which are apparently strong but by resisting the storm are broken by it. The supple willow yields to the storm and springs back once it has passed and in this way survives. The willow's resilient nature makes it popular for use in crafts, constructing fencing and making cricket bats. The willow tree has also been associated with sadness and grief, mourning and loss and has been depicted in poetry many times conjuring up pictures of loss and sadness. Shakespeare starts to describe Ophelia's death saying:

'there is a willow grows aslant a brook
that shows his hoar leaves in the glassy stream.'

Being under the dominion of the moon, which itself represents fickle and changeable emotions ('Someone is said to be inconstant as the moon'), the willow has become a symbol of forsaken love, the emblem of the rejected love, in the language of flowers.

According to the Doctrine of Signatures (see Chapter Six), willow has been used since medieval times to treat arthritis and rheumatism, as it grew in damp places where rheumatism was rife. It was also popular as a remedy to bring down fevers and for treating intermittent fevers as in malaria. In 1763 the Reverend Edward Stone noted the connection between low marshy regions of England, the white willow and rheumatism. After trying out decoctions of the bark on sufferers of rheumatic complaints and noting its success, he was responsible for the later isolation of the active component salicin from the willow and the production of salicylic acid, which is the basis of the common aspirin. So willow bark is excellent for treating all the symptoms which are relieved by aspirin – it reduces inflammation of arthritic and gouty joints and eases the pain of rheumatism, aching muscles and headaches, head colds, flu and fevers. It contains tannins with astringent properties which make it a good remedy for diarrhoea and dysentery, and to check bleeding as in heavy periods, and externally from cuts and wounds. It can be used as gargles for sore throats, and mouthwashes for mouth ulcers and bleeding gums. It has diuretic properties which reduce fluid retention and help to eliminate toxins from the body via the urinary system.

Because it also has the ability to prevent the blood clotting too quickly, aspirin has been used recently

to help prevent heart attacks and strokes. The willow is likely to be a good remedy therefore for circulatory problems. One of the drawbacks of using aspirin (which is salicylic acid isolated from the other constituents of the willow bark) on a long-term basis is that it irritates the lining of the stomach and can cause bleeding from the stomach wall. When the herbal remedy is used, the tannins in the bark astringe the lining of the stomach wall and protect it from irritation so willow is itself a good remedy for the digestion, strengthening a debilitated digestive system and relieving heartburn and acidity.

Witch Hazel: *Hamamelis virginiana*

(Spotted alder, snapping hazel nut, winter bloom)

Parts used: Leaves, bark and twigs, and distilled water

*'The wreathing odours of a thousand trees
And the flowers faint gleaming presences,
And over the clearings and the still waters
Soft indigo and hanging stars.'*

J. C. Squire

Witch hazel is an attractive deciduous small tree or shrub reaching up to 15 feet high, native to the eastern states of North America and Canada. It is popular in gardens for its bright yellow, spidery flowers which bloom in the middle of winter. Its red and yellow leaves in autumn are also lovely. One of its common names, snapping hazel, derives from its seeds being ejected violently when ripe.

The bark, twigs and leaves all have a similar astringent action and were long used medicinally by the Native Americans. Taken dried, it was popular as snuff to stop nosebleeds and to relieve sore muscles and soothe inflamed eyes. The branches of witch hazel were valued in America as divining rods for detecting underground water and metals, in the same way as water diviners in Britain have used hazel twigs to find water – perhaps why it became known as 'witch' hazel.

Witch hazel extracts were famed in the days of our grandmothers as a first-aid remedy for scalds and burns, bruises and swelling and to stop bleeding. Distilled witch hazel is still available in almost all pharmacies for the same purposes. The high levels of tannin that occur in the plant give it an effective astringent action, making an excellent remedy for bleeding. It can soothe the pain and irritation of insect bites and stings, ease aching muscles, tone the skin and reduce broken capillaries. As a decoction, a tincture or a distillation, witch hazel is useful as a douche for vaginal infection and irritation; it can be applied to cuts and wounds, used as a mouthwash for mouth ulcers and infections, and included in a lotion or an ointment for bleeding haemorrhoids.

Witch hazel also helps to speed healing, to reduce pain, inflammation and swelling, and to provide a protective coating on wounds. It has a contracting action on the muscles and blood vessels, giving it a wonderfully wide variety of therapeutic uses. As a lotion or ointment it can be applied to soothe the pain, irritation and swelling of varicose veins and phlebitis, and the itching of haemorrhoids, and also to speed healing of varicose ulcers. In a poultice or compress it relieves scalds and burns, inflammatory skin conditions, sprains and strains, bruises, engorged breasts and bed sores. Mixed with rosewater and used as an eye bath, or to soak eye pads, it helps sore, tired or inflamed eyes, including conjunctivitis. It can also be used as a gargle for sore throats, tonsilitis and laryngitis.

ABOVE: *The strange spidery flowers of witch hazel make a welcome splash of colour in the depths of winter, and the sweet scent is equally appealing.*

Yew: *Taxus baccata*

Parts used: Fruit and leaves

The name *Taxus* comes from *toxas*, meaning an arrow, since the Celts apparently used arrows poisoned with the juice of yew which acts as a nerve poison. Yew leaves contain a toxic alkaloid, taxine, which is dangerous to cattle and horses, though sheep, goats and deer seem to be immune to their toxic effect on the heart and circulation. The seeds of the fruit are also poisonous, although the fruit, which is succulent and sweet and surrounds the seed, is not.

The wood is hard and yet elastic and was traditionally used to make bows. Shakespeare was well aware of this and the poisonous nature of yew when he said:

*'Thy very beadsmen learn to bend their bows
of double-fatal yew against thy state.'*

There appears to be a variety of different reasons why the yew is so commonly associated with churchyards. Some say it was to stop archers from obtaining branches to make into bows, as cutting trees in churchyards was a punishable offence. Others say that yews were planted in churchyards to keep them well away from horses and cattle, who might otherwise come to grief if they were closer to grazing land. It may have been to discourage farmers from allowing their animals to graze on church land. Apparently the yew was sacred to the Druids and was planted by their temples. With the coming of Christianity, churches were often built on old temple sites, and the association between places of worship and yew trees was perpetuated. In the Christian tradition, the yew tree, with its evergreen leaves and long life, was a symbol of immortality. Its heartwood is red while the sapwood is white and this represented the blood and body of

Christ. Yew was said to keep away evil forces and to protect the church from high winds. Some of our ancestors regarded the yew with some suspicion, however, believing that its roots preyed on the dead bodies lying in graves lying below it. The yew was often cut on Palm Sunday and taken into the church to be given out to the congregation, who would take it back to their houses for good luck. It was traditionally burned to make ash for Ash Wednesday.

The yew, despite the dangers associated with its use, was at one time prepared medicinally. The juice of the berries was used for chronic bronchitis, and the leaves have been used in homoeopathic medicine for epilepsy, nausea, dizziness and Ménière's disease, rheumatism and arthritis, liver and urinary problems. Recently taxol, a chemical occurring in yew leaves, has been found to be effective against some kinds of cancer. It is licensed as a medicine now for patients with ovarian and breast cancer and is administered intravenously.

Yew can give the power of love divination. A maid must pick a yew sprig from a graveyard in which she has never before set foot. If she sleeps with this beneath her pillow then she will dream of her heart's desire.

It is said that if yew is brought into a house, a death in the family will occur within 12 months. Also, cutting down or damaging a yew will cause death within 12 months. Witches, ghosts and demons are supposed to hide amidst the dark foliage, but on the other hand yews have been planted close to houses not only to protect them from storms, but also to keep the evil spirits at bay.

Many a schoolboy has related tales of surviving eating yew berries and spitting out the seeds. The sixteenth-century English apothecary Gerard said:

'When I was young, and went to schoole, divers of my school fellows and likewise myself, did eat our fils of the berries of this tree, and have not only slept under the shadow thereof, but among the branches also, without any hurt at all.'

The witches in Shakespeare's *Macbeth* included it in the brew:

'liver of blaspheming Jew
Gall of goat and slips of yew.'

RIGHT: *High above the Wye valley, near Tintern, grows an ancient yew in a small earthy hollow behind a tall pillar of rock known locally as The Devil's Pulpit. However, this tortured yew on its rocky podium appears eminently more suited to the title. The rock was almost certainly once covered by soil, but several hundred years of wind and rain have eroded it away, leaving this weird old sentinel clinging bravely to life.*

OPPOSITE: *Church yards – the most familiar haunt of the yew.*

Woodcrafts and Industries

S everal woodcrafts and industries have been described at some length in this section. Some, such as wheelwrighting and coopering, belong to a bygone era and, although there are still a few individuals who practise these skills, their commercial importance has well nigh disappeared. Others, including timber-framed building and basketry, either still have a lively and sustainable future or, as in the case of charcoal, are making a comeback. All have contributed in their time to the vitality of the British woodlands.

WOOD CRAFTS OF A PAST AGE

The twentieth century has seen the virtual demise of the woodcrafts once employed to construct the bodies and wheels of wagons, both the rustic forms which were the mainstay of rural transport and the numerous carriage styles, from the humble hansom cab to the elegant landau and state coach. Craftsmen do exist who can build such vehicles, but much of their time is now spent in restoration work.

The building of farm-wagons for haulage by horses or oxen, which were the universal form of rural carrier until the advent of the petrol engine, was a conjunction of several different craftsmen's skills. Carpenters built the wagon bodies and frames, wheelwrights constructed the wheels, blacksmiths forged all the fastenings, brackets, brakes and wheel tyres, and painters decked the finished cart in distinctive regional livery.

As all the counties of Britain evolved their customary colours for wagons – Prussian blue in Sussex, buff in Kent, ochre in Essex, bright yellow in Oxford – the styles of carts varied as well. Some had deeper bodies, some a pronounced boat shape; some were relatively plain affairs while others had fretted and chamfered decoration to the bodywork. Earlier wagons had four large wheels, but the later trend of having smaller wheels at the front gave wagons greater manoeuvrability.

Different timber was chosen for the various constituent elements of the vehicle, and all of it had to be well seasoned, for the stresses and strains upon

it were extreme. Ash and oak were used for the framework, ash with its resilience, was ideal for rails and ladders. Elm was the usual choice for the panelling and flooring of the body, and for the two undercarriage assemblies (front and rear), the connecting poles between which were also of ash, as were the shafts for the horses or oxen.

In his *Country Crafts*, James Arnold provides a comprehensive description of wagon building and wheelwrighting. He includes an inventory of specialised tools required to make a wheel. With strange names such as samson, jarvis, buz and spoke dog, all were vital for specific jobs.

Construction of the wheels brought together three different timbers. The central boss of the wheel, known as the nave, was always of elm, for this was the only wood that could withstand the cutting of ten or more mortis holes for spokes plus the tapered axle hole through the middle, not to mention the rigours of years of daily use. Oak was used for the spokes, and it was crucial for these to be cleft rather than sawn to reveal any natural weaknesses or faults in the grain. The heartwood was always kept at the back of the finished spoke to maintain the greatest strength where needed. The wheel rims, or felloes, again for optimum durability, were of ash. Close study of any cart-wheel reveals the dished shape of the spokes within the rim – that is, the spokes were not set at exact right angles to the nave and felloes. The dishing, coupled with a commensurate canting of the axle ends, greatly reduces the stresses on the whole wheel structure. The felloes were cut in sections that would each accept the outer ends of two spokes, and as these were fitted around the wheel they were pinned together with dowels to hold the rim true. The final part of wheel construction was the application of an iron tyre. In early wheels these were in sections called strakes, which were nailed to the felloes, but latterly continuous hoop tyres were used. A previously measured tyre would be heated in the forge, dropped around the wooden wheel, hammered home, and then allowed to cool; as the metal contracted, it pulled the whole structure together with a vice-like grip, and the extreme pressure would soon reveal any weaknesses in the timber.

In *Wild Life in a Southern County*, Richard Jefferies spoke of how wagons, from their construction through to their 'voyages' along distant roads and the goodly 'crews' who steered them forth, resembled sailing ships. He related that the builders of Chinese junks

RIGHT: *A stunning gipsy caravan recalls the skills of the old wagon builders.*

always used timber that had grown naturally into the required bends and knees, rather than manipulating the timber by steam. In the same way the wagon builders selected appropriate timber, storing it in their yards to season. For, like a ship, the true old-fashioned wagon is full of curves, and there is scarcely a straight piece of wood about it. Nothing is angular or square; and each piece of timber, too, is carved in some degree, bevelled at the edges, the sharp outline relieved in one way or the other, and the whole structure like a ship, seeming buoyant, and floating, as it were, easily on the wheels.

The Craft of the Cooper

Oak casks, or barrels, crafted by coopers, have been used for centuries to transport all manner of alcoholic beverages, from beer and wine to brandy and whisky. The oak not only made a tough, leakproof container, but also contributed to the flavour of the liquor within. Dry casks were also once used for a wide variety of foodstuffs. Casks have traditionally been made in the following sizes: hogshead (54 gallons), barrel (36 gallons), kilderkin (18 gallons), firkin (9 gallons), pin (4 gallons).

Coopers were either employed in large permanent workshops within breweries and distilleries or they worked as freelance tradesmen travelling the countryside to wherever their services were required. Making a cask can be divided into four processes, dressing, raising, heading and gathering. Dressing was the preparation of the cask staves, which must be precision-cut and planed so that they will all butt perfectly together. Raising was the assembling of all the staves, using ash trusses to form the familiar shape. Small fires on cressets were lit inside the cask in order to get the staves to yield to the trusses and bend and contract into a tight fit. The ends were then worked before the steel hoops were applied. Heading was the making and fitting of the two cask ends, which fitted into slots previously cut inside the open ends. Gathering was the final hooping and finishing of the cask, after a bung hole had been bored in a slightly wider stave and lined with a metal bush.

Master coopers have virtually disappeared, since the carriage of beer and cider is now almost exclusively in steel casks. The romance may have gone, and even the contributory flavour of the old oak, but the reality is that steel casks are remarkably low in maintenance and cheaper to manufacture.

Shipbuilding

John Evelyn introduced his *Silva: or A Discourse of Forest-trees* with an expression of deep concern at the plight of the British navy in the shadow of what he considered to be almost rampant vandalism of the nation's woodland by irresponsible landowners. He berated them for their 'disproportionate spreading of tillage' and their accompanying demolition of

'goodly woods and forests' and added the following warning:

'...there is nothing which seems more fatally to threaten a weakening, if not a dissolution, of the strength of this famous and flourishing nation, than the sensible and notorious decay of her wooden walls, when, either through time, negligence, or other accident, the present navy shall be worn out and impaired'.

Dr. Hunter in his 1776 edition of *Silva*, pointed out that Evelyn's worries were nothing new. He recalled that Tusser, writing in 1562, had expressed exactly the same concerns, 'that men were more studious to cut down than to plant trees'. Hunter himself recommended that wealthy landowners look to their own estates and the potential for tree planting.

Until the early years of the nineteenth century, perennial panic about the lack of suitable oaks for the shipbuilding industry was highly misguided. Ships were relatively small until the mid-eighteenth century,

TOP: *Raising the cask. The cooper and his mate fit the ash trusses on to the cask prior to firing. The photograph was taken at the Reading brewery of H.&G. Simmonds Ltd. in 1955, a time when most large breweries still had their own cooper's shop.*

ABOVE: *This photograph dating from 1902 shows Samuel and William Wilson of Breachwood Green, Hertfordshire, retyring a cart-wheel. These men were most probably general blacksmiths, rather than wheelwrights, judging by the rather makeshift support instead of a proper tyring platform.*

RIGHT: *Clinker-boats at Dungeness.*

RIGHT: *Clinker-boats at Dungeness.*

OPPOSITE: *Built in 1869, the graceful* Cutty Sark *was built for speed as a tea clipper for the trade with China.*

RIGHT: *Peter Faulkner of Leintwardine in Shropshire at work on one of his coracles. The framework is made of hazel, the woven flooring and the braided gunwale of willow, and the seat of ash.*

CORACLE BUILDING

One of the earliest types of boat known in Britain, the coracle, is effectively a woven boat. Dating from well before the arrival of the Romans, its name is derived from the Welsh 'corwgl'. These rather amusing little oval boats have been traditionally associated with the rivers Teifi, Towy and Wye and certain stretches of the Severn, and each area has its own slight regional variation in shape. They are designed for a single person, and their extreme light weight (about 30 pounds) makes them very portable. The design is always the same in principle: cleft willow laths are interwoven to make the framework, the ends being pulled upwards and secured in a woven gunwale of willow laths or, alternatively, unsplit hazel rods. This rather large basket is then covered with hide or calico, which is painted with pitch. A bench seat is secured inside. A piece of ash is formed into a short, large-bladed paddle, and all is ready to venture upon the water. The craft will last for only two or three years, but it is a simple enough task to make another – probably much simpler than learning how on earth to steer the coracle in the desired direction!

and Britain had plenty of suitable trees amongst its hedges and parkland to supply the shipyards, for it was these trees which supplied the timber correctly shaped for the knees and ribs, rather than the tall standards which were better suited to building construction. If the naval shipyards had a problem, it was that they were constantly starved of funds and thus couldn't afford the best timber. Apparently merchant shipyards had no difficulty obtaining the right timber when they required it, perhaps because they had access to the returns from increasing international trade.

The Napoleonic Wars saw the greatest upsurge in naval shipbuilding, which continued along with the expansion of merchant shipping until about 1860, when ironclads began to eclipse wooden ships. The price of oak rose steadily during this period, not because of its scarcity, but due to the dramatic increase in demand for oak tan bark. So much did the tanning industry influence woodland management, that great tracts of oak wood were simply coppiced rather than allowed to grow to maturity, and a great many oaks were planted during the early nineteenth century. However, with the huge increase of iron and steel construction, and then chemical substitutes for oak bark tannin, the demand for oak soon slipped away. The legacy from this period is evident now in the stands of quite closely packed oaks, with sparse understorey, creating slightly unnatural and rather unremarkable woodlands.

English oak was always the best timber for shipbuilding, although for many smaller clinker-built boats, larch became a popular alternative for the planking. By the late nineteenth century the larger ships were all built of iron, but as late as the 1960s small fishing vessels and pleasure craft were still regularly clinker-built. Then, just as iron had displaced timber in the previous century, the advent of glass fibre for hulls saw the gradual decline of wooden boats. Never again would so many regional variations of boat and ship designs grace the British coastline. Gone were the Cornish luggers, the Whitby yawls, the Essex sailing barges, the Scottish fifies and the cobles, ketches and mules, along with the elegant pleasure craft of the Lakeland and the chunky canal barges and butties which plied their trade along the great canal network.

As with so many splendidly crafted vessels of the past, they are now the fodder of museums and preservation societies, and yet they are not all in mothballs. A few sailing barges still venture out along the lower Thames reaches and the Essex/Suffolk coast, occasionally having a bit of a race with one another. There are also many enthusiasts racing vintage clinker-built yachts and dinghies at regattas up and down the country.

Building with Wood

From the Neolithic period onwards, people began to settle down in permanent communities, largely as a result of management of farmland and woodland which provided the necessary sustainable infrastructures. Whatever natural materials that came to hand were pressed into service. The first houses are thought to have been round in shape and constructed from palisades of poles, with roofs woven from branches or reeds in a rudimentary thatch. Large centre posts would often act as an anchor for the whole structure and wattle-and-daub would have been used to interweave and bind it together.

The principles of timber framing dominated construction across much of England right up to the sixteenth century, and in several regions well into the nineteenth. Both the Anglo-Saxons and the Normans had skilled stonemasons, but they were usually employed upon grander constructions such as cathedrals, churches and palaces. Many of these splendid buildings have survived to this day, where houses, farms and barns built with timber have long since crumbled away: Britain's earliest surviving vernacular architecture stems from the fifteenth and sixteenth centuries. All buildings, however, required a certain percentage of timber construction for the roof members and the floors, doors and windows: a

study of some of the oldest-surviving examples reveals that relatively small-scale wood was used wherever possible. Even in medieval times there was great movement of timber around the country, particularly when huge pieces were needed for special projects, such as the gigantic roof members of cathedrals. Oliver Rackham also found that there was a brisk import trade of timber as far back as the thirteenth century, with documentary evidence confirming that pine from Norway was used in the building of Ely Cathedral.

From the sixteenth century onwards, according to date stones, the increased use of locally quarried stone and the widespread manufacture of bricks appear to have ousted a lot of pure timber buildings. The timber-framed style persisted in areas where substantial amounts of suitable timber, predominantly oak, were available, and local building stone unavailable or unsuitable, though there were exceptions. As Rackham points out, Cambridge and its environs were once predominantly composed of timber-framed buildings, yet there was a plentiful supply of good stone.

R.W. Brunskill has divided Britain into eighteen regions which display distinctive vernacular styles. Five of these have a significant quota of timber-framed buildings to this day, each with its own particular characteristics of design and decoration. With the exception of East Anglia, they all lie in the same parts of Britain as the major concentrations of native woodland. The other principal regions are the southeast, including Kent, Sussex, Surrey and Hampshire; the Home Counties, to the west and north of London; the West Midlands, mainly Herefordshire and Shropshire, along with the eastern edge of Wales; and Cheshire and Lancashire to the north where a great deal of early timbered architecture has undoubtedly been eradicated in the past to make way for industrial sprawl and its associated dwellings. The long band of Cotswold limestone, which gives this great swathe of central England its distinctive honey-coloured stone buildings, splits the five regions of timber-framed buildings.

East Anglian timber-framed buildings are typified by their closely spaced vertical studding and tall pointed roofs, often thatched from Norfolk reed, which is the very best of thatching material and plentiful thereabouts. Coloured washes to the rendering between the studding are a popular decorative addition which, when a whole row of buildings receives the treatment, can generate some stunning effects. The oak studding is usually left unpainted, so that the weathered timbers and soft colours create a much mellower appearance than the stark black and white of the west.

In the southeast, Kent and Sussex hold some of the finest examples of wealden houses. Essentially these are built around a central hall, which rises to the full height of the building. Either end of the house is on two storeys, typically with jettied first floors to the front and sometimes the ends of the house. Curved

NEXT PAGE:
HIGHFIELD

Timber construction of buildings has become more fashionable in the past ten years, with many people opting to build their own house in wood because of its longevity, strength and natural beauty. This huge tithe barn was constructed in the depths of Warwickshire in 1998 and is the largest green oak structure to be built in Britain for over 300 years.

The timber frame was erected and secured using traditional construction methods where wooden dowels (pegs) are hammered into basic mortice and tenon joints. No nails or screws whatsoever are used in the timber frame.

It took two years, 21,000 man hours and 350 oak trees to build the aisled barn. The owner planted nearly 10,000 new oaks to replace those used in construction.

Used solely as a private entertainment complex, Highfield houses a stunning swimming pool, cinema, sauna and gymnasium.

RIGHT: *The longest-surviving hammerbeam roof in Britain, in Westminster Hall, London. Completed in 1401 by the carpenter Hugh Herland, the Royal Commission describes it as 'probably the finest timber-roofed building in Europe'. The quantity and dimensions of oak used in the construction are awesome, the hammer-posts alone being 39 by 25 inches and 21 feet long.*

braces in the studding are often a typical feature. Although some buildings have rendered panels, it is equally common to find them of natural red brick, usually laid in a stretcher bond, but sometimes, particularly in older buildings, a herringbone pattern has been adopted. In the southeast, timber is also used for weather boarding; the horizontal lapped timber cladding is applied as the outer skin of the walls. This was easy to apply to the outside of a timber-framed building and cheaper than, though not as durable partial brick or stone construction.

The Home Counties to the north and west of London present a smaller proportion of half-timbered properties than the other regions, but with the understandable influences coming from the wealden homes of the southeast hinterland as well as East Anglia.

To the west lie the West Midlands and the eastern borders of Wales. This region is best known for its striking black and white buildings, of which some villages are almost totally comprised. The most common style is known as box-frame which, as its name suggests, is identified by the square sections between the timber studding. Decoration to these panels may be with diagonals or chevrons, which might also add strength to the structure. The visual emphasis is always upon the striking black and white paintwork.

Northwards, in Cheshire and Lancashire, one finds

yet another distinctive style of timber-frame building. Here the studding is often closer than anywhere else, and the small panels it forms frequently contain geometric patterns utilising stars, crosses, diagonals or quatrefoils. Most of the surviving timber-framed properties in this part of the world tend to be on the grand scale, typically ancestral homes, large merchants' houses or great galleries of commercial buildings such as those to be found in the centre of Chester. Much

ABOVE: Typical fifteenth-century dwelling in Lavenham, Suffolk. The four-centred arched door frame exposed with the studding shows the changes the building has undergone over the centuries.

BELOW: Lower Brockhampton is a splendid late-fourteenth-century moated manor house with a detached fifteenth-century gatehouse.

JOHN ABEL

Very few of the early builders of half-timbered structures are known by name, it is only from surviving contract documents or distinctive design features that they can usually be defined. This building, known today as The Grange, in Leominster, was built as the Market Hall by John Abel in 1633/4. In its original form the ground floor would have been an open arched structure in which the weekly market transactions would have taken place. The building's recent history has been quite eventful. In 1855 it was taken down and offered for sale, being bought by a Mr. Francis Davis for the princely sum of £95. Shortly afterwards the local MP, John Arkwright, bought it for the figure from Davis and promptly offered it back to the council if they would undertake to re-erect it. They declined, and the building lay in pieces in a builder's yard until Arkwright rebuilt it on the present site as his family home. During the 1930s there was a risk that the building would be sold and exported to America. The council stepped in, paying £3,000 for it, and brought it back into civic use.

like the timber-framing of the West Midlands, great emphasis is placed on pitch-black studding and white panel infills.

Timber-framed buildings can be divided into two basic types of construction. Cruck-framing is characterised by pairs of huge cruck blades which arc gently from the base of the walls to meet at the ridge purlin. There will be a pair at each end of the building with one or more intermediate pairs, depending upon the length of the building. The ridge purlin, side purlins and wall plates tie the frames together along the length of the building, whilst tie beams pegged between the cruck blades stop them from spreading under the roof load. The technique dates back at least to medieval times. Walls would originally have been turf or wattle-and-daub, while

further developments would have utilised timber-framed walls and, latterly, stone or brick. Cruck frames are usually concealed beneath the fabric of buildings and are thus not immediately discernible. The giant cruck blades were usually cut from one large tree. Until recently this was always assumed to have been oak but some early crucks have now been identified as black poplar: its slightly arced form would have made it an obvious candidate. The immense size of some of these crucks provokes wonderment about how they were manoeuvred to the building site and erected. Box-framed buildings, which were a slightly later development but almost certainly overlapped with cruck-frame, were comprised of vertical and horizontal timbers joined in squares or rectangles to make a rigid box structure. Triangular trusses formed the roof, braced by purlins lengthways and then again by rafters laid on top. The joints, and there were many of them, were invariably secured by wooden pegs.

The expression 'half-timbered' refers to buildings with exposed studding on the exterior. The panels were traditionally filled with wattle-and-daub, while more recent infills are of brick. Oak has always been the most sought-after timber for this type of building for it is the strongest and most durable wood; but elm has also been used in many buildings, usually for roof trusses and purlins and for floorboards. Richens states that elm was used principally in regions such as East Anglia, where oak was relatively scarce, though, even where oak has always been prolific, as in Herefordshire, elm appears with some regularity as roof members. So much remnant suckering elm still exists in Hereford's hedgerows that large trees must have been plentiful and convenient to use. Even so, when they become worm-eaten, as both timbers eventually will, oak is still the superior timber. It may appear badly infested on the outside whilst the inside is still hard as iron,

while elm tends to become affected right through the timber and will probably fail well before a similar oak member. Timber-framed buildings were invariably built from green timber, unless materials were being recycled from other buildings (it is a popular misconception that old ship's timbers were used to build houses), and since all green timber will move and settle as it dries, this has resulted in some bends and bows appearing in old properties. Elm will warp even more than oak, so a noticeably dipping ridge line indicates its presence.

Close inspection of even the oldest of timber-framed buildings reveals the remarkable skills of the carpenters of centuries long past. Joints were cut precisely, though with rudimentary tools; carpenters marked codes on the joints of trusses which had been assembled in workshops and then broken down and carted, thus enabling them to be correctly erected on site; huge sawn and hewn timbers bear distinctive cross-cut and adze marks; and on some of the more affluent residences are marvellous carvings and panellings inside and out.

Wood Turning

Whether for bowls from which to eat, legs of chairs on which to sit, handles of tools to till the land or shafts of weapons to harry the foe, wood has been shaped into all manner of rounded items for hundreds of years. Though wood grows into cylindrical and spherical shapes, sticks and smaller branches can never offer the uniform and flawless shapes for enduring articles, and young wood is highly prone to warping and splitting. Seasoned wood is superior for turnery, being taken in the cube from the main trunk of certain trees.

The principle of turning is that, rather than having a static piece of work around which the craftsman moves, the piece is rotated across a static tool in order to make the required form. One of the very earliest methods of turning was the pole lathe, for which the spring of a long wooden pole, combined with a foot treadle, supplied the simplest of power sources. The piece to be turned was set in a horizontal bed, secured at either end between poppet-heads, at a convenient height for the operator, who usually stood. Then a string was run from the treadle, looped around the piece, and tied to the end of the long pole, the other end of which was set in the ground a few feet away. Pushing down with the treadle revolved the piece one way, while the spring of the pole revolved it in the reverse direction. The turning chisel was only applied on the forward stroke, being withdrawn as the piece revolved the opposite way; a variety of chisel types and sizes were used to obtain the different shapes required.

Pole lathes were extremely simple and quick to construct and were widely used in woodlands all over Britain. One of the greatest concentrations of such

ABOVE: *Mr. J. Davies of Abercych, Cardiganshire, turning a sycamore platter on his pole lathe in 1932. Some turners developed the remarkable ability to turn nests of three or four bowls out of one piece of wood.*

LEFT: *A sumptuous array of bowls and cups turned by Chris Robertson from a wide variety of woods.*

ABOVE: *Mike Abbott, pole lathe turner. The overall view shows how the interaction between the treadle and the springing pole powers the lathe. Although, in true bodging tradition, Mike makes plenty of chair legs, he has also become something of a specialist making babies' rattles, such as this one which he has almost completed.*

turneries was in the Chiltern beech woods, where dozens of chair bodgers set up their hovels and lathes in amongst the coppice wood which they had contracted to work. The chair legs they turned were strongest when taken from a cleft of a mature beech log, thus retaining the natural direction of the grain. After the rough clefts had been cut to the correct length, the corners were trimmed back and the taper of the ends was worked with a draw knife; the pole lathe was then used to turn the finished leg. The beech was worked green, and the finished legs were stacked to season.

As the twentieth century progressed, fewer and fewer turners worked with pole lathes. The harnessing of water power enabled large industrial turneries to produce massive quantities of items such as bobbins and reels for the cotton industry. Great ranks of belt-driven lathes could be powered by a single water-wheel. Many of the mills were in Cumberland and Westmorland where there was a ready supply of silver-birch coppice well suited for these items, though, unlike beech, the birch had to be seasoned for some time in the wood before it was ready for turning. Later, as electrically powered lathes became commonplace and made the turning process radically faster, the demand for mass-produced turned items began, perversely, to decline. Shortly after World War II the entire textile industry began a downturn as well as an increased use of plastic components where wood had once been the only choice.

For many years vessels such as bowls, platters, beakers and boxes were turned for use in ordinary homes. Some of these aged utensils, often made of elm or oak, have survived and are avidly sought by collectors of treen, who admire their simple shapes and their centuries-old patina. In the nineteenth century a great interest arose in decorative wares of all descriptions. Craftsmen produced beautiful and intricate wooden items upon the lathe: candlesticks, goblets, tobacco jars and an infinite variety of exquisite boxes were but a small selection of their output. The Victorian desire for lavish architectural decoration brought a demand for all manner of turnery for balustrades and bannisters as well as intricate furniture patterns.

The commercial side of wood turning has moved to computer-controlled lathes which produce endless identical components for the furniture industry, often made of oriental hardwoods from forest sites of dubious sustainability. There are, at the other end of the craft, still plenty of individual craftsmen out there making beautiful things in garden sheds or tiny studios, although few of today's turners fill an indispensable niche in any industry. There are even a few exponents of the pole lathe getting back into the woodlands once again. For the fine art turners, who may well also make functional pieces, the main interest is interpreting and developing the intrinsic qualities of an individual piece of wood. The thrill for these turners is to go out and rummage through woodland and hedgerow in search of wonderful natural lumps of wood, keeping a weather eye open for unusual trees such as yews or fruit trees which have been wind-blown or felled; for a while, after thousands of elms were smitten by Dutch elm disease, there were copious quantities of elm wood to be scavenged for turning, but this supply is now virtually exhausted. The finds are hauled back to studios where they can dry out or finish

seasoning, sometimes with the end grain coated in wax to inhibit splitting along the medullary rays. As the artist lives with these natural forms they begin to suggest how they should be turned and what they will eventually become. Even something as simple as a bowl can have an infinite variety of forms, and some turners produce such thin and delicate ones that they take on a transparency akin to china. Other turners revel in the knots, figures and even holes in the wood. Strangely enough, marks in the wood made by disease or fungal attack can often add to the visual interest of a piece.

Several woods have traditionally been used for specific turning purposes. Birchwood was used for bobbins, barrels, bungs, clothes pegs, bowls, spoons and butter pats. Sycamore, which was noted for its property of not tainting foodstuffs, was ideal for bowls and platters, butter pats and rolling pins; since its clean white wood was not prone to staining, it was also used for rollers in laundries and the textile industry. Beech has always been one of the most universal woods for turning. As Edlin puts it: 'Turners employ beech for all the hundred-and-one odds-and-ends of household use where a "bit of wood" is needed.'

Ash has been, and still is, the most popular wood for handles of all manner of tools, while more specialised applications were billiard cues, cricket stumps, parallel bars for gymnastics and rowing oars. Elm was commonly used for bowls, candlesticks, egg cups and a variety of household items. Oak was never the most popular of woods for the turner, for unless exceptionally well seasoned it tends to warp and split. The splendid woods of walnut, cherry, plum, pear and apple have all been used to make decorative pieces. Some of the harder, closer-grained woods such as box, holly and yew have also been used for their decorative attributes. Box has a delightful creamy yellow colour and, because it seldom grows to any great size, it was used for small and delicate items such as lace bobbins, chessmen and draughtsmen, as well as small pulleys, rollers and knobs for furniture. Holly has similar qualities and uses to box, but was more receptive to staining, thus it was often stained black by way of a substitute for ebony. Yew has perhaps the most mouth-watering array of reds and purples within its wood, although the frequent voids and hollows of its more ancient boles can make it a tricky wood to work into larger pieces.

Wood Carving

Widespread evidence of wood carving in Britain dates from the medieval period, when the great majority of carved wood was in oak. Tools were rudimentary and only as fine as the blacksmiths who wrought them. Much carving of this time is distinguished by the simplicity of forms and, to the modern eye, an endearing naivety in the execution. Some depictions were allegorical while

OPPOSITE: *'The Forest Man' by Len Clatworthy. This 20-foot figure is to be found near the Dean Heritage Centre in Gloucestershire. Constructed largely of hazel and willow, he is being allowed to erode naturally.*

others, which today seem to defy all explanation, might now be interpreted as humour or ribaldry. Maybe they were simply that.

Churches are by far the richest repositories for the art of carvers of both stone and wood. In these sanctuaries their work has weathered the various caprices of fashion, religion and politics, including the successive ravages of Henry VIII and Oliver Cromwell, better than anywhere else. While depictions of religious themes are to be expected in a church, it is fascinating to find imagery of a decidedly pagan origin frequently in close proximity with Christian iconography. Little faces and figures representing the Green Man are surreptitiously located under seats, behind pillars or high up on roof bosses. Perhaps they were pictorial reminders of the Old Ways, which were to be set aside in favour of Christian worship, or astute compromises on the part of the Christian authorities in deference to the deep-rooted alternative sensibilities amongst their flock.

Little is known about the identity of the vast majority of wood carvers, but Grinling Gibbons (1648-1721) is still regarded as one of the finest artists of all time. His work is found to this day in many a stately home and church up and down the country. John Evelyn recorded with some pride his acquaintance with Gibbons's work:

'...the festoons, fruitages, and other sculptures, of admirable invention and performance, to be seen about the choir of St. Paul's and other churches, royal palaces, and noble houses in the city and country, all of them the works and invention of our Lysippus, Mr. Gibbons, comparable, and for ought appears, equal to anything of the antients. Having had the honour to be the first who rec-ommended this great artist to his Majesty Charles II, I mention it on this occasion with much satisfaction.'

Gibbons was one of the finest exponents of the architectural genre, one of the three broad divisions of wood carving. His work was mostly confined to interiors. Up to the seventeenth century there had been a long-standing tradition of carved oak on the outside of buildings, in particular on timber-framed

constructions, but these were usually the preserve of the more opulent buildings or those of some public significance. Oak was the timber most suited for these exterior works; four or five centuries on, it is still amazing to see how many carvings have withstood the batterings of the elements. Carving within buildings was usually decorative panelling, friezes, fire surrounds, staircases and the like. From the Jacobean period onwards, furniture made for the wealthier families and their splendid homes was carved with figures, motifs and patterns such as linelfold on the panels. By the nineteenth century machines had been developed

TOP: *Sunlight through a stained glass window illuminates a medieval carved frieze.*

ABOVE LEFT: *Medieval carving of two little pigs with their heads in a trough on the choir stalls of Little Malvern Priory. The connotation of such imagery would surely be appreciated today, but maybe the privileged monks were oblivious.*

ABOVE RIGHT: *This little chap from the fourteenth century was rediscovered during recent renovation in a Herefordshire church. If his existence was known, then all the previous guidebooks chose to ignore his presence. He is still proving controversial since some people want to publicise him as part of the church's interest, while others think that it's not in good taste.*

LEFT: *Love spoons: Sycamore (opposite) was the most popular wood for these, since it is easy to carve, stable will not taint if used for food or drink. Each spoon is carved from a single piece of wood and some of the more intricate designs defy belief. Typical patterns involve balls within balls, or within boxed stems, interlinked chains and beads on spindles; hearts and initials are frequently included.*

RIGHT: *Outstanding detailed carving in lime wood at Kirlington Park in Oxfordshire, by the great master Grinling Gibbons.*

BELOW RIGHT:
This sculpture in elm by Dame Barbera Hepworth (1959-1960) is called Nyanga.

BELOW: *A timber horse, by Dave Johnson, in the depths of Brockhampton woods near Bromyard, which was inspired by the horses once used in woodlands to extract the larger timber.*

BELOW, RIGHT: *'Rapunzel' by Dave Johnson. All the above wood sculptors were active at a recent Sculptree event at Westonbirt Arboretum. In this group event sculptors banded together for a week to exhibit their skills and to create pieces which would be auctioned for the charity Tree Aid on the final day. Tree Aid's purpose is to restore tree cover around the villages of the Sahel, in Africa, where they are imperative for food, fuel, shelter, medicines and animal fodder, as well as protection of top soil to enable the cultivation of food crops. The huge pieces of timber that the sculptors work upon are all wind-thrown or diseased trees that have been saved specifically for this event.*

which could cut out regular patterns, so that much of the flamboyant carved woodwork found in Gothic-style interiors has been turned or wrought in some way by machine. Late-twentieth-century wood carvers may seldom have the chance to design and execute whole interiors, but their skills are constantly in demand for restoration purposes.

Carvers of small and personalised items were a separate group unto themselves. Many such carvers were enthusiastic amateurs or were compact enough in their needs and products to set up as cottage industries, and quite often these craftsmen turned wood as well as carving it. They used woods that had fine figuring, such as burr walnut or yew, or those, such as the fruit woods, laburnum or box, which provided interesting colour and were also very stable. Box was in fact the most sought-after wood for the comb makers, from Roman times through to the seventeenth century. The Romans developed an ingenious saw with two blades slightly staggered, so that as one blade finished its cut the second blade was starting the next, thus allowing progressively equidistant teeth to be cut. During the Tudor period it was fashionable for young men to give splendid carved and decorated box combs to their ladies. Another token of affection has long been the love-spoon, particularly in Wales, where they were carved by young men for their intended.

The fine-artist wood carvers are more easily identified as a separate group in the twentieth century. The delineation may be subtle, but these are carvers who often work on quite a grand scale. The end product may serve no practical purpose, but be purely a statement of vision or emotion, either representational or abstract. These are the sculptors of wood.

Basket Weaving with Willow

There are literally hundreds of different varieties and hybrids of willow which can be used for basketry. Some will split or peel better than others, while some also offer the options of different colours, such as golden willow or purple willow. Probably the archetypal species is the osier (*Salix viminalis*), although almond willow (*Salix triandra*) is also favoured by many weavers, who aver that it makes the best-quality baskets.

As willows grow most successfully in wet or marshy ground, they have traditionally been cultivated commercially in the flat, low-lying areas of Britain, most typically the fens of East Anglia and the Somerset Levels. However, willows have the capacity to do well in almost any damp situation: a cut twig simply pushed into the earth will almost immediately begin to throw out roots. Sets are usually cut and planted in the winter, before the sap has begun to rise, and may be as dense as 25,000 plants per acre. After two years they have grown enough to be harvested, and thereafter may be cut annually for about 30 years, although some osier beds are said to be more than 50 years old. The willow stems, usually called withies, are cut as close to ground level as possible, between November and April. The one-year-old growth, which is thinner and hence more pliable, is known as a rod, while the older and thicker product is called a stick.

There are three principal types of rod – brown, white and buff – not to be confused with the multitude of different natural colours and names. Brown is the general term applied to rods with their bark still intact, and, as a rule, these have been used for the less refined applications of basketry, such as log baskets and agricultural containers. Baskets made of brown rods are not as well seasoned as the better-quality baskets, which are made from the white rods, and so will have a shorter life. White rods are cut and peeled in the spring, when the sap is rising; the bundles of rods are frequently stood upright in running water to prolong the peeling season, a process known as pitting. In the past the peeling would appear to have been largely performed by women, using a tool called a break, which comprised two iron blades in an upright wooden frame which pinched the withy and then scraped off the bark as it was drawn through. Now a machine with large abrasive rollers speeds up the process. Buff rods are the ones of a pleasing golden-brown colour most often used for domestic basketry. Buffs are brown rods which have been boiled for a few hours, during which the tannin in the bark is released and migrates into the white wood, staining it brown; they are then peeled in the same way as whites. The use of buff rods has attained its greatest popularity in relatively recent times.

The method of constructing an article of basketry is always based on the principle of weaving thin sticks

ABOVE: *Stripped willow being bundled for transportation, somewhere in the Cambridgeshire fenland, in the early twentieth century.*

LEFT: *Willow hampers and baskets piled high in an improbable fashion, doubtless for the benefit of the photographer, form a backdrop for these two basket weavers in the doorway of their workshop.*

LEFT: *Simon Cameron, one of the new breed of basket weavers, begins a piece in his workshop.*

BELOW: *Osier willows used for basketry.*

around or between thicker sticks, known in the trade as stake and strand baskets. Forms are usually of a round or oval nature, the exceptions being items such as fishermen's creels or picnic hampers which tend to be rectangular boxes. The willow thickness is matched to the strength demanded by the container, as are such details as types of handle. Sharp butt ends are carefully kept to the outside of baskets that may be used for laundry, so that clothing isn't snagged. The most important foundation for a sound basket is a sturdy base or slath, which is made in a domed form, thus making it stronger, in larger baskets designed to carry considerable weights. Once woven to the desired width the spokes of the slath are turned up to begin the sides; this is the upset. Thicker stakes are slid into the edge and the main body of the basket is woven. There are several technical terms for the different types of weave. A single strand is called randing, grouped rows of two or more strands is slewing, two rods intertwined prior to weaving is called pairing and more rods twined together is known as three (or more) rod waling. Siding with intermittent gaps is known as fitching.

At one time baskets were a universal necessity, though the art of the basket weaver has been in a good deal less demand since the rise of plastic, about the time of the last war. Miners used baskets for coal and ore (yet these may have been the stronger oak swill type rather than willow), fishermen needed baskets for their harvest, fruit and vegetable growers used them to transport their produce, jars and bottles were protected with basketwork, and before the plastic carrier bag virtually every housewife owned a willow shopping basket. In 1949, Edlin catalogued not only over 100 different kinds of baskets, but also babies' cribs, lobster pots, eel traps, birdcages and dovecots, umbrella stands, cushioning for fragile loads dropped from aircraft and various types of hats. A larger and increasingly important use of woven willow is either as soundproofing screens alongside busy roads and motorways, or as willow mattresses for stabilising reclaimed land or new landscape schemes. As with several other woodcrafts, basket weaving has had a small revival over recent years, but the craft is largely the preserve of creative individuals producing a variety of basketry for an essentially leisure market. This in no way devalues the excellence of skill employed in their production, and in many respects the humble basket has been elevated to the status of art, yet the plain truth is that cheaper mass-produced containers

fulfil the utilitarian role. Beautifully produced baskets, often made with different-coloured willows, have become designer accessories for the upper bracket to dangle on the arm and use for flowers gathered from the garden or champers and foie gras for Ascot.

Charcoal Burning

How on earth can lighting up the barbecue and enjoying the delights of an alfresco repast revitalise our woodlands and generate growth in the rural economy? The answer is charcoal.

The original demand for charcoal was primarily derived from the intense heat used in the metallurgical industries of Britain which, though they existed prior to the Roman occupation, were considerably expanded by that new order. There is plentiful documentation of the various early centres of the iron industry and its attendant demands upon local forestry in the Kent/Sussex Weald, the Forests of Dean and Wyre and the Furness district of Lancashire. It is a common misconception that the ironmasters were responsible for denuding the wooded landscape, particularly during the Industrial Revolution. In fact, the reality was quite the contrary, since the ability to maintain sustainable fuel supplies in close proximity to the iron industry was fundamental to its survival. The last remaining charcoal-fired furnace at Backbarrow, in Cumbria, was in production into the early years of the twentieth century. Charcoal was also once the principal fuel required for the making of steel and glass, although, as with the iron industry, it was superseded by the use of coke and even more recently oil- or gas-fired blast furnaces.

During the sixteenth century the consumption of Britain's wood for charcoal in the iron industry caused periodic panic. Royal Commissions recommended the prohibition of felling mature trees, in order to

RIGHT: *Charcoal burners in Epping Forest. This 1879 engraving depicts a typical encampment with the little hut which was used as a shelter for the round-the-clock vigil, necessary in case of flare-ups. The screens to the right of the kiln are placed there for the same reason, as they deflect any wind over the stack.*

keep the timber for shipbuilding, and both the manufacture of iron and the establishment of new ironworks were widely curtailed, as there was concern that cannon might be exported to potential adversaries such as Spain. Observers with a more aesthetic leaning denounced the despoliation of the landscape.

The traditional mainstay of wood supply for charcoal has been from regularly coppiced hardwoods, usually on a rotation of between seven and 15 years. Ash, hazel (which regenerates the most readily), oak, sweet chestnut and hornbeam are all excellent for making suitable fuel charcoal, while willow produces the best artist's charcoal. The charcoal made from alder buckthorn has traditionally been considered the finest type for gunpowder, when crushed and mixed with sulphur and saltpetre. Besides heat for industry, gunpowder is the major use for charcoal.

With the decline in demand for charcoal as an industrial fuel, and the demise of other applications such as filtration, insulation and various chemical constituent purposes, the whole charcoal-burning industry seemed doomed by the mid-twentieth century. Hard upon this decline came a commensurate abandonment of coppicing practice, which heralded the loss of woodland vitality on a nationwide scale. Happily, however, recent years have seen a dramatic turnaround.

For centuries the customary method of producing charcoal was from within an earth-covered stack or kiln. Although there were many minor regional variations in their style, the principle was always the same. The charcoal burners would set up camp for a few months at a stretch, sometimes with their whole families, in a large tract of woodland, where cordwood had been cut and stacked several months before to season. A large circular hearth would be cleared on level ground, preferably protected from prevailing winds. A central flue was constructed, with a lattice of split logs or branches, usually in square or triangular section. Around this flue the cordwood was stacked in regular and tight radial formation, with the thicker pieces to the centre of the kiln. Once the large hemispherical kiln was raised, it was covered with grass, bracken or turves, before the final layer of earth was carefully packed into place all over the top, in order to exclude any air. The kiln was then lit by dropping red-hot embers or charcoal down the central flue, with a little suitable firewood. Once the fire was underway the top of the flue was capped off with more turf and soil. The whole kiln was then fine-tuned with appropriately pierced vent holes in the sides, which caused the burning to be even throughout. Finally, the vents would also be closed off, so the whole kiln could gently char over a period as long as ten days, depending on the size of the kiln. The charcoal burners had to be on station 24 hours a day, in case of a flare-up which might ruin the whole kiln. When the smoke turned from white to blue and finally disappeared to

MODERN CHARCOAL BURNING

By the 1930s the making of charcoal with earth kilns had virtually disappeared, with the last recorded commercial burner working up to 1948. In recent times there has been a renewed interest in the process, which has encouraged countryside museums to put on demonstrations, such as this one at the Dean Heritage Centre in the Forest of Dean, in the expert hands of Pete Ralph and John Reed. The sequence shows the construction of the square flue for the centre of the kiln. The kiln is built of neatly stacked cordwood, leant towards the flue, and covered with turf. Hot embers are put down into the flue along with small billets of wood to start the fire. When this has taken hold and smoke is beginning to appear through the kiln sides, the flue is capped with more turf and earth is scattered over the whole stack to seal any air holes. This quite small kiln will then be left and watched for two to three days. It will be dowsed with water if there should be any untoward flare-ups which would consume the charcoal. Eventually, little or no smoke will be seen, and only a shimmering heat haze above the kiln betokens the completion of the process.

a heat haze, the charcoal was ready. The kiln would then be carefully dismantled, so as not to break up the sticks of charcoal, with plenty of water available to damp down possible flare-ups. Depending upon the type of wood, and the speed at which it was charred, the average yield from six tons of wood might be one ton of charcoal.

The earth-burn method of production is now obsolete, and only performed occasionally for demonstration purposes at such locations as the Dean Heritage Centre and the Weald & Downland Museum.

The increasingly familiar metal kilns now used by commercial charcoal burners are widely thought to have come into favour about 60 years ago, yet there is documentation as far back as 1808 which records their use on Lord Egremont's estate, at Petworth in Sussex. Steel kilns are infinitely more user-friendly than the old earth kilns, as they are not at the mercy of the weather and once fired up can be safely left to burn to a satisfactory conclusion.

Some 95 percent of Britain's 60,000-ton charcoal consumption is currently imported, much of it being derived from unsustainable sources such as tropical rain forests. An organisation called the Bioregional Development Group has set up the Bioregional Charcoal Company (BRCC), a co-ordinated network of around 120 charcoal makers who market their product to major national retailers. Since the available quantities of coppice wood far outweigh the demand

WHAT EXACTLY IS CHARCOAL?

H. L. Edlin, in his comprehensive Woodland Crafts in Britain, *describes it thus:*

'*Charcoal is made by heating wood in the absence of enough air for complete combustion. The water contained in the wood is first driven out, followed by all the volatile compounds, including creosote and tar. This leaves behind a residue of solid black carbon, together with a little mineral ash. The process of wood decomposition becomes, at one stage, exothermic, producing more heat than it absorbs; this fact explains how a small fire, lit at the heart of a heap of wood, can cause the whole to become charred, although only a small fraction is actually burnt up.*'

as far back as the medieval period, although the industry does not appear to have achieved boom proportions until around 1780. The demand for leather for the following 70 years turned a minor woodland by-product into a lucrative industry. Great tracts of oak wood were given over specifically to bark production from the coppice rotation. Since the mid-nineteenth century was a period of increased importation of conifers for building construction, Oliver Rackham surmises that the massive planting of oaks carried out by landowners, keen to cash in on the oak-underwood bonanza, was largely responsible for the legacy still growing today. With both iron and coal supplanting timber for shipbuilding and fuel respectively, this period might otherwise have seen a dramatic decline in the planting of oak.

Bark is taken from oaks in two different ways: either stripped from large timber trees as they stand (a practice peculiar to the Forest of Dean) or after they have been felled; or, most commonly, through the regular cutting of oak coppice. The latter practice prevails today, usually on about a 25-year rotation. Whereas the bark-peeling crews of a century ago might have had as many as 30 or 40 workers, today small groups of two or three people in one or two last outposts of the industry are all that remain.

The Wyre Forest and the neighbouring town of Bewdley, near the Worcestershire/Shropshire border, supported, until 1928 when the last tannery closed down, one of Britain's most prosperous tanning industries. Bewdley was perfectly sited with plentiful supplies of oak bark, and the soft water of the adjacent

ABOVE: *The use of steel kilns for charcoal making makes life much easier. The kiln is loaded with wood, before being lit from beneath. At first the lid is propped slightly open to encourage the fire to take hold. Once this is satisfactory, the lid is firmly shut, four metal chimneys are installed around the kiln, all seams and vents are sealed with earth and it is allowed to burn for about 24 hours. Unlike the old earth kilns, which had to be watched day and night, these can be left on their own to burn to a conclusion.*

ABOVE: *After the burn. In a woodland clearing Colin Gardiner examines some fine charcoal made from coppice wood, which he bags up and sells for barbecues, using the special rib shovel.*

for charcoal, it is clearly a sustainable industry, with the added advantage that its success will bring back vigour to previously neglected and mismanaged woodlands. Take an overcrowded tangled woodland and start to selectively coppice: suddenly fine displays of long-dormant flowers emerge once again, encouraging an attendant population of butterflies, and the remaining trees have more space to develop freely and vigorously.

More than 2,000 old charcoal hearths have been located within the bounds of the Forest of Dean. Although there is little likelihood of charcoal production returning to such a scale, as more charcoal burners take up the trade each year it may well become an important element of localised rural economies. The need to purchase a stand of coppice is a drawback for some enthusiastic woodsmen. What they require is rough woodland that is probably of little interest or use to landowners. In a perfect world, the landowners might appreciate the revitalising role of a coppice worker, who is almost certainly working for decidedly minimal profit margins and might perhaps wish to enter into some kind of management and profit-sharing agreement. If rural crafts and trades are to remain viable, there must be a prospect of profitability to hold the workers to the cause; otherwise it will only be a passing fad.

Bark Peeling

Until the beginning of the twentieth century, when chemicals superseded the traditional use of bark for tanning leather, the bark harvest was an important contribution to the income of the woodlander. Although bark has been stripped from a variety of trees, such as alder, birch, sweet chestnut, rowan, willow, spruce, larch and hemlock, by far the best tree for the purpose is the oak, which has the highest tannin content.

The use of oak bark for tanning leather was recorded

OAK BARK STRIPPING CALCULATIONS

By 1894, James Brown, in his woodland management tome The Forester, bemoaned the arrival of cheap imported oak bark from the Continent and the resulting depression in the English market. Nevertheless, he still managed to devote a considerable section to the cultivation of oak trees and the harvesting of the bark. Following his calculations of potential yields from a given area of coppice, Brown went into the logistics of ascertaining the relevant size of workforce required 'in order to strip one ton of bark each day, including all the necessary operations pertaining thereto'.

1 lad stripping lower parts of stems, at 1/6d. per day	£0/1/6
1 man 'laying in' these, at 3/- per day	£0/3/0
2 cutters, at 3/- per day	£0/6/0
2 pruners of trees, at 2/6d. per day	£0/5/0
3 lads pruning branches, at 1/6d. per day	£0/4/6
2 boys carrying branches, at 1/- per day	£0/2/0
1 man putting up stages, and attending to bark on them, at 2/6d. per day	£0/2/6
2 lads carrying bark to stages, at 1/6d. per day	£0/3/0
12 women or boys stripping, at 1/6d. per day	£0/18/0
Total wages bill	**£2/5/6**

river Severn. Bill Doolittle of Far Forest Works is now the only bark peeler still working the Wyre, but his peel harvest finds a buyer in Cornwall, since there are a few tanneries who still believe that oak bark makes superior leather.

Even though the industry has become a mere vestige of its former glory, the principles are still the same. The prime time for bark peeling is the period when the new leaves are emerging and the sap is rising, generally from late April until mid-June. An area of standing woodland is purchased for the year from the landowner (in the Wyre this is the Forestry Commission) prior to felling. When the coppice poles have been cut they are taken to a clearing, trimmed of any small twigs, and laid on X- or Y- shaped frames. A small hand tool known as a barking iron, a type of chisel with a sharpened spoon-like blade, is then used to slit the length of the bark on the pole before being levered to and fro beneath the bark in order to remove it as one cylindrical piece. The peeled bark is then stacked on racks to dry, with the inner surfaces out of the rain, as this could leach out the tannic acid. The peeled boughs, known as black-poles, were once used for charcoal burning, firewood and hurdle making. These days, in Bill Doolittle's works, they end up as rustic furniture, arbours and garden fencing.

Sadly, the continuing existence of bark peeling depends upon there being a market demand for the produce and a worthwhile return for those who peel.

Cricket Bat Willows

Ah, the sound of leather on willow, the polite ripple of applause as the ball skids sweetly over the boundary for a four, teacups politely chinking in the pavilion and towering plates of neatly cut sandwiches awaiting 22 hungry faces from the fray upon the green. Few things can be more quintessentially English than the village cricket match. And whether it's the thrash of the enthusiastic amateur or the calculated stroke of the professional, every man on the field must eventually try his hand with the bat.

All cricket bats are made from the same wood – willow. However, not just any old willow will do. There are several hundred species and varieties, but the best one for bats is a variety of *Salix alba* called 'Caerulea', which produces wood with the important qualities of lightness combined with great resilience and strength.

In the early years of cricket the selection of the right sort of willow for bats was largely a hit-and-miss affair, but over the last 100 years there has been widespread, closely controlled cultivation of Caerulea in Norfolk, Suffolk and, most particularly Essex. The vast majority of cricket-bat willows are now grown in plantations, rather than harvested from the wild. Even though they are planted in fairly regular formations, such trees have formed a distinctive landscape feature where they have been planted

alongside streams and ditches of the Essex countryside.

In earlier times it was common practice to plant rooted sets, but nowadays a nursery bed is laid out with parent willow sets which are allowed to produce several leaders each. When these have attained the prescribed height of about 12 feet 6 inches, they are cut and taken to a suitable site where they are pushed at least 2 feet 6 inches into the ground. Roots will soon develop and take the tree forward. The side shoots are rubbed from the main trunk on a regular basis, so that a long straight stem will develop without knots. The trees will make the required size for harvesting after 15 to 20 years, when a circumference of 4 feet 8 inches can be measured at 4 feet 8 inches from the ground. The trees are felled and transported to the processing works.

Usually a clean trunk of about 10 feet in length will cut into four rolls of 2 feet 4 inches each. Each roll is then split, using an axe and wedges, into eight pieces or clefts. In the summer it is important to process the timber within a few days, or it tends to darken in colour as the natural tannins in the bark stain the wood, and most cricket-bat manufacturers prefer a lighter colour. This is purely a cosmetic aspect and does not affect the playing quality of the finished bat. The quarter-cutting of the rolls is important as it produces clefts with the grain running parallel to the front of the projected bat, a necessary quality for performance. Each cleft is then sawn into the rough blade shape, before being waxed on the ends to prevent shakes appearing and then stacked to dry. Once seasoned and before despatch to the manufacturers, the blades are graded according to grain and colour, straighter and lighter being the most desirable. The grading is very important, since the best possible bats will be reserved for the top-flight professionals.

There are only a handful of cricket-bat manufacturers in Britain, and even though mechanisation has greatly increased productivity, there is still an overriding element of craftsmanship involved, based on sensitivity

ABOVE: *Husband and wife team Bill and Geri Doolittle are usually hard at work bark-peeling for about two months in the spring. In the foreground are the coppice poles newly arrived from the woodland, whilst the peeled or black poles are stacked behind to season.*

ABOVE: *The barking iron is used to split the bark and then to prise it from the pole. The tannin in the bark soon blackens the hands.*

CRICKET BAT MANUFACTURE

Recently felled cricket-bat willow logs in the yard.

The cleaving of the roll is done with special plastic wedges.

All blades are graded before despatch to the bat manufacturers.

Duncan Fearnley hand-finishes a cricket bat destined for a county cricketer client.

OPPOSITE: *Cricket-bat willows planted alongside a drainage ditch in Essex.*

RIGHT: *Colin Gardiner at work on a hazel hurdle in his Welsh woodland. The secret of good hurdle making lies in the mastery of the simultaneous turn and twist at each end.*

heavily weighted rollers that give the face of the bat the added strength and power required to strike an exceptionally hard cricket ball a long way without smashing to pieces. Meanwhile split cane which has been bonded into a sandwich with three pieces of rubber (known as springs) are cut and turned into bat handles. The end of the handle to be set into the bat is trimmed to a point, while the bat blade has a matching V-cut to receive the handle; the two parts are glued and tapped firmly together. The bat is then brought up to its final finished state with the shoulders and end correctly shaped, a process which is still very much a hands-on skill.

Finally, handle sheathing is applied to the cricket bat. In recent years, the trend is to cover the face also with an adhesive transparent plastic coating and thus reduce the risk of splitting, as it moves with the surface of the bat. The process has now superseded the traditional application of linseed oil, once used by every batsman to maintain his bat.

Hazel-hurdle Making

The making of hurdles, or short sections of woven wooden fencing, is a woodland craft once practised in almost every corner of Britain. The hazel coppice woodland which supplied the raw material was, and still is, widely available. Yet the late twentieth-century agricultural supplies of cheap wire fences and tubular steel gates and hurdles have all but seen the demise of hazel hurdles.

The earliest uses of hazel hurdles were for folding sheep on temporary or winter grazing land or confining them for lambing or shearing. The hurdles used by the shepherds were typically 4 feet 6 inches long by 3 feet high. Hurdles subsequently utilised in gardens have tended to be larger, to provide shelter and keep out leaping animals such as deer. Woven hazel was used in the wattle panels of building construction: when coated liberally with cow manure, horsehair and lime, it became wattle and daub.

A similar concoction, but without the cow manure, was also plastered on to hazel laths in interior walls and ceilings. The existence today of so much of the old lath and plaster is a fine testament to its long-term durability.

Hazel is traditionally cut from the coppice at about seven years of age, when it has usually attained the appropriate thickness and length for hurdle making. The bundles of cut sticks, which are cut in the winter months are then stored vertically (so that they will not rot on the ground) before hurdle making begins in earnest in the late spring.

First a row of slightly stouter rods are pushed into a wooden beam, known as a break, placed on the ground, with holes about 6 to 9 inches apart; these will be the uprights of the hurdle, and are known as sails. The thinner rods of hazel, sometimes in the round and sometimes having been cleft through the

to the raw material, an eye for detail and the intuition that can only be acquired from years of experience. The manufacturer's first task is to cut the blank blades to the familiar shape of the finished bat, with its very slightly convex face and domed back. The most important process is then the pressing, or compaction, of the bat's face, which is done by passing it through

middle, are then woven between these sails to make the hurdle. The secret of good hurdle making is to get the weave tight and even, with all the loose ends wreathed into the structure. The hardest part to master is the ends where the rod being woven must be turned around the end sail and twisted over at the same time, forming a firm and strong end to the whole hurdle. Such wrenching and twisting would break almost any other wood in two, yet hazel has the amazing capacity to be literally tied in knots without snapping. Usually the first and last rods of the weave are braided together, which again adds strength to the top and bottom of the hurdle.

Even though the resurgence of interest in hurdle making is largely to cater for a leisure-based market, it still also provides the structural basis for a few specialised applications. Where habitation is close to busy roads and motorways, hazel hurdles provide extremely effective sound screens. Hazel woven into mattresses and stuffed with reed is being used for stabilising river banks. Contrary to the 'out with the old, in with the new' building of recent decades, there has recently arisen an increasing interest in accurate conservation and restoration of historic buildings; for the first time in many a year, builders are relearning the old recipes of wattle and daub.

A few individuals are now making hurdles as part of a widespread, but localised, coppice woodland industry which, if it is part of a diversified business strategy, can actually contribute to a viable rural living. However, it is tough going! Both the hurdle maker and the customer have to balance the price of a craftsman-made hurdle against the off-the-peg garden-centre lap-fencing section.

Field Gates

The favoured wood for constructing field gates has always been oak. It will weather the elements better than even the most carefully treated softwood gate. In times past it was customary for many such field gates to be made of cleft timber; the main reason being that cleft timber tends to be stronger, as the cleaving process will follow the natural full length of the grain. However, most gates made in the last 50 years are inevitably constructed from sawn timber.

The most important elements of a gate are the upright post, called the harr or heel, where the gate will be hinged, and the top rail, from which the whole weight of the gate hangs. The top rail usually tapers away from the heel, thus keeping more of the weight at the hinge end. The key to a finely constructed gate is the strength of this important joint of the two pieces. The next most important part of the structure is a cross brace between the base of the harr and a point usually midway along the top rail. The rest of the gate is composed of parallel rails, a thinner upright at the opening end of the gate and a variable pattern of cross braces to complete the structure.

The two posts between which the gate is set are ideally of oak, although it is not unknown for yew to be used instead. On old gates it was customary to fit hinges of a ring-and-pin type, which would have been wrought by the local blacksmith. The secret of hanging the gate is that, while it should sit vertically between the posts when closed, the bottom pin is set further out from the heel, thus, as the gate opens, the bottom leading edge rises. This stops it grounding on uneven earth and allows the weight of the gate to work positively in swinging back to the closed position.

Part of the fun of travelling through the countryside is spotting the different regional variations in gate patterns. Sometimes a particular style proliferates on one estate or in one village where an individual craftsman has adopted his own personal pattern.

Sad to say, the all too familiar tubular steel has ousted many a wooden field gate, where frequent use and the ravages of time have seen them slowly disintegrate. Wooden gates don't seem to stand up too well to severe biffings from heavy agricultural machinery, but in the backwaters of rural Britain, among many of the old hill farms, an abundance of ancient gates continue to serve well.

OPPOSITE: *There is a long standing tradition of using oak timber to make gates and fence posts. In the past the custom of planting oaks in hedgerows provided a ready source of construction material on site. Today, the increased use of steel gates has negated the need to plant such hedgerow oaks.*

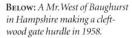

LEFT: *This is the most common form of five-bar field gate, with opposed V braces, although this one appears to be masquerading as a four-bar gate.*

BELOW: *A Mr. West of Baughurst in Hampshire making a cleft-wood gate hurdle in 1958.*

CHAPTER TEN

Trees and Woods in British Art

Trees and woods flourish in British art. They often seem to have at least as much, life and personality as the human figures in the landscape, reflecting a culture in which they are a focus of strong feelings for the natural world. No other part of the vegetable kingdom inspires such affection and attachment. Their longevity and appearance make them adaptable as human symbols, and British artists have used them to portray and amplify human attributes and emotions. Oaks have traditionally been esteemed as masculine trees, sturdy, and patriarchal; elms as feminine trees, graceful and alluring. Grand parkland trees have signified the aristocrats who owned them; humble hedgrows the peasants who worked them.

Some painters and illustrators, engravers and photographers create botanical or silvicultural illustration, others incline to human portraiture or to devotional icons. In any one picture there may be an interplay of meanings and associations. We could describe this interplay as the cultural ecology of art, which includes not only its beauty and variety but also its conservationist concern about the depletion of trees and woodlands on the ground. Whether the artistic response has been dappled nostalgia for the fabled greenwood or a clear-eyed admiration for freshly planted, well-managed groves and hedges, artists have celebrated Britain's woodland heritage as an economic, ecological and emotional resource.

OPPOSITE:
Ovenden, 'Tree At Painter's Bridge'

LEFT: *Miriam MacGregor An Apple Tree 1996*

WOODS: ANCIENT AND MODERN

By Stephen Daniels

The greenwood has exerted a powerful hold on the English imagination. It is not the dark forest of howling beasts and tribal bloodshed that haunts the folklore and literature of mainland Europe, but rather a golden world of sunny glades and leafy trees where the living is free and easy, a happy, hospitable realm, Merrie England. The Robin Hood ballads portray the greenwood as a bright, blossoming place of flowers, fruit and songbirds. It represents fertility and renewal; it belongs to all; it evokes a realm beyond unjust laws or grasping landlords, or whatever forms oppression and exploitation might now take, from motorways to acid rain. English artists have drawn on both the mythological associations of the greenwood and its nature as a working world used by herdsmen and huntsmen, woodmen and wayfarers.

As a boy in Suffolk, Thomas Gainsborough (1727–88) persuaded his father to write notes excusing him from school so that he could go sketching in the local countryside. Gainsborough later recalled that he 'had no idea of becoming a painter then, yet there was not a picturesque clump of trees, nor even a single tree of beauty, nor hedge row... that he had not perfectly in his mind's eye'. While 'nature was his teacher and the woods of Suffolk his academy', he learned the language of art in London by studying the works of Old Masters. He found in the work of Dutch painters like Ruisdael and Hobbema a landscape of woods, hedgerows, rivulets and rutted lanes that provided a format for portraying his own native countryside.

Cornard Wood displays Gainsborough's skill in depicting the fine detail of forest scenery, the foliage and bark of different species of tree, the plants and grasses growing in the glades, the subtle light and colour that pervade the wood. The artist is no less intent on portraying the wood as a place of human resort. In the foreground a woodman ties a bundle of firewood, while across a track another woodman leans on his axe to chat with a young woman. Donkeys graze beneath the trees. Two travellers make their way along the track to a village. It is a scene which celebrates the commonplace, the greenwood as a landscape where people work, rest and make their way. We are not deep in the forest; we are, as so often in English art, at its edge, looking through the trees to the village beyond. Here is that borderland where people were able to make an independent living, grazing cattle and pigs, cutting furze, perhaps poaching and purloining timber.

Such scenes reflect the taste for "the picturesque" in Georgian England. The Reverend William Gilpin, high priest of the picturesque cult, searched out suitable scenery on his many tours around the British Isles and he found it fully realised in the New Forest, where he lived as a parson. Here were a variety of noble trees, of a kind which graced the work of the Old Masters. Like many admirers of woodland scenery, Gilpin felt in two minds about the woodlanders who lived and worked there. Their freedom from the discipline field labourers were subject to made woodlanders seem both appealing and subversive. Wandering by night, stealing deer and timber, they appeared dangerous outlaws. Grazing their animals, gathering firewood, they appeared pastoral innocents; Gainsborough painted many such scenes, of woodlanders returning to a cottage brimful of bonny children.

Nineteenth-century artists portrayed woodland as both a workaday world reflecting the rhythms of the more modern age and as a place apart, a relic of a former time, a relaxation from modern life. J. M. W. Turner (1775–1851) and John Linnell (1792–1882) portrayed woodmen felling, barking and transporting timber, much as they portrayed field labourers harvesting, quarrymen digging or masons building. Their work is shown as integral to economic and cultural development, to the making of the landscape. Linnell's pictures of woodcutting in Windsor Forest also declare an interest in natural history. The barked trees and cut boughs are studied with the meticulous

BELOW: *J. M. W. Turner, 'The Forest of Bere', 1808. Petworth House.*

ABOVE: *John William North,*
'Gypsy Encampment,' 1873.
V & A Museum, London.

attention other artists devoted to specimens of dead and dissected wildlife.

However, Victorian artists found it difficult to invest the figure of the woodcutter with the dutiful heroism they visited on other workers. Even Thomas Hardy, who wrote a subtle account of forest work in his novel *The Woodlanders,* later endorsed a negative view of woodcutting in his poem 'Throwing a Tree: New Forest':

> *'Two executioners stalk along over the knolls,*
> *Bearing two axes with heavy heads shining and*
> *wide,*
> *And a long limp two-handled saw toothed for*
> *cutting great boles,*
> *And so they approach the proud tree that bears*
> *the death-mark on its side.'*

A sentimental view of nature is evident in Victorian paintings of picturesquely neglected woodland, thick with undergrowth, strewn with fallen boughs. Such woodlands were often the resort of gipsies, more picturesque figures than woodcutters, living what seemed to polite viewers an appealingly free life. The greenwood continued to be a source of inspiration and drew forth the distinctive Victorian combination of scientific inquiry and religious reverence. John Ruskin, whose writings promoted a theology of woodland, found the act of drawing trees to be one of revelation. In his autobiography, Ruskin recalled lying on a bank in Fontainebleau forest, languidly drawing an small aspen against a blue sky:

> *'and as I drew, the languor passed away: the beautiful lines insisted on being traced, – without weariness. More and more beautiful they became, as each rose out of the rest, and took its place in the air. With wonder increasing every instant, I saw they "composed" themselves by finer laws than any known to men...the woods, which I had only looked on as wildness, fulfilled I then saw, in their beauty, the same laws which guided the clouds, divided the light, and balanced the wave...and I returned along the wood-road feeling that it had led me far'.*

In his *Modern Painters,* Ruskin explored the laws of development which he discerned in the form of trees, with detailed sketches and diagrams of buds and blossom, leaves and branches. Budding artists reading Ruskin would have seen an entire architecture in the form of trees. Here was a habitat, strong and sacred, which had lessons for an industrial society busy stripping the world of its natural foliage and its poetic enchantment. Its significance could be discerned in its details. Ruskin once gave a lecture on "tree leaves as physiological, pictorial, moral and symbolic subjects".

Victorian artists extended the variety of trees in British art and the range of places in which they grew. They helped to move woodland appreciation in landscape art beyond the sunny glades and hedgerows of southern England to mountain and moorland scenery. In 1866, the Scottish painter Peter Graham took the Royal Academy by storm with his spate in the Highlands and it helped to fix a typical idea of Scotland and its scenery. 'The pines,' noted one critic, 'relics no doubt of a primeval forest such as one sees in Scotland, are like the shaggy tawny cattle at large on the moor or in the glen. In this picture, a clump of pines, in the middle distance, stands like a remnant of an army making a last, desperate defence. In the foreground, the roots of one that has fallen before the gale, gleam ghastly white, like bleaching bones.' Graham's Scottish contemporary James MacWhirter toured Norway and Switzerland as well as Scotland in search of mountainous scenery and its vegetation. His picture of a birch grove, 'The Three Graces', became one of the most reproduced Victorian pictures. 'In our islands the birch is most commonly associated with Scotland,' noted our critic, 'where it often stands out in marked contrast to a background of moor and mountain; or we look from the mountain-side, between the gleaming stems and network of branches, at the still dark waters of the loch far below.'

Eighteenth-century woodland art is usually set in summer or early autumn, the trees in full foliage; but Victorian artists looked at woodland scenery in all seasons, in the chill of early spring and approaching winter, in deep mid-winter with snow on the ground and hoar frost on the branches. Trees in art had traditionally signified dramatic events, notably the storm-blasted trees of sublime landscapes, but Victorian artists greatly extended the range to capture a variety of nature's moods. John William Inchbold's 'Study in March, Early Spring' shows a grove of trees surrounded by signs of spring, primroses on the bank, a ewe and two lambs beyond. It was first exhibited with a line from Wordsworth: 'When the primrose flower peeped forth to give an earnest of the spring'. Bare-branched, the trees themselves will soon burst into life.

Early in this century, in response to fears about the depletion of woodland reserves, the British government, in the shape of the Forestry Commission, planted large tracts of countryside. By 1939 the Commission had created 230 new forests, many of coniferous softwoods on mountain and moorland terrain. From something of a folkcraft, forestry was transformed into a scientific profession. For lovers of picturesque landscape, the new forests with their regimented stands of pine and spruce seemed an alien intrusion. Campaigners against afforestation in the Lake District enlisted Wordsworth's complaints about coniferous planting a century before. The Commission attempted to make the forests more attractive, by increasing the range of species, contouring the planting and encouraging public access with footpaths and campgrounds. They also promoted a coniferous aesthetic, which looked back to such Victorian views of sublime grandeur as the crag, loch and pinewood image of highland Britain, and also looked forward with modernist views of a bright, functional, well-designed countryside.

Below: *Francis Danby, 'View of the Avon Gorge', 1822. Bristol Museum & Art Gallery.*

Some nineteenth-century woodland was carefully maintained as an amenity for polite visitors. Leigh Woods, clothing the Avon Gorge (and, incidentally, a refuge of large-leaved lime), were open to the people of Bristol and Clifton to take their recreation. Paintings by Francis Danby (1793–1861) show the citizens in their Sunday best, reading, conversing, making music, as naturally at home in their habitat as the nymphs and dryads of the mythological scenes on which the pictures are modelled. Here is that civic, democratic view of nature which inspired the building of public parks in the nineteenth century and the designation of national parks in the twentieth.

British artists simultaneously retreated into the picturesque and stepped out into the modern age. Taking its inspiration from Thomas Bewick, there was a revival of woodcut, often to illustrate classic books on country life by William Cobbett and Gilbert White, but also modern ones which described contemporary life on the land. Woodcuts by artists such as Eric Ravilious, Clare Leighton and John Nash are filled with traditional images – of old pollards, apple gathering, coppices – but there is a freshly modernist, abstract quality to the compositions. Symbols of the modern age are incorporated, too: telegraph poles in Gertrude Hermes's views of the woodside verge in the headlights of a car, new plantations and woodstacks in works by Paul Nash. The woodcuts in Forestry Commission guides promote forest parks as progressive versions of a traditional scene and with an almost sculptural attention to natural materials such as wood, stone and water.

The ecological concerns of recent times have brought British artists into the woods again. The sculptor Andy Goldsworthy made his first outdoor works in a small wood near his boyhood home in a suburban housing estate on the outskirts of Leeds: 'It is typical of many planted and managed woods, composed of separate areas of different kinds of tree; pine is succeeded by larch and again by beech. Exploring it involves walking through a series of separate worlds.' While doing holiday work on a local farm, Goldsworthy also learned to look upon the countryside as a crafted landscape. There is a strong sense of woodmanship in his sculpture; he works every aspect of trees, from trunks to branches and leaves, and he carefully places them in different environments. At one extreme are massive, outdoor pieces, like the woodland sculptures commissioned for Grizedale Forest in the Lake District. Grizedale is working woodland run by the Forestry Commission, open to the public for recreation. Goldsworthy's 'Seven Spires', seven sets of clustered pine trunks grouped in a tall pyramid made from surplus timber, is off the main track. To experience the sculptures in their gothic grandeur the visitor must walk among the trees and look up the spires from the base, from where one can share Goldsworthy's sense that there exists 'within a pine wood ... an almost desperate growth and energy driving upward'.

Goldsworthy is well known for his leafworks, small ephemeral constructions created in woods, hedgerows, fields and streams and recorded in colour photographs. Each work grows, stays, decays – integral parts of a cycle which the photograph shows at its height, marking the moment when the work is most alive. The photograph does not need to shrivel and fall to the ground for change to be part of the purpose. It is an outdoor experience expressed in an indoor place. It is as appropriate to that space as it would be inappropriate to hang a framed photograph from a tree in a wood.

Garden and Grove

If the greenwood is a common place, a realm of natural and social freedom, the plantations of parks and gardens signify the privacy and propriety which are no less characteristic of the English countryside.

Trees, and the planting of trees, were central to the prestige of the landed gentry of Georgian England. Plantations produced heavy yields of profits (timber often yielded more money than farmers' rents), pleasure (ridings for hunting, coverts for game, groves for garden parties), and patriotism (the resources for building ships for the navy). In the hundreds of paintings of estates and parks commissioned by the gentry, trees and plantations are as conspicuous as country houses. Indeed there is a strong tradition of estate portraiture in which woodland is the definitive feature. Great avenues mark extensive influence, and great trees – suggest deeply rooted authority. Woodland provides a setting for the fauna of the park, well-groomed pedigree cattle and horses posed under healthy pruned groves, and watchful deer at the forest margin, their antlers echoing the stag heads of old trees.

Landowners themselves assumed the mantle of Knights of the Greenwood. The aristocracy were, in the words of the eighteenth-century statesman Edmund Burke, 'the great oaks which shade a country'. Old oaks not planted by aristocratic families, often former

ABOVE: *John Ruskin, 'Spanish Chestnut', 1845. Art Institute of Chicago.*

John Ruskin exercised a powerful influence on all Victorian artists, and in particular those of the so-called Pre-Raphaelite Brotherhood. Artists like Edward Burne Jones and William Holman Hunt strove to revive a minutely detailed, enamelled style of art practised before the time of the Renaissance painter Raphael: Their art was a vehicle for reviving what they imagined to be the more pious, devotional view of the world in the Middle Ages, but also for a very modern botanical analysis of the natural world. There was a moral seriousness in the very detail with which pre-Raphaelite artists painted woodland scenes. Much of the subject matter, from toiling woodcutters to courting couples, is heavy with fateful significance. Millais specialised in chill, dreary landscapes full of foreboding. Even his garden scenes, of children picnicking under orchard blossom, or gathering autumn leaves, are melancholy, with intimations of mortality. There is not much merriness in the pre-Raphaelite greenwood, even in the many chivalrous scenes, from Shakespeare, cavalier history or Arthurian legend.

woodland or hedgerow trees, were frequently incorporated in parkland. They feature as subjects themselves in parkland views, or to frame various views. A popular style of portraiture placed the family beneath an oak, with the man of the house aligned to the trunk. Gainsborough, who made his money portraying the wealthy, was as adept at this style of picture as he was with more vernacular woodland scenes. Gainsborough's portrait of Mr. and Mrs. Andrews (see next page) shows the newly wedded couple on their Suffolk estate. They are posed beneath an oak tree, dressed in the latest London fashions for country life, he with his shooting jacket, she with her milkmaid-style cap. In the distance are the newly laid hedges enclosing paddocks for sheep and cattle and well-tended plantations for timber and game. No wood-gatherers or wayfarers would dare to trespass upon a scene such as this!

Many eighteenth-century aristocratic garden scenes show fine details of planting design and management, from espaliers to parterres. Balthasar Nebot's views of the gardens of Hartwell House portray the razor-sharp angles of the topiary garden and the no less calculated groves of the wilderness garden. Various temples and statues, arches and lakes are framed through the trees. Gardeners are shown clipping, rolling, scything and sweeping, along with courtly-looking family and friends promenading. This consciously regimented, even geometrical, view of nature, found expression in topiary-yew and privet clipped into cones, pyramids, spheres and the figures of animals and humans.

The clipped, formal style passed out of fashion by the late eighteenth century. Later, picturesque styles of planting and gardening were regarded as more natural, even more English. Such gardens were often labour-intensive, but gardeners seldom appear prominently in paintings of them. The trees and groves appear to grow spontaneously. In contrast to the sociable, even boisterous scenes in formal gardens – music-making here, bowls there – pictures of picturesque gardens are relatively underpopulated. The contemplation of nature was a quieter, more meditative pursuit. Garden buildings are fewer, and relatively unobtrusive, and some, like root houses constructed from unbarked timber and branches, reflected the new cult of nature. It was a short step for landowners to go out of the garden into the woods to appreciate the virtues of nature, and to be shown as such in portraits. Joseph Wright shows Brooke Boothby as a Man of Nature, lying down full-length in the woods, holding his volume of Rousseau.

The leading promoters of picturesque gardening were two Herefordshire squires, Richard Payne Knight and Uvedale Price. Their own well-wooded estates were exemplars of the picturesque style. Knight would allow decayed branches and storm-blasted trunks, but 'banish the formal fir's unsocial shade, and crop the aspiring larch's saucy head'. Knight modelled his planting plans for his estate at Downton on scenes of the Roman campagna in paintings by Old Masters, such as Claude Lorraine and Salvator Rosa.

Knight's fellow picturesque connoisseur Uvedale Price had more of a taste for Dutch art and for Gainsborough, a family friend who had sketched the woods at Foxley. 'It is in the arrangement and management of trees, that the great art of improvement consists,' Price declared. 'They alone form a canopy over us, and a varied frame to all other objects; which they admit, exclude, and group with their beauty is complete and perfect in itself; while that of almost every other object requires their assistance.' As Michelangelo thought the human figure was latent in marble, the sculptor only having to remove the stone to release it, so Price considered there were pictures in every wood, the improver only having to remove branches and undergrowth to reveal them.

Thomas Hearne was one of a number of English artists who specialised in painting the woodland glades of picturesque parkland. He produced a set of commemorative views of Richard Payne Knight's planting at Downton, focusing on the variegated foliage at different distances from the eye and describing how the trees articulated the space of the landscape as natural architecture. Hearn was a painter of antiquarian buildings as well as of woodland, and his art echoes the idea that Gothic architecture evolved from the avenues of forest trees. His portraits of great, gnarled parkland oaks and chestnuts have 'a monumentality enlivened with intricacy and variety that is comparable with his treatment of architecture'. Hearne's view of the great oak in Moccas Park was published as an etching in 1798, when the war with France revived the associations of ships and naval power, Britain's 'wooded walls'. 'The oak is the first class of deciduous trees,' he ran accompanying letterpress; 'and it is a happiness to the lovers of the picturesque, that it is as useful as it is beautiful ... no one who is lover of his country, but, in addition to the pleasure which he has in contemplating this noble plant, must feel his heart glow on reflecting, that from its produce springs the British Navy, which gives our Island so honourable a distinction among surrounding nations.'

The landscape gardener Humphrey Repton tempted his clients with enchanting watercolours of the designs he proposed. This might involve pruning as well as planting to create parkland views, and uprooting trees like firs, larches and Lombardy poplars. Such species were regarded as unbecoming for traditionally minded gentry, as flashy, fast-growing trees, like the precocious parvenus who favoured them.

Repton and the poet William Wordsworth compared the benevolent, deciduous ways of ancestral landowners with the harsh, coniferous ways of the nouveaux riches. In one of his illustrations, Repton

contrasted an old England, in which the parkland groves of the gentry blended with the greenwood of the commoners, with a new England in which thick plantations of larches and ploughed commonland signified a harsher country. An old stile by an old tree marking a pathway through the wood was replaced by a high fence and a sign warning of mantraps and spring guns.

The modern-minded Victorians, however, welcomed conifers along with a range of evergreen trees and shrubs as a way of both enriching parks and gardens and enlivening the English landscape at large. Around smart villas and restyled country houses near the main roads and railways, on newly emparked estates of northern England, they planted their grounds with a range of new species. New shapes and forms and, above all, strong colours were celebrated. 'Strong colours are like the voice of Nature,' pronounced one plantsman, 'speaking to mankind in tones that all can understand.' The palette of the planter expanded to match that of the painter. Victorian gardens rivalled the chromatic register of Victorian art.

The range of available garden trees and shrubs expanded enormously, and private and public pleasure grounds exhibited species from around the globe.

ABOVE: *John Everett Millais, 'Autumn Leaves', 1855-6. City Art Gallery, Manchester*

OPPOSITE, TOP: *John William Inchbold, 'Study in March', 1855. Ashmolean Museum, Oxford*

OPPOSITE, BOTTOM: *Andy Goldsworthy, 'Torn Sycamore Leaf Trail Floating On A River Pool'. Leaves for Goldsworthy represent 'part of a wider approach to the land – leaf and tree, light and space, all form part of the leaf. There is a whole world in a single leaf'.*

Victorian albums of watercolours and coloured prints revel in the exotic variety of plants and the splendour of individual specimens. Visitors behold towering palm trees in glasshouses, magnificent pines on the lawns, rhododendrons running riotously down hillsides. So called 'pictorial trees' came in many shapes and forms, most strangely the monkey puzzle. Plantsmen were keen to colonise the countryside at large with new species, with an eye on the old forests preserved for public use. One proposal for Epping Forest found that 'the colour of the trees now existing in the Forest are, for the most part, a uniform green' and planned to send in an international brigade of scarlet oaks, Norway maples, Douglas firs, cypresses, cedars, wellingtonias and rhododendrons to enliven the place.

There was reaction to such schemes, indeed to exotic developments generally in parks and gardens. Old England was envisioned in a variety of period styles. Topiary was revived, in plantings meant to recreate the courtliness of Tudor times. Yew hedges were integral to the design of cottage gardens, as a frame for climbers and wild flowers. In each case the arts of painting and gardening intersected. Courtly gardens featured in the historical scenes favoured by Victorian artists, from David Cox to Burne-Jones. Cottage gardens were made famous in the paintings of Helen Allingham, along with their sunbonneted children, free-ranging hens and ducks. They evoked a country life that city dwellers yearned for and, from the enduring popularity of Allingham's art, still do.

Nineteenth-century nurseries supplied an astonishing variety of trees and shrubs, both exotic and indigenous, adapted and improved for a variety of sites and situations. There was a veritable explosion in varieties of orchard fruits, both for commercial orchards and for domestic gardens. Orchards had long been valued for enhancing the scenery of both country and city and for marking the distinctive character of the countryside of Kent and Herefordshire. The efforts of horticulturists like Thomas Andrew Knight, grafting and experimenting, drew attention to the beauty as well as the botany of the fruit. Knight's lavish folio *Pomona Herefordiensis* (1811) is one of a number of fine albums which elevated English horticultural illustration to a fine art.

The planting of individual, sometimes small gardens collectively enriched the landscape at large. Victorian artists were drawn to the leafy suburbs, where flowering trees both softened new developments and gave them brighter tones. Poplars glittered in the breeze. In exile with Claude Monet and Alfred Sisley from the Franco-Prussian War, Camille Pissarro realised his Impressionist vision in the south London settlements of Norwood and Sydenham. Ford Maddox Brown's 'An English Autumn Afternoon' looks from his house in Hampstead across the russet villa gardens to Highgate and the Heath. It stands in an enduring tradition of English suburban art: viewed from the upper storeys of houses, sometimes framed by windows with pot plants on the sills, a multitude of garden trees – ornamental trees and shrubs, the remains of hedgerows and orchards – compose the landscape as a whole or form a foreground for views to woody hills in the distance.

Hedgerow and Copses

In much British landscape art, trees articulate a wider agrarian landscape of fields and pastures through hedgerows and patches of woodland, covert and copse. The bosky nature of this landscape is enhanced in views along lanes or streams overhung with trees, or in long views where hedges, woods and plantations blend, to give the impression of continuous woodland, even, it seems, the verge of a forest. The pre-eminent painter of this kind of countryside is the most popular English landscape artist, John Constable (1776–1837), who was attentive not just to the singularity, even personality of trees, which he portrayed with loving care, but to their role in the look and management of the countryside.

Constable was brought up in the valley of the river Stour, between Suffolk and Essex, where his family ran a milling and boating business, and riverside scenes came to form a main subject of his art. In their scale and vantage point, many of Constable's riverside scenes combine a child's view of the world, the landscape looming large, with an adult appreciation of its form and function. The viewer is also offered a number of ways through the landscape, out to the brightness of sunny fields and pasture, into the shady depths of hedgerow and grove.

In one of Constable's best-loved paintings, 'The Haywain', the wain is poised in the middle of the millstream (carts were stood in the water to tighten up the iron wheels), to cross to the bright hayfield to collect its load, but our eye is drawn into the cool depths by the cottage, a woodland dell. 'Flatford Mill' picks out the elms, ashes, willows and poplars which line the river. Constable could be sentimental about trees, but he recognised their role as a resource. 'Boatbuilding', set on the bank of the river Stour at the family's drydock, shows in a sequence the planing and shaping of timber, and how it fitted to construct the hull of the barge on the stocks. Constable country is not just a pretty picture, a rustic dream, but a living and working region.

Country lanes were a favourite subject of nineteenth-century art, the deeper and greener the better. Uvedale Price, a connoisseur of the picturesque, loved the lanes of his native Herefordshire, 'discovering the different strata of earth, and the shaggy roots of trees: the hollows are frequently overgrown with wild roses, with honeysuckles, periwinkles, and other trailing plants…in the summer time these little caverns afford a cool retreat for sheep'. Artists like Myles Birket Foster and David Cox made careers out of painting the green lanes of rural Britain, deep in shady dells, or where they emerged at stiles, gates and stepping stones under old boundary trees, or in more open country where clumps of hawthorn, yews or pine marked the way for drovers and a pasturing place for their herds. The green lanes are the settings in Victorian art for rural life, with milkmaids, shepherds, and children nutting, blackberrying, and flower-picking; they are also a projection of the rustic dreams of the city-dwellers.

Twentieth-century artists have continued down country lanes in search of a more rooted world: Graham Sutherland among the gnarled oaks of the twisted

LEFT AND BELOW LEFT: *Humphrey Repton, 'Improvements', from* Fragments on Landscape Gardening, *1816.*

BELOW: *Thomas Hearne, 'An Oak Tree', water-colour, 1786. British Museum.*

lanes of Pembrokeshire, Hamish Fulton in the hollow ways of the North Downs on his walk between Winchester and Canterbury. John Hubbard's 'Sunken Lane' is a colourfield image, a stylistic mixture of French Impressionism and American abstract expressionism portraying the nooks and crannies, the twisting and trailing growth of a lane in Dorset.

When Constable, the countryman, moved to London to live in Hampstead, then a semi-rural suburb, he continued to study trees intensively, with the scientific attention he devoted to the drawing of clouds. He often showed trees against the sky, counterpointing their liveliness, their changing form and light, with that of the clouds. He contributed to the image of Hampstead as a leafy village of London. His feeling for trees was intense. 'I have seen him admire a fine tree,' wrote his biographer, 'with an ecstasy of delight like that with which he would catch up a beautiful child in his arms. The ash was his favourite, and all who are acquainted with his pictures cannot fail to have observed how frequently it is introduced as a near object, and how beautifully its distinguishing peculiarities are marked.' In a lecture given at Hampstead, Constable illustrated some hints for

drawing trees from his own tree studies. One was of a 'tall and elegant ash', of which he said:

'Many of my Hampstead friends may remember this young lady at the entrance to the village. Her fate was distressing, for it is scarcely too much to say that she died of a broken heart. I made this drawing when she was in full health and beauty; on passing some time afterwards, I saw to my grief, that a wretched board had been nailed to her side, on which was written in large letters "All vagrants and beggars will be dealt with according to the law". The tree seems to have felt the disgrace, for even then some of the top branches had withered. Two long spike nails had been driven far into her side. In another year one half became paralysed, and not long after the other shared the same fate, and this beautiful creature was cut down to a stump, just high enough to hold the board.'

Twentieth-century artists have charted the patterns of neatly hedged and wooded landscapes in all their regional variety. Rowland Hillier portrayed the 'garden of England' in Kent, its orchards and oasthouses, James Bateman the haystacks seeming to emerge from woodland clearings in the Cotswolds, James McIntosh Patrick the windbreaks and copses on the hills of Angus.

Tree Spirits and Sacred Groves

In folklore and literature, trees and woodland often possess magical power. Tree spirits, sacred groves, the maypole, the Green Man form an enchanted realm which is echoed in a visionary tradition of British art.

Poet, painter and engraver William Blake (1757–1827) saw a spiritual world in 'England's green and pleasant land'. His vision was often urban, a New Jerusalem arising in his native London, but he took it with him on excursions to the surrounding countryside. 'The tree which moves some to tears of joy is in the Eyes of others only a Green thing that stands in the way,' he declared. He was proud to possess 'double vision', both an 'inward eye' and an 'outward eye' which enabled him to perceive the inner life of the natural world.

Blake's doubled spiritual vision also had a social dimension. In his graphic poetical works, the natural world symbolised both human happiness and suffering, contentment and enslavement. His depictions of trees mobilised the traditional symbolism of the communal tree, the maypole, and that of the 'accursed tree', the tree of execution, Christ's cross. In 'The Echoing Green' from the *Songs of Innocence*, an oak is a symbol of communal happiness, a spreading tree on the village green under which old folk and young mothers sit smiling and around which children joyfully play, yet in the 'Prelude to America', a poem on spiritual and social regeneration, the oak is a symbol of repression,

RIGHT: *John Constable,*
'The Cornfield', 1826.
National Gallery, London.

with Orc, a figure of revolutionary energy chained to its trunk in an attitude of crucifixion, before his resurrection in the image of a vine.

Blake inspired a group of artists, the self-styled Ancients, to seek the spiritual world in the English countryside. The most notable was Samuel Palmer (1805–81), who found his 'valley of vision' in the valley of the river Darent, around Shoreham in Kent. Here Palmer and his fellow Ancients read poetry, the psalms and visionary books of the Bible, made music and undertook midnight walks to feed their artistic imagination. Literary Londoners, they found rather less inspiration in the working countryside around them. Its tasks and travails had no room in their pastoral reverie.

The world of Palmer's pictures during his sojourn at Shoreham is a luminous, luxuriant one. Palmer's trees are magical images of natural richness, loaded with enormous leaves, luscious fruit and clotted blossom. 'In a Shoreham Garden' shows a tree with branches heavy with pink and white, surrounded by

a profusion of plants in a riot of colour. His magical trees included newly fashionable species. 'A Pastoral with a Horse-chestnut' shows a piping shepherd and his flock beneath a tree that had hitherto scarcely figured in the gallery of popular English species.

Inspired by some 'monumental oaks' in Milton's *Il Pensoroso*, Palmer went in search of some ancient oaks in Lullingstone Park. He found that 'the Poet's tree is huger than any in the park: there, the moss, and rifts, and barky furrows and the mouldering grey, tho' that adds majesty to the lord of the forests, mostly catch the eye before the grasp and grapple of roots; the muscular belly and shoulders the twisted sinews'. His watercolours of the Lullingstone oaks depict the details of their surface texture within a vigorous anatomy that recalls Blake's muscular figures. Palmer's words also echo Blake: 'Sometimes trees are seen as men,' he told a friend. 'I saw one, a princess walking stately with majestic train.'

Some writers found a robust pagan spirit in the woods. In Richard Jeffries's *Wood Magic* a squirrel tells the boy Bevis of the treachery of elms: 'I have known a tree, when it could not drop a bough, fall down altogether when there was not a breath of wind, nor any lightning, just to kill a cow or sheep, out of sheer bad temper.' Pan figures strongly in the literature of the turn of the century, in *Peter Pan* and as the genius loci of *The Wind in the Willows*, the guardian of 'small waifs in the woodland wet'. Inspired by the antiquarian Alfred Watkins, ley-hunters roamed the countryside, noting the associations of sacred trees with sacred stones and clumps of pines on ancient earthworks. Old pollards and stately avenues were drawn into the alignments, in that ancient, perhaps mystical network charted across the countryside. J. Ernest Phythian's *Trees in Nature, Myth and Art* moves from pre-Raphaelite painting to woodland management to the Tree of Life: 'the source of all things, including time and space, the upholder of the universe'.

James Frazer's book *The Golden Bough*, first published in 1890, invigorated the visionary tradition in British art and culture by connecting indigenous rituals and folktales throughout the world. Frazer took the title of his book from Turner's painting 'The Golden Bough', which shows a scene from Virgil's *Aeneid:* set around Lake Nemi, near Rome, a sacred grove stands at the entrance to the underworld.

> 'A mighty tree, that bears a golden bough,
> Grows in a vale surrounded by a grove,
> And sacred to the queen of Stygian Jove.
> Her nether world no mortals can behold,
> Till from the bole they strip the blooming gold.'

Frazer was fascinated by the relationship between the picture and a recently discovered classical commentary on the lines in Virgil's epic; the commentary described a cult of woodland priests that had continued for centuries near Rome, and some excavations around Lake Nemi which had revealed a shrine. In travel narratives from other parts of the globe, Frazer came across evidence of similar cults still practising. Were there religious and magical constancies in human culture across time and space? In *The Golden Bough*, the community of tree worshippers extends from the Mayday customs of Berkshire and Bavaria to sacrificial rites in Cambodia and Wakamba, East Africa.

A spirit of place illuminates neo-Romantic art of this century. Artists like Graham Sutherland, John Craxton, John Minton and Cecil Collins took their inspiration from earlier English mystical traditions, rediscovering the work of Samuel Palmer and searching out their own valleys of vision, for Sutherland the countryside of Pembrokeshire, for Collins that of Cornwall and Devon. In Collins's landscapes the trees appear to grow not upon the land but inside it, swelling the hills. It is a world of natural power reminiscent of that conjured up by Dylan Thomas: 'the force that through the green fuse drives the flower'. Other neo-

Romantics like John Piper followed a more picturesque tradition, in the overgrown avenues and glades of half-forgotten country houses.

The new mysticism in British art was also influenced by modern, international currents of art and thought, by surrealism, biology, psychoanalysis and anthropology. Paul Nash used a modernist style to probe primordial powers. Trees in his art are totemic symbols of the spirit of place, even, like the pylons of the new age of electricity, conductors of the currents of the anima mundi. In painting, wood engraving and photography, Nash depicted trees and woods of many kinds, in every state of life and death, at every stage of human use: ancient woodland, modern orchards, budding branches, weathered stumps, neat woodstacks on a farm, blasted trunks on battlefields. His landscapes are largely devoid of other kinds of vegetation, and almost empty of animal life except, significantly, birds. He found animation enough. 'The country round about is marvellous,' he noted in a letter, 'grey hollowed (hallowd) hills crowned by old trees of Pan-ish places down by the river. Wonderful to think on. Full of strange enchantment. On every hand it seemed a beautiful legendary country haunted by old Gods long forgotten.'

The woodland scenes which most enchanted Nash in England were on chalk downland, where trees

BELOW: *John Hubbard, 'The Sunken Lane', 1982.*

amplify the form of the land. Wittenham Clumps, two copse-crowned hills near Wallingford, above Iron Age earthworks and the remains of an ancient forest, recur, like a personal motif, in a number of Nash's pictures. 'I have haunted them often,' he confessed, 'and now am going to try & interpret some of their secrets.' He described them as 'pyramids of my small world', an indication of how formally, even abstractly, Nash discerned the world, and of the power he perceived in such forms and their correspondences.

The English visionary tradition in the visual arts underwent another revival in the 1970s, as part of a new fashion for country life among the young, expressed in a variety of forms from folk-rock music to Laura Ashley dresses. In 1975 a group of painters, notably Peter Blake, David Inshaw, and Graham and Annie Ovenden, formed the Brotherhood of Ruralists, in a conscious echo of the Pre-Raphaelites and the Ancients. William Blake and Samuel Palmer were the inspiration, along with Stanley Spencer, Edward Elgar, Thomas

Hardy and Lewis Carroll's *Alice in Wonderland*. Some ruralists, like Graham Ovenden, went in search of the mystical landscape of ley lines and Druids' groves re-presented for a New-Age readership by their friend John Michell. Others, like Blake, drew the fairies at the bottom of the garden.

David Inshaw cast a bright, sometimes surreal light on familiar country scenes, a cricket match in parkland, an orchard, an oak grove. Inshaw's 'The Badminton Game' is still one of the most popular pictures in London's Tate Gallery in terms of sales of reproductions, although it seldom now hangs on the walls. It is a dreamy, even visionary scene. Two young women in the Victorian style fashionable in the 1970s – flowing hair and smocks – play badminton on the lawn of a lush garden on a brilliant summer day, the shuttlecock poised in mid-flight, the women fixed in statuesque poses. Behind them tower topiary bushes, backed by rhododendron bushes and a giant monkey puzzle, and beyond by wooded downs. The original title of the painting was a line from a Thomas Hardy poem, 'Remembering mine the loss is, not the blame'. The picture is, according to the artist, an attempt to hold the complex memory of a poignant love affair in a moment. It is a scene of 'happy thoughts as well as sad, full of waking dreams and erotic fantasies'. The picture gives the traditional association of love and gardens an erotic charge with the shimmering heat of a summer afternoon and that tumescent topiary. 'Everything in this picture is taken from near my house in Devizes and rearranged into its right place. I changed everything I used in the picture in order to increase the mystery and wonder I felt all around me in the magic place.'

Photographer John Blakemore's woodland vision has roots in his childhood enthusiasms of birdwatching and leaf-collecting, his experience as a wartime evacuee, a spell working as a farmhand in the Warwickshire countryside, and Victorian romanticism. 'Glory be to God for dappled things,' runs the line from Gerard Manley Hopkins's poem with which Blakemore glosses a recent album titled *Inscape* after Hopkins's keyword. 'I seek to make images which function both as fact and as metaphor, reflecting both the external world and my inner response to it, and connection with it,' Blakemore explains. Photographs taken in North Wales in the winter of 1968 following the breakup of his marriage reflect the turbulence of his emotional life.

Blakemore went on to photograph the energy of the landscape in a less personal way: with 'the source of (photography's) power, its reliance upon reality, the camera must be pointed away from self'. Like a Victorian artist he traced the life cycle of a fern to establish 'the energy of growth itself' and used the changing form of trees and woodland to chart climatic, atmospheric conditions. In his 'Wind Series' he uses multiple exposures to show the ferns and leaves of

Below: *William Blake, 'The Echoing Green', from Songs of Innocence, 1789. Fitzwilliam Museum, Cambridge*

LEFT: *Samuel Palmer,
'Old Trees, Lullingstone Park,'
1828. National Gallery of
Canada, Ottawa.*
*Palmer's picture echoes a passage
from Milton's Paradise Lost:*
  *'Of goodliest Trees load'n with
  fairest Fruit,*
  *Blossoms and Fruits at once of
  gold'n hue*
  *Appeared, with gay enameld
  colours mixt:*
  *On which the sun more glad
  impressd his beams.'*

Derbyshire's Ambergate woods shimmering with an otherwise invisible power. Blakemore is drawn to 'the dynamic of the landscape, its spiritual and physical energy, its livingness'.

By the 1980s the woodland vision of British artists had recovered the global register of 'The Golden Bough'. The destruction of tropical rain forests was found to threaten not only the livelihood of peoples in Amazonia or Africa but life on the planet as a whole. Trees are often symbolic of life, fertility, growth and regeneration, wisdom and knowledge. They are close to the centre of the mythology or belief of almost every culture and religion. The 1989 Tree of Life exhibition found 'new images for an ancient symbol' in a variety of media, from drawing to sculpture.

LEFT: *Paul Nash, 'Landscape of
the Vernal Equinox (III)', 1944.
Scottish National Gallery of
Modern Art.*
  *Nash's picture of a beech wood
on Ivinghoe Beacon is an almost
sculptural rendition of trees,
echoing the shape of the downs
and radiant with the latent
power of nature in early spring,
around the vernal equinox. It
was, Nash said, 'a synthesis of a
scene and mood of nature at a
particular time of the year –
March. The red 'foliage' is the
effect of massed buds of beeches
before they break.'*

NEXT PAGE: *J.M.W. Turner,
'The Golden Bough,' 1834.
Clore Gallery, London.*

WOOD ENGRAVING

By Simon Brett

A wood engraving can only be produced by its own particular process and tools, which account exactly for the way it looks. It may be distinguished from other forms of woodcut by the use of carefully prepared, mirror-smooth blocks of end-grain wood and fine tools which provide a way of making pictures with marks at a size on or just below the threshold of the eye, so that a particularly rich and subtle tonal range can be produced in print. The medium's aesthetic characteristics derive directly from its physical properties and from the fact that, in engraving marks which will come out as white when printed, the artist is, in effect, drawing with light.

Engraving, it has been alleged, began with the dawn of mankind, for there are incised marks and images alongside painted ones in the earliest caves. The shallow relief carvings and incised decorations of early civilisations also represent a kind of engraving, and in some early cultures engraved seals were impressed into writing tablets. But engraving for print required the idea of ink or pigment being transferred from one surface to the other and a surface to print on, which the clay and papyrus-based civilisations of the Mediterranean and Near East did not arrive at. In China, seals appear to have been printed with pigment, as individualising marks, on to wood or fabric even before AD 100; but with the invention of paper about that time, printing as we know it began, and its potential for multiplication was quickly realised. As print spread with Buddhism from China to Japan and then westward, it came with missionary and democratic intents of one sort after another, even while governments sought to license or forbid its power. The image preceded the word. In Europe, the block-book (where an entire page was cut onto a single piece of wood) preceded the introduction, also from China, of movable type. In pre-Reformation and Reformation Germany, it has been observed, the painted altarpiece was supplanted by the printed image, and then the printed image by the printed word: a theological movement from altar to book, a democratic movement from a communal, pictorial focus to numerous individual interpretations of a text.

The wood used had to withstand increasingly complex demands. The earliest European woodcuts brought in a drawing style of thick, strap-like black lines, simple in themselves however sophisticated the design; these were mostly used for devotional images which might then be hand-coloured. The surge of renewal Albrecht Dürer (1471–1528) brought to the woodcut saw the development of a finer line, textural rendering, shading which followed the contours of the form and much more complicated subject matter. In the latter part of his career, Dürer introduced tonal groupings into his compositions and a real sense of light. In his hands alone, the woodcut's status travelled from that of a playing card to that of a Renaissance painting. In all these developments, the design was conceived as if done with a pen – at first a broad pen and later, in Renaissance times, a finer one used with a lot of cross-hatching and shading. The woodcutter cut out the spaces between these lines, at first simple wide areas, then increasingly tiny chips of space between the cross-hatched lines of the drawing. The skill of the woodcutters who cut Dürer's designs was astonishing.

OPPOSITE: *John Farleigh, 'In My End is My Beginning,' 1943.*

BELOW: *Paul Nash, 'Black Poplar Pond', 1921.*

The wood had to be strong enough to accept such cutting, to stand up as a black line without crumbling and to cut cleanly rather than to tear or crush because of being too fibrous. Throughout most of history, the choice of wood was indigenous and not artistically critical. The Japanese used wild cherry. The earliest European block that survives is walnut. The German engravers of the sixteenth century used the limewood that their wood-carver colleagues employed for sculpture. Sycamore and maple are also very suitable.

In Europe the woodcut went in and out of favour, while in Japan it remained a consistent, slowly evolving tradition unrivalled by other media. When the European woodcut made its astonishing return in the late nineteenth century at the hands of Gauguin and Munch, due partly to the example of the Japanese prints that influenced the Impressionist painters and partly to the impact of 'primitive' art, the use of the wood was essentially the same to print either upstanding black lines or solid areas of colour. Though the grain and characteristics of the wood used naturally affected how it could be worked, they were not allowed to dictate the final appearance of the print. In our own century, as a variation, it has been part of the artist's

purpose to let the wood have its own way, to allow its woodiness and the character of the tools which cut it to show. There is a torn, hacked look to much expressionist printmaking. From the time of Munch onwards, artists responded to and printed the grain of the wood, as indeed the Japanese had already done, incorporating its flow and pattern into their expressive purposes. They have used the wood not as an inert or 'invisible' material but as a 'found object' whose own character contributes to the work's identity.

Wood engraving represents the one real exception in this 2,000-year history, for the wood is used the other way up, on the 'end-grain', the tree having been sawn into sections rather than into planks. When the wood is used in this way and when it is a close-grained wood such as box, a new sort of drawing results, equally from the tree itself (it could not be done on any wood) and from a change of artistic perception.

The perception occurred slowly. There is evidence of end-grain engraving being practised around 1700 in Armenia and Constantinople and around 1750 in France, but it is not always possible to know just which engravings were or were not end-grain when they were done in the existing style of 'black-line' cutting – rather as early motor-cars were made to look like horse-drawn carriages, no one having yet decided what a car should look like. Thomas Bewick of Newcastle (1753–1828) is the first who can be securely credited with doing wood engraving as we understand it today – indeed, as he defined it for us. Trained as a metal engraver of spoons and clock-faces, Bewick used the fine, single-shank metal-engraver's tools on end-grain wood for printing. He recognised that when a single spiral of wood was removed, a white line resulted before a black one was achieved and that therefore the drawing proper to 'wood engraving' was done in white.

Bewick was a natural observer in the manner of his age. In 1790 he published the first edition of his *General History of Quadrupeds*, followed over the turn of the century by the *History of British Birds*. These were the books, in particular their humorous and imaginative tailpieces, which fired the imagination of the young Jane Eyre, hidden behind the curtains in the first chapter of the eponymous novel. He was a defender of a bold peasantry, and his illustrations to various works of literature and natural history are vignettes of vernacular life. He prized those 'neglected spots, bounded and bordered with a background of ivy covered ancient hollow oaks, with elms, willow and birch... an undergrowth of hazel-whips, broom, juniper and heather with the wild rose and woodbine and brambles.' Bewick was the ideal illustrator for Joseph Ritson's anthology of songs and ballads about Robin Hood, which, as the book declared, show 'a spirit of freedom and independence which has endeared him to the common people'.

Bewick's apprentices took the engraver's skills to London and spread them throughout the industrialised world. For the nineteenth-century printing industry, wood engraving meant that detailed, high-quality images could be printed alongside text, because both type and picture could now be printed in relief. In one of history's greatest democratic disseminations of knowledge, popular newspapers, periodicals, catalogues and encyclopaedias broadcast increased awareness of the world; meanwhile the world was ransacked for the raw materials with which to do it.

The particular wood on which the new drawing depended was box, a slow-growing tree; hence the closeness of its rings and the even perfection of its grain. Centuries, indeed millennia, of growth must have been cut down in Europe, Turkey and the West Indies to provide printing blocks, in some cases for fine works of graphic art and astonishing works of graphic skill by artisan engravers, but also for the trivia of commercial publishing, journalism and mail-order catalogues. Boxwood of the maturity and girth seen in nineteenth-century blocks is simply not available now. Even ancient box trees do not grow very thick, and blocks of 3 or 4 inches or more were and are constructed from several pieces of wood, matched for consistency in cutting, flawlessly jointed, and polished to a silky, uniform surface. The slightest hairline opening in a joint will show up in the printing. Like a large and perfectly stretched blank canvas, a large block is a thing of rather daunting beauty. It is made 'type-high' – that is, the same height as printing type.

Box remains the best wood to use for wood engraving. Other woods need to have a similar consistency of grain even if they do not have the same closeness. 'Lemonwood' (genaro), pear and, in America, maple are widely used. Most of these woods grow to a larger diameter than box, and it is common to find a 6-inch block made from just one slice of the trunk. The differences for the engraver are between the firm sweetness of box and relative grittiness or fibrousness in the cutting of other woods, and in the varying capacity of the woods to sustain a fine line.

The handmade woodcut or metal engraving, besides providing a way to produce images, was also the primary, if not the only, way to reproduce a painting or drawing before the invention of photography. Until then, when a work by Michelangelo or Titian or any recognised artist was to be reproduced, it was simply copied by another, lesser artist, with all the margin for misinterpretation that implies, directly on to the wood and cut by craftsmen in wood-engraving firms. For most of history the woodcut artist was an artisan – not an artist at all, but a tradesman. 'Engravers', as printmakers were generally known, were regarded as the servants of art and were not admitted to academies on an equal standing with painters. By the end of the nineteenth century and under pressure from photography, they began to chafe about this.

W. J. Linton wrote in his manual on wood engraving (1884) that 'only the artist-engraver, while he upholds the dignity, can ensure the future of engraving', explaining that since the best drawings are not made in line but in tone with a brush the engraver must 'draw with his graver such lines as shall represent colour, texture and form… In such work he is an artist

ABOVE: '*Tree Felling by Machinery,*' The Graphic , 1878

LEFT: *Peter Smith, 'Fallen Tree',* 1990.

BELOW: '*Bird Garden' by Betty Pennell.*

(though only "copying" the painter or draughtsman) in exactly the same degree in which the translator of poetry is a poet.'

History has obliterated these distinctions. Many prints of past centuries, made for functional reasons as reproductions in their day, are now prized and traded as art; for where there was room for misin-

terpretation, there was also room for excellence and variety, and originality is now perceived where only fidelity may have been intended. Perhaps the greatest of the reproductive wood engravers was Timothy Cole, an American who spent most of his life, up to the brink of the First World War, in the great galleries of Europe, engraving their collections for *Scribner's* (later *The Century*) Magazine in New York.

The status of the engraver as artist began to be changed in the 1890s in Britain by Charles Ricketts and Charles Shannon, two young artists who met in the trade-oriented wood engraving class of the City and Guilds Schools in London. Rather than finding jobs as trade engravers, which was the probable ambition of most of their working-class fellow students, they had the means to set up the Vale Press in James McNeill Whistler's former house in Chiswick, where they printed their own pages and engraved blocks of their own designs. Even the revolutionary master-craftsman William Morris had not done that; he employed (he also designed the typography) engravers to engrave and pressmen to do his printing. Stylistically, the work of Ricketts and Shannon was attenuated, exquisite and *fin de* siécle, and the historian of wood engraving Douglas Percy Bliss commented, 'It is difficult to see what they could have gained by making facsimiles of their own drawings, which the excellent Hooper [Morris's employee] or any other professional engraver could have done as well if not better.' They had, in other words, reverted, as revolutionaries often do. From the rich, realist tonality of Victorian commercial engraving they turned to an austere, linear style based on Italian Renaissance models, which were, of course, woodcuts. But they had made a new beginning, albeit more in the field of book illustration than in that of independent prints. It would take the new artist-wood engravers some time to work through the truth-to-materials idea (as it would later be expressed) and come to terms again with the full range of greys which is one of the glories of the medium; but very soon they began to create again in the terms characteristic of wood engraving.

Wood engraving, then, like much printmaking a late entrant among the fine arts, shares with the woodcut a sturdy history in trade. The great woodcut artists of the Renaissance tended to arise from the trade rather than approach the technique as artists, though from the fifteenth century onwards, in the West, there were artist-printmakers (most of whom were not relief printmakers), such as Rembrandt, Callot, Hogarth and Goya. A rediscovery of relief printing in the creation of the nineteenth-century wood engraving industry prepared the way for the revival of both forms (woodcut and wood engraving) as artists' media in the twentieth. But not until the advent of mechanical rather than manual reproduction of images, well into the twentieth century, was printmaking finally divided from trade and freed as art. The artist-printmaker continues a

flirtatious and experimental relationship with the latest technologies – it is part of being a modern artist to do so – but the divide between the print as art and printing as a highly mechanised, commercial industry is now complete.

The twentieth-century medium of wood engraving that is used to make prints for their own sake, for illustration and for applications in graphic design, is a medium of kitchen-table simplicity. A block can be engraved with a couple of engraving tools and can be printed by using an old teaspoon to burnish the back of the paper, when it is laid on the inked block. Many of the best artists never did anything else. It can also be aesthetically simple. After turn-of-the-century development at the hands of William Nicholson and Edward Gordon Craig, following on from Ricketts and Shannon, a flowering of the art took place in the 1920s and 30s, when, in the wake of World War I and under the impulse of modernism, simplicity and directness answered the needs of the day. The direct, black-and-white woodcut qualities of these prints are given extra refinement by their being done on wood-engraving blocks.

In Britain, in the 1920s and 30s, when wood engraving was a widely practised, avant-garde medium, its artists made some of the finest prints of the whole era in wood: angular, Vorticist prints made during World War I by Bomberg, Wadsworth or Gaudier-Brzeska are mainly catalogued as woodcuts but may have been done, like Nicholson's, on end-grain blocks. Paul Nash made an influential sequence of landscape and still-life engravings; and, between 1932 and 1938, large partly abstract figure compositions by Gertrude Hermes and Blair Hughes-Stanton married the formal insights of modernism to an awareness of psychological forces, with a grandeur that makes surrealism look flippant.

Towards the end of the period, a decline in the market for prints (associated with the Depression) and the rise of a number of fine private presses led to engraving being deflected more into book illustration again. Hermes, Hughes-Stanton and Agnes Miller-Parker, in association with the Gregynog Press, and Robert Gibbings and Eric Gill at the Golden Cockerel Press produced the great collectors' books of the time. Many of the same artists were employed on the Penguin Illustrated Classics, and wood engraving was widely used at every level of publishing. These are but a handful of the names involved.

Though wood engraving no longer occupies the limelight of the avant garde, the revival of the Society of Wood Engravers in 1984 and the later formation of American associations focused a renewal of interest in both countries. The quality of current work is as high as, and probably more varied than, ever before – the latter, curiously, because it is not in fashion; there is therefore no fashion to follow. Whereas in the 1920s nothing was more immediate than the hand of the

LEFT: *Edwina Ellis, 'Parterre, Llanerchaeron'.*

OPPOSITE: *Guy Malet, 'A Sussex Lane,' 1930.*

artist on the material, today the hand of the advanced artist is on the button, the computer mouse or the concept. Wood engraving therefore reflects the advanced art of the time less than it did. In the age of installations

BELOW: *George Mackley, 'Broken Willow', 1946.*

Above: *Simon Brett,*
'King Willow', 1990.

Below: *Gwen Raverat,*
'Elms by a Pond', 1917.

a collector's market of its own.

The essential and abiding characteristic of wood engraving is that the artist begins with a black rectangle, the block, and works into it with white. This drawing with light will always give a wood engraving a look like no other sort of work. Black lines can, of course, be made to carry the drawing too, and usually are to some degree, but the directness with which an artist has latched on to the principle of light will have much to do with the success and impact of a print. The best pictures contain a few carefully selected and grouped tones, whether the selection is designed to be stark or to be rich. Many turn-of-the-century and 1920s engravings were strongly simplified, quite black-and-white and woodcut-like, in reaction against the Victorians' use of the whole 'photographic' keyboard of tones, from black through every possible grey to white. Today's work may be graphic or richly tonal again, silvery and pewtery in the textures unique to the medium. Its images tend to be complex, as though the slow and contemplative nature of engraving encourages mulled-over ideas.

Some of these points are evident in the portfolio of wood engravings which accompanies this book, even in the varied treatment of the foliage in the prints of trees. What cannot be picked up from a selection on a single subject is the diversity of theme and character found in wood engraving overall, including the strong traditions in America, Russia, eastern Europe, and the Low Countries, as well as in Britain.

Then there is the wider context of art to consider, and the trees that are the subject of this book. A collection of prints which deals with trees may, by inference, be seen to be dealing with the perennial state of the world. An art which is energy-simple, using resources that in a properly managed world should be renewable, and using very little of them at that, would seem to be the ideal for our time. Compared with the timber consumed by the construction, furniture or paper industries, the wood engraver's needs are a model of environmental tact.

Art is not about how it is done, what it is done on or whether its methods are newfangled or old. In wood engraving, concentric rings of boxwood are jointed together to make a flawless printing block. The wood shows the slow and steady growth that makes fine engraving possible and yet makes human existence seem like a breath upon the wind. Staring into that surface, the artists see neither clockwork nor carpentry, not wood but the mirror of their dreams. Like any other art, the only thing that matters is what is done, the high-risk possibility that just here, just once, the stuff of the world is transmuted into spirit. The environmentalist and conservationist are rightly concerned with matter, lest we all perish. The task of the artist is to put meaning into matter, lest we never truly live.

and video, when painting is nevertheless far from being dead, and when the woodcut on big, new sheets of manufactured timber plays a vivid part in the contemporary scene, wood engraving is one element in a multifarious art world. A minority art, as, in fact, it always was, it has a literature, a following and

PRINTMAKING WITH WOOD

Wood Engraving

Engraving a block. One hand holds the tool which engraves the block; the other turns the block against the tool, continually moving and presenting it at the right angle for the cut. The block is made of a section of the tree – the 'end-grain' – shaped to a rectangle. Larger blocks are made of several pieces of wood flowlessly joined together. Before beginning, most engravers darken the surface of the wood so the engraving will appear light, and draw onto it the outlines of the design with a pen, pencil or brush. A wood engraver thinks of a design or picture in terms of white marks on a dark background, drawing 'with light'. Some of the tools have a broad cutting edge like tiny chisels, while others have a more pointed cutting tip in various shapes.

Making a single knife cut in a plank of wood does not remove any wood at all nor leave a white line. Two cuts are required to do that – or the double cut made by a gouge. That is the essential technical difference between 'cutting' and 'engraving'. The artistic difference is that in most woodcuts the background has been cut away to leave a black drawing line standing proud, whereas in most wood engravings the drawing is done with the white line which the tools most naturally produce.

Printing a Block

The block has been rolled up with ink and printed off on to paper in a press. When the paper is carefully removed, the image engraved in the wood has been transferred to the paper. The image is of course reversed, like a mirror-image, from block to print; but what the engraver saw as the dark surface of the block is still dark; what appeared as light wood-colour as it was cut now reveals the more intense contrast of the white of the paper. The press is not essential. Printing can be done by hand, by burnishing the back of the paper, which makes wood engraving an attractively simple, kitchen-table medium.

In Japanese prints and the European chiaroscuro woodcuts of the sixteenth century, blocks

Simon Brett carefully engraves a 'block'.

The engraved image has been transferred to the paper.

were used to print flat areas of colour. The coloured areas filled the linear outlines of the main design. The main design and each colour would be cut on separate blocks.

The Etching revival

The movement known as the Etching Revival (which refers to two bursts of interest in etching, around the 1830s and the 1860s) led in 1880 to the foundation of the Society of Painter-Etchers for the promotion of original printmaking because of the Royal Academy's continued refusal of printmakers. The word 'original' distinguishes between printmaking which produces an image and commercial printing which reproduces one. The paradox is that an original print may be a simple little woodcut in one colour only, but (for what it's worth) from the artist's own

hand; while a reproduction may be the product of sophisticated, multi-colour technology and legitimately more costly.

'Originality' therefore also becomes linked to the idea of the 'limited edition', – that is, a specific number of prints an artist chooses to produce from a block as opposed to the limitless number of reproductions that could, in principle, be mechanically run off on demand. Such distinctions correspond to the social roles of the artisans or artists who produce them. Victorian industrial wood engravers worked ('unlimited') for a mass public, in newspapers and illustrated magazines; and for an informed and specialist book-buying public too.

The work of the artist-illustrator-wood engraver today may similarly be seen in journalism; in graphic design on supermarket packaging, where it probably won't be recognised; and in illustrated books, where it will – the informed reader will look for the illustrator's name.

Except in gallery editions, engravings are always reproduced nowadays (as here), printed from the block only in books in 'fine print' publications, which are themselves produced in limited editions. In this case the illustrator has become more like an artist-printmaker, making signed works as freestanding artistic expressions for a public which specifically appreciates his identity and message. In practice, few artists will have confined themselves to only one of these overlapping roles; all are aspects of printmaking today.

In the complex sociology of all art practice, the purchaser shades into the enthusiastic amateur and the part-time professional.

The Continuing Tradition

In the post-war era, wood engraving was no longer the first medium of the modernist print, as lithography seemed to evoke more vividly the anguished scratchings of the time, and the bright, commercial brashness of Pop Art in the later 1960s found its perfect, impersonal medium in the screenprint. Nevertheless, the significant

earlier sequences of work by Hermes, Hughes-Stanton and Nash were matched by the work of John Farleigh, Clifford Webb, and Geoffrey Wales; and, a little later, by George Tute and Monica Poole, Still, in 1997, at the height of their powers. Tute's series of visionary landscapes was done in the 1970s at the time of the medium's lowest ebb, when wood engraving was very nearly extinguished simply by the scattering of its forces by other fashions in art.

Poole's engagement with the rocks and trees of her native Kent, in images at once austere and richly complex, began in the 1950s and shows no abatement yet. Among illustrators, David Gentleman, has designed many postage stamps, and his blown-up wood engravings decorate a London Underground station. Both he and John Lawrence, prolific in books, continue the lighthearted, decorative tradition of Eric Ravilious in the 1930s and Derrick Harris, whose Radio Times work and Air India advertising campaigns typified the 1950s.

The Art of Craftsmen

Most designers of pictures were tradesmen too; but even when they achieved personal fame (Dürer or Burgkmair in the West, Hokusai or Hiroshige in the East) or were already famous painters (Titian, Pontormo or Rubens) their designs were cut in the wood by craftsmen. The more complicated the projects, the bigger the workshop. In Japan, an artist such as Hokusai made a drawing on paper. A craftsman copied the drawing on to thin rice paper which was pasted down on to the block, so that the block-cutter literally cut through the paper around the lines of a drawing conceived and made with a brush. Further workers printed the blocks, all by hand. In the workshops of sixteenth-century Nuremberg and Venice or nineteenth-century London and New York, the same applied, though without the rice paper and with the printing done in the West on presses, which during the nineteenth century became mechanised.

CHAPTER ELEVEN

Conservation and Tree Planting

Trees and forests have been used and misused by people throughout the history of the human race. Tree cover has increasingly been removed to provide timber and to make way for agriculture and dwelling places. As the human population has increased so, too, has the pressure on trees. Public awareness of the need for conservation and sustainability on a world scale is increasing. There is considerable pressure to save the rain forest, but where are the wood products the rain forest supplied, or the things we made from them, to come from? Sustainable forests are accepted as a good idea, but how should they be planned and managed?

Diversity of species must be the key to success. Greater site diversity is another factor that will increasingly be available to us as farming practices change. It will enable new species to be tried out in a range of different conditions. In addition the opportunities to strengthen non-tree benefits such as recreation and environmental enhancement fit in very well alongside increased species diversity. The future for tree planting has never been as interesting as it is now and the potential for success never greater.

CONSERVATION AND REFORESTRY

Increasingly, European and worldwide organisations are talking seriously about the conservation of wild plants. It has become clear that if the plant resources of the planet are used up, people will not survive. So in order to ensure our own future we must not jeopardise the future of plants. In the long run, however, it is not enough simply to preserve only the plants we need and use every day. Whole plant communities must be protected from exploitation. In this way the biological diversity necessary for healthy development is safeguarded. Genetic diversity, too, must be kept at a satisfactory level: an insufficient number of genetically distinct individuals in a species population reduces vigour, health and disease resistance. The IUCN (International Union for Conservation of Nature and Natural Resources), founded in 1948, has taken a leading role in the world conservation of plants. Discussions are in progress about innumerable aspects of plant conservation. Recently the Convention on Biological Diversity (Rio de Janeiro, 1992) set out to define biological diversity, and lay down rigorous rules and general guidelines for the conservation of all living organisms.

Legally the rules only apply to countries which are contracting parties to the convention, so trees and forests elsewhere are not yet totally protected. Individual states each have to define the conservation importance of their own flora, and inevitably this is done in the light of politically driven land-use policies. Logging and agriculture, on which many communities rely exclusively, can obviously not be abandoned instantly or willingly while some trees remain standing. However, the exploitation of forests has progressed to such an extent now that sustainable management must be implemented, otherwise timber reserves will soon be exhausted. Beneficial plants will continue to be lost; some of them will not even be discovered or classified before they become extinct – their potential as 'useful' plants will never be realised. Furthermore, if substantial areas of forest are not reinstated or conserved, the climate will be considerably altered in many areas. The reduction of transpired moisture from foliage will have a profound effect. In addition the modifying characteristics of trees on the climate will be lost to whole regions of the world. Droughts, storms and erosion will prevail.

Exploitation of natural resources is driven largely by customer attitudes. Trees are affected directly by the human requirement for wood of every kind, drugs, food and paper products. They are affected indirectly by environmental damage associated with a whole host of other, often unrelated, commercial activities. Consumers of wood products are mostly quite detached from trees and forests, and they have little idea of the implications of their demands. If attitudes can be gradually changed so that non-essential requirements at least are reduced, for instance through public awareness campaigns, the eventual benefits would be considerable. Trade in non-essential luxury items, or in raw materials for dubious purposes, is a serious threat to trees. Increasingly, it is subjected to official condemnation, but catching the perpetrators of illegal traffic is another matter.

Trade in rare and threatened plants has become a major cause of species loss and habitat degradation. It particularly affects tropical plant communities. In this respect trees are less at risk than exotic flowering plants which are greatly sought after by collectors and gardeners in affluent societies. International monitoring and control of the trade in vulnerable plants is co-ordinated through CITES (Convention on International Trade in Endangered Species of Wild Fauna and Flora). The convention was signed in Washington, D.C., in 1973 and came into force in 1975; 120 states had become members by 1993, each responsible for implementation through its own local scientific authorities. Lists of species have been drawn up in

PREVIOUS PAGE: *The old and the new at Collyweston Woods in Northamptonshire; an old pollard small-leaved lime is still tended while young lime coppice is also encouraged.*

BELOW: *Careful planting will ensure the continuing survival of many tree species.*

which international commercial trade in wild grown specimens is not permitted, and other plants that are regarded as likely to become threatened are also strictly regulated by import and export permits. The countries that have agreed to be bound by the provisions of the convention have become an effective global conservation network. The implementation of CITES is overseen by a small secretariat in Geneva. Working closely with CITES and based in Britain is the World Conservation Monitoring Centre, funded mainly by WWF (World Wide Fund for Nature); its officers monitor trade and the use of wild plants and animals worldwide. Even scientific bodies and botanic gardens that wish to obtain endangered wild plant material from abroad must first have an export permit and prior permission from the country of origin. Artificially produced material, which is increasingly grown under controlled conditions from seeds, or vegetatively propagated, should obviate the need to collect from the wild at all; meanwhile permits showing proof of origin are required to import or export such material. By 1995 botanic gardens round the world held secure collections of some 80,000 species of higher plants.

Plants in botanic gardens and arboreta are usually out of their natural habitat (*ex situ*). Very often their habitats have been completely destroyed, and artificial conservation is the only option left for them. Many trees can survive now only in an arboretum, which is a 'tree zoo', or in a gene bank. It is tempting to suppose that one day all *ex situ* species might be returned to the wild, but, unfortunately, that will not happen. Re-introduction is a complex procedure: for many plants there is no 'wild' left to return to, and if they are 'released' into another apparently similar habitat they could endanger other vulnerable species that are already there. One possible step up from the zoo environment is the micro-habitat, which is also a contrived *ex situ* setting but one that is specially adapted for a range of ecologically related flora and fauna to live in.

Although many named species of trees are endangered, which means they are very rare, and many more have already become extinct, trees as a whole are probably not threatened on a world scale at all. The past activities of volcanoes, colliding asteroids and other celestial debris, continental drift, uncertain sea levels and ice ages, have all failed to annihilate trees completely. Most of these phenomena far outweigh the destructive consequences of modern human activities, even when devastating chainsaws, mobile logging machines, herbicides, toxic fumes and matches are brought to bear.

Trees have a long history of survival: they have recovered completely and come back in abundance following apparently total destruction. In Britain the last complete clearance of trees occurred during the most recent ice age, which only started to relinquish its freezing grip between 10,000 and 15,000 years ago, during the period known by geologists as the Late-

OPPOSITE: *Trees like the sycamore will quickly take the place of a felled oak.*

BELOW: *Tiny sycamore seedlings thrust their way towards the sunlight in their battle for survival.*

BOTTOM: *A devastated beech clump after the 1987 gales.*

glacial. This was like yesterday in a geological sense, and a mere fraction of the time that trees have been on the planet. Before that, in the Full-glacial period, woody plants only survived in refugia to the immediate south and possibly in what is now Cornwall. Sub-fossil remains of dwarf birch and alpine willow have been found in Full-glacial deposits there.

The recovery of trees after present-day flooding or forest fire is similar in many respects to the recolonisation after the ice age: wherever there is exposed earth, seeds of some sort will soon grow on it. Where the seeds come from, if not simply from adjacent land, is of great interest – they may be blown on the wind or carried by birds. Some are remarkably watertight and will cross oceans. In some species, seeds, and even the trees themselves, are almost fireproof. For them, hot, sterile volcanic slopes and brush-fire sites appear to be more of a challenge than a difficulty. Eucalyptus can be, and often is, burned to a cinder above ground in Australian bush fires, but in a single growing season most of the charred stumps will recover and sprout green shoots again. Fire is actually an essential aid to the reproduction of some trees. The 'closed-cone' pines of California and Mexico, and the American redwoods, will not usually open their cones to shed any seed until they feel the heat of a forest fire. Cleverly, the fire itself only breaks a resin seal, and the cone will not open fully to release the seeds for some hours, until the fire is out and the bare, ash-enriched ground has cooled.

We would be deluding ourselves if we really thought that we could wipe trees off the face of the earth: they are better equipped to survive adversity than we are. They probably regard our puny efforts to destroy them as a tiresome irritation and invariably seem to turn hardship to their own advantage. For many a big tree that is destroyed, a hundred new

ones will spring up. In southern Britain these may be elder, birch or sycamore, to replace a fine oak, but they are trees all the same. The threats posed by insects, and the long-running contest with fungal and bacterial disease, must be of far more concern to trees than what people are up to.

There are several thousand diverse sorts of tree. Each of them can tolerate slightly different conditions. Since the first Carboniferous trees appeared 300 million years ago, there has always been some tree that can adapt to every changed geographic and climatic situation. The first trees (tree ferns) inherited a world with far higher levels of carbon dioxide in the atmosphere than modern carbon-dioxide enrichment caused by human proliferation and fuel consumption: over millions of years, tree ferns purified and oxygenated the atmosphere, packing surplus carbon into their stems and branches, where much of it has remained in fossil form until this century, thus making the world a suitable place for us to live in. While increasing global temperatures cause us considerable concern at present, for trees a rise in temperature simply means that species which are tender in Britain at this moment will be in their proper element in the future. It is true that a few British sub-Alpine species will probably die out, but that is the nature of evolution. The harsh reality is that a species which can adapt to new situations will thrive and a species that is less able to do so will perish. Two and a half million years ago the British climate was similar to that of Florida. Magnolia and redwood trees flourished here as Pre-glacial native species. Perhaps they will again sometime. They already have a foothold in British arboreta and gardens so they will not need to migrate here naturally next time.

Even physical destruction, which they frequently have had to endure at the hands of man, is nothing new for trees. There are species that are specially adapted to being smashed to pieces, such as some crack willows which have abandoned sexual reproduction completely in favour of vegetative proliferation relying on breakages to reproduce. The smallest broken twig or slither of shattered stem will root instantly if it falls on to damp earth or into shallow water. The small brittle branches are specially designed to 'crack' off easily in the slightest wind. Such a plant has little use for seeds, and nothing whatever to fear from us.

Though trees can prevail on earth despite cataclysmic adversity, habitats, on the other hand, cannot. They are fragile and can easily be destroyed or diverted away from their preordained evolutionary path. They are created and modified by the life forms within them and do not remain static even if they are left alone. This presents us with a problem because it is generally agreed that saving plants *in situ*, in their existing habitats with all of the associated biodiversity, is crucial to us. However, habitats are in a constant state of evolutionary change. Each

ABOVE: *A black poplar at the entrance to an old farmstead. Regular loppings have kept this old tree in good shape.*

component part is a living, developing organism, some of which are in the process of becoming more successful, and others of becoming weak and failing. All of them are actively competing with each other for light and nutrition. The effects of introducing a new element into an existing habitat can be dramatic. The whole evolutionary direction of the community might be immediately changed so that a different kind of habitat is soon produced. Alternatively, very little might happen and the infiltrator might soon be suppressed. Physical change, such as felling mature trees, will change a habitat, in this instance by altering the physical environment which regulates the lives of the other inhabitants. This will be to the advantage of some organisms and to the detriment of others.

Most tree habitats in Britain have a long history of human interference. Very few British trees, for instance, are actually native trees, which means they are growing in their present location entirely naturally. Almost every tree, even if it is a native species, has been introduced from elsewhere. As such, they are, in the strict sense, alien to the habitat they are in. They will have interacted with whatever was already there. It could be argued that there are no natural habitats left in Britain. This presents a problem for conservationists who are concerned to preserve the integrity of the countryside in its present state. It is impossible to stop the clock and preserve everything in a habitat exactly as it is, or as it used to be. Only one passing animal with seeds attached to its fur is needed to have a profound effect. More seriously, the number of naturalised plant species in Britain that have originated from garden rubbish dumped in country ditches is incredible. The wind plays its part in this respect, too, as it blows seeds from alien plants in parks and gardens. The spread of Oxford ragwort, which escaped from a botanic garden in 1794, is a well-documented example; the seeds germinated alarmingly quickly beside roads and canals, and later along the railways. In the same way modern motorways support a whole new crop of introduced plants wafted along in the wind generated by traffic, and seaside plants in particular thrive on de-icing salt hundreds of miles inland. These habitat modifications are all fairly obvious while they involve species that are readily distinguishable from the local native vegetation.

A far more complex problem arises when the introduction is of the same species as the existing vegetation but is from another source. What should be done, for example, to resolve the dilemma presented by a 300-year-old oak wood which is genetically or historically shown to have originated from East European or French acorns? It supports a diverse web of life and looks perfectly natural in the British landscape. It is not natural, though, and it will be evolving differently to a truly wild wood (if there is such a thing). Shutting the gate on it and doing nothing will never make it into a native wildwood again. Cutting it down will not do much good either. Alien pollen will have spread from it to British oaks outside, and acorns will have regenerated in adjacent areas wherever jays could take them. As it stands, the wood makes a valuable contribution to the environment by retaining water, fixing carbon dioxide and softening the impact of physical forces. It is home and larder to numerous insects and invertebrates. The lengthy time-scale involved has blurred the circumstances of its origin so much that the full consequences can never be sorted out. The only practical thing to do is accept the wood exactly as it is and regard past human activities as a valid biological contribution to what is there.

Conservation of wildlife is often concerned with species survival for its own sake. Trees are more likely to be preserved to support wildlife or for human use, enjoyment or prosperity. Huge numbers of trees are protected just for amenity. Campaigns to save these usually focus on individual notable specimens, and they attract the kind of enthusiastic support you would expect to see given to saving the local pub or village school from closure. It is obviously a good thing to save individual notable trees, but it is a pity they are generally regarded only as a means of providing comforts and benefits for people. In some respects they are reduced to the lowly status of the domestic cow. The whole object of their conservation is the end product and not the provider.

Trees in Britain are seldom conserved, or preserved, for their own sake; their conservation usually depends upon their contributions to the conservation of something else. It might be a landscape, or the shelter they provide for birdlife, animals or insects. It might be the fungi and invertebrates they directly support. Likewise, tree habitats are not usually managed just for the benefit of the trees. The trees simply form a framework for everything else.

Except in forests, habitat management often actually works against trees. They are cut out of wetlands, where the maintenance of open water is the objective.

They are rooted out of heathlands, where their presence, it is claimed, would eventually cause the heath to evolve into forest: a process prevented, certainly since the Stone Age, by the incidence of brush fires. They are even cleared out of woodlands to create glades for butterflies and other insects. These kinds of habitat require constant manipulation and stimulate evolutionary retrogression so long as they are kept up, but they can only be temporary. Left to nature, trees will eventually return and dominate once again.

Specialist conservationists often regard trees as being there simply for a bird to sit in or for a particular beetle to chew. For them trees usually carry less weight than the butterflies and animals that live amongst them. Trees have to be artificially manipulated in order to benefit particular organisms. Habitats that have been blatantly altered in order to preserve an animal or invertebrate, or a small number of them, are always going to be difficult to maintain to prevent evolution from reinstating itself. One only has to look at a derelict industrial site or an immaculately manicured garden after five years of neglect to see this happening.

There is a completely different destructive force silently at work, which seriously threatens tree species as we know them at present. Although it is usually caused by human activities, it generally goes unnoticed by most people. It is genetic pollution, and it has the power to obliterate completely the identity of species.

NEXT PAGE: Willows hybridise so successfully that it is often quite difficult to identify them.

BELOW: A dead oak in Windsor Great Park is stage managed into a 'natural' habitat. With the aid of hawsers and winches great boughs are ripped from the old tree to create shapes that are similar to the aftermath of a gale rather than the abrupt and unnatural cuts of the tree surgeon. These tears and breaks also make much more inviting nooks and crannies for bird and insect habitation.

ABOVE: *One of the finest wild service trees in Britain, outside the Snooty Fox public house at Lowick, Northamptonshire. In the foreground is just one clump of the abundant suckers which the tree has pushed up.*

Genetic pollution is exacerbated by the artificial movement of plants from one place to another, something which gardeners, foresters and agriculturalists now do all of the time. It is of particular concern to systematic botanists, taxonomists and historians. There are several forms of genetic pollution. Some threaten the integrity of true species or provenances *in situ*, and others threaten the historical integrity and succession of heritage trees. The latter type is so advanced in Britain that most original British native trees are already adulterated by imported stock to such an extent that further mixing of genes is not likely to be of any real consequence. Many endemic populations are no longer of exclusively British native origin, and they are only vaguely related to our original postglacial trees. Commercially there is often some advantage to be gained from artificially breeding selected individuals to produce 'improved trees'. But little is known about the impact of these improved trees when they mix with local native stock nearby and new hybrid races start to evolve.

A handful of British natives, however, still have some authenticated status and distribution. These remnants of our tree heritage and their progeny deserve special protection from genetic pollution. Probably the most researched species in this category at the present time is the small-leaved lime. It is an ancient tree which has always been reluctant to spread naturally over a very wide area. Isolated coppice stools are known that date back over 2,000 years. For hundreds of years the climate has not suited small-leaved lime particularly well and seeding has been infrequent. Recent warm spring and summer days, however, have encouraged the species to fruit again. By coincidence this has unfortunately occurred simultaneously with an upsurge in popularity of lime as an amenity tree. The demand for plants is greater than availability in British nurseries so imported European stock is widely used. If foreign trees are planted within pollen range of ancient native trees, they will destroy the authenticity of any seed that is collected or any natural regeneration that might occur in the future. While small-leaved lime continues to be coppiced, it will live almost for ever, but old stored coppice and fully grown trees do eventually die. If coppicing stops, and regeneration is genetically polluted, the authenticity of British native small-leaved lime will be lost.

Another very rare tree in Britain is the native black poplar. This subspecies of the European black poplar has been subjected to systematic, but quite unintentional, genetic pollution for over 200 years, since the introduction of straighter, more productive hybrid timber trees. As a result of this, the native tree is on the verge of extinction. No spontaneous seedlings anywhere in Britain can now be regarded as true to type. Pollen from cross-fertile alien hybrid poplars can travel up to 11 miles, enough to pollute every true black poplar in the land. Fortunately, the subspecies grows perfectly well from cuttings. These are, of course, true to type, but because hybrids are so much more productive, very few of the original native species have been grown, from seed or cuttings, since 1750. Poplars are short-lived, and many old specimens, which may have been genetically significant, have already been lost for ever. The genetic base of the last few remaining native trees, females in particular, is now alarmingly small.

The native wild service tree in Britain has been subjected to genetic pollution on two distinct occasions separated by several hundred years. It was formerly cultivated for its fruit and artificially grown in orchards. No doubt this involved a degree of selection for the best fruit, and the inevitable transportation of plants from one place to another. At the same time small populations of wild trees appear to have survived in woodlands. These are considered to be an indicator of ancient woodlands, but it has never been a common tree. The taste for the edible fruit, and the drink fermented from it, was superseded eventually by improved cultivated apples, and orchards were gradually abandoned. Whether surviving trees are of natural origin or descended from cultivated stock is often unknown. Now, after hundreds of years, wild service is popular once again, this time as an amenity and urban street tree. Seed is usually predated by a wasp so British plant supplies are scarce. Imported material certainly makes up the shortfall. Authentic British trees could now be finally eliminated because germination is so poor. Few seedlings are ever produced, and those which are could be contaminated by imported pollen. Fortunately, wild service trees also reproduce themselves from root suckers, which in some circumstances ensures a longer genetically pure life for some existing plants. It would still be prudent, however, not to plant imported stock next to ancient woodland specimens where they might at the very least be confused with the 'real thing' in the future.

A species can also be adulterated by promiscuity. Willow and eucalyptus, for example, can hybridise to such an extent that the identity of the original parents may be lost in a confusing mass of intermediate hybrids, but for two quite different reasons they have developed species individuality. Willows, and their close relatives poplars, often remain distinct because of the differences in their flowering times. A plant that flowers in March can not cross with a plant that flowers in May, even if they are within the same sexually compatible sub-genus: all the flowers are dead and gone by then. If, however, a compatible 'third party' species that flowers in April is planted nearby, it could easily cross with either or both of them. Later on, the resulting progeny might re-cross and back-cross until all the surviving plants lose their original identity and indeed their species status. Once this happens the genes can never be sorted out again, and new plants will go on

to bridge the 'time gap' between themselves and yet more species. In eucalyptus, which flowers over a much longer period, the development of distinct species is often due to distance. Plants which are sexually compatible have become distinct through geographic isolation from each other for thousands of years. If at any time they are artificially moved so that they come together again they will cross-pollinate, and a new hybrid mix of genes will occur. In this way the genetic integrity of many original eucalyptus species in cultivation has been lost.

Although the risks of genetic pollution in Britain are well understood by the forest industry, there remains a need to grow trees commercially to supply wood products and paper pulp. The best options that are likely to meet these two distinct requirements appear to be the use of local native stock in or near old woodlands where it is possible to do so, or the use of alien and highly productive species in plantations away from old woodland sites. These should not be related in any way to British native species so that it is impossible for them to hybridise with local trees or cause any pollen contamination. Any unwanted natural regeneration that may spring up in sensitive areas can be immediately recognised and quickly pulled out. Imported hardwood species such as oak, beech and cherry are most likely to interbreed with native trees because they are so widely used. In Britain,

BELOW: *Imported hardwood species like beech are most likely to interbreed with native trees.*

however, these are already so mixed up with previously imported material that further genetic pollution is unlikely to be recognisable.

Conserving Your Own Trees

The one aspect of tree conservation which is most important to ordinary people is how to establish a tree and keep it alive and healthy. Many trees are planted these days by non-professional people in their own gardens or their local community. Success depends upon a few very basic principles: the right choice of tree is paramount, and some professional advice is always useful at this stage; good planting is also extremely important, and then the lengthy business of protecting the vulnerable young sapling begins. There is great satisfaction to be had from watching a tree that you have planted yourself grow (see Appendix B). This can only be achieved when the initial planning and groundwork have been done properly.

The Sustainable Forest

Tree planting in Britain as an objective is accepted in principle by most people as a good idea. The arguments in favour of it are overwhelming. Now the question increasingly being asked is not 'Should we plant trees?' but 'Which trees should we plant?' The answer is fairly complex and should be determined on the basis of a combination of the following criteria:

The needs:
- To enrich the poor species coverage we have at present in urban situations and in forestry.
- To make up the deficit in the wood that will be available to us (as a result of depleted foreign sources).
- To maintain a regular supply of wood and wood products.
- To enhance landscape planting with good alternative amenity trees.

The problems:
- The length of time trees take to create an impression and become mature.
- The uncertain market requirements for wood in the distant future.
- The possibility of climatic change.
- The opinions and aspirations of the general public.

ABOVE: *The Norway maple has been used as a fine street and park tree for many years. It does well in urban areas and grows quickly. This is the ordinary green-leaved variety, but there are also numerous cultivars with different-coloured and variegated foliage. The vivid green display in early spring gives the appearance of young foliage, but it is, in fact, the great clusters of flowers which emerge first. Although the tree may have been somewhat overused by urban landscapers in recent years, it still plays a vital role.*

The answer:

∾ Diversification, in terms of actual species, and in terms of the number of different species used in each place.

What is needed

It is significant that 90 percent of our existing British forest area consists of only 30 species, and the majority of forests actually contain fewer than five planted species. The decline in species diversity, particularly in the twentieth century, is probably due to the considerable quantity of foreign timbers that have been imported from overseas. There has been little need for home-produced wood, or incentive to grow high-quality timber. Most town and street trees these days are also drawn from just a handful of species, all chosen for their reliability and ease of establishment. It is difficult to imagine how many Norway maple and wild cherry trees have been planted in towns during the past 20 years.

The way ahead in theory is relatively easy. In Britain we enjoy one of the best tree-growing climates in the world, and, if anything, the climate changes that are predicted for us will enhance this. Some 2,500 different trees will grow here, so there are plenty of alternatives to choose from for every situation. Obscure trees, seldom seen now outside botanic gardens and arboreta, deserve wider use and re-evaluation. There is at present

an insatiable worldwide demand for wood products, especially paper pulp, which shows no sign of abating. This has mostly been supplied over the past two centuries by foreign forests, which are now becoming exhausted or no longer available to us.

In Britain there is a surplus of agricultural land at the present time. If this situation continues then trees as a farm crop on former fields, especially arable fields, might become part of the solution. The environmental impact of this is likely to be minimal. In most circumstances plantations would be more favourable to wildlife than a cornfield, especially for small birds during the thicket stage. Public enthusiasm for trees near their homes and places of relaxation or work is strong, but these are clearly not potential sites for vast sustainable forests. There is considerable demand for improved landscape planting too, but cutting down the 'landscape' to make furniture or paper would appear to be be less enthusiastically received. In order just to maintain our present levels of wood consumption in Britain we would need to increase our output of home-grown timber by about 900 percent. Furthermore, a large part of this would have to be high-quality hardwood and veneer timber, which will be in particularly short supply. It would be unrealistic to attempt to be self-sufficient in timber in Britain, because of the present size of the population relative to the land area and the unlikely prospect that consumption of wood products will drop significantly in the future.

There is substantial potential for planting more quality hardwoods and alternative species that will yield veneer-grade logs. These are mostly broadleaved trees that grow to great size and incidentally look good in the landscape while they are doing so. They would do very well as small areas of woodland or even in field corners. For them to succeed, high standards of silviculture and protection would need to be maintained. Fencing to prevent browsing by domestic stock and deer is essential. Rabbits and grey squirrels, if they are present, would have to be rigidly controlled. Once the stem of a tree is physically damaged by any of these it can never recover enough to produce top-quality wood. The timber will always carry a blemish. Shotgun pellets, especially steel ones, and any other metal objects hammered into trees also spoil the wood for ever. Such things can cause serious damage to saws and injury to milling machine operators too.

A good deal of quality timber could be, and indeed is, grown in urban situations. It seems to be a good idea to enjoy trees while they are standing, and then enjoy the wood they produce when they are cut down. The public perception of this is seldom so idealistic, but with good planning the present general reluctance to see anything cut down might be relaxed. Amenity trees, planted purely for the pleasure they give, regardless of any timber or end product, are already pretty diverse. Although recent plantings have tended to be somewhat monotonous, we do have a rich but elderly legacy of exciting and curious survivors from Victorian and Edwardian times. Already some local-authority tree officers and arborists are looking at these critically as alternative modern species. The diversification of urban tree planting is probably an easier concept than diversified forestry to sell to funding authorities.

Growing amenity trees is more of an art than a science. Like most art, it has a long history of patronage from wealthy landowners. Amenity trees are not 'working trees' as such – they were never expected to produce any financial return. They simply had to perform well and please the eye. The landscape they contributed to was all that mattered. Through trial and error since the seventeenth century the most successful varieties have prevailed and become popular. Only in the late twentieth century has the range of species for planting been reduced to include nothing but the most reliable, hardy, drought-tolerant types. By using these, the cost of replacements is practically eliminated.

The way forward with amenity trees must now be to select better examples that will meet new and changing circumstances. Space may be more restricted, air may be more polluted and water in many localities is becoming scarce in summer. Trees that do not disrupt underground services, walls or structures are increasingly needed as building sites proliferate.

The Problems

The length of time a tree takes to develop and reach maturity is a problem that we all have to face up to. Our expected life span and rate of growth bears no relation to that of a tree. Responsibility for the management of trees has to be passed on from one generation of people to the next. History shows that we do not excel in this. Except for the fastest-growing species, planting a tree is an act of faith. The person who cuts it down is probably not born yet, and the people who will begin to enjoy its shade are probably still children. This situation is both a challenge and a responsibility. Two things are certain: if we don't plant any trees we shall have failed, and if we plant a lot of different trees some of them will succeed and future generations will thank us for it.

For commercial growers, investing in trees and forests now, the quandary is how to predict what will be in greatest demand in 50 to 100 years' time. If this were easy, funding would be instantly available, and forests would be appearing in every available space. History shows us that predictions can go very wrong, because of the pace of change. Who could have possibly predicted in the early 1900s, for example, when much of our existing conifer forest was planted to provide mining timber, that the coal-mining industry would virtually disappear in Britain by the 1990s. Growers tend to be cautious, and such events do not boost confidence. It is difficult to try to persuade the people who are funding tree planting, and forestry in particular, to switch from what seems to have been safe in the past, to something that appears to be very speculative. This is an understandable and by no means new situation. When Sitka spruce was introduced from America in 1831, very few foresters would have rushed out to plant it. By the mid-twentieth century, however, this species outnumbered all the other forest trees being

BELOW: *Conifer diversity makes commercial forestry much more appealing to the eye and creates superior wildlife habitats. This hillside in Dunnerdale, Cumbria, bears a healthy mix of firs, spruces, larch and broadleaf trees.*

planted in Britain. In wet upland areas there is still nothing to beat it for productivity, though public dislike of evergreen monocultures, or its insect predators, may prove to be its downfall. Present-day timber prices are also unhelpful as a guide to the future. They are notorious for fluctuating wildly, often because of disastrous storms in recent years, and supply and demand consequently often get out of step. Landowners may have potentially valuable fallen timber selling for firewood, while consumers continue to use highly priced imported material for cabinet-making. Finished wood products command astronomically high prices while at the same time trees of similar species cut in the forest are often sold cheaply. Treated peeled fence posts, for example, have become so expensive now that on occasions the fence is worth more than the forest

it encloses. If one fence post is used as a tree stake, for a long time it is worth more than the tree.

Predictions about possible changes in the climate of the British Isles are now becoming less speculative; it is evident that there will be some warming up. Many trees will benefit from this, while a few others might suffer. It remains to be seen how many predating insects will benefit from a milder environment. Some bacterial disorders and fungal diseases will certainly thrive. New species that were once regarded as risky because they are tender can reasonably be considered to be hardy in some areas; and several of these are rapid-growing timber producers. The strengthening heat and intensity of summer sunshine is exacerbating drought conditions, and is also causing stem scorch and foliage damage on some trees. Air pollution is increasing in built-up areas, so pollution-tolerant species, of which there are many, will need to be increasingly used in urban localities. Many new community forests are being established in exactly these situations.

In Britain public opinion is at present very much in favour of trees, though selective about the tree types and where they grow. Large and veteran trees are loved the most. There are favourable reactions to most tree planting so long as it is not extensive, coniferous or of only one species. In general people like to see trees but not young trees in a tangle of weeds, in plastic tree shelters, or as a forest of wooden stakes. Thicket-

stage plantations generally only interest naturalists and bird watchers. Trees in rows still sound a discordant note, although many more people do now appreciate that there is no other way to plant small saplings if you need to find them again. Pollarding, to increase longevity and ensure safety, is beginning to be accepted; considerable efforts have recently gone into explaining this ancient management system to people. In general, people do not like to see trees cut down: a tree that blocks one person's window usually contributes to another's landscape. A tree today is very likely to be seen as a kind of ornament, and not in any way related to planks of wood, which, in some respects like meat, are regarded by many as a product of the superstore rather than of a tree.

There are two market benefits that will certainly come to investors in the future. Firstly, wood of every kind will soon only be available from sustainable planted stocks. There is at present a price premium if products are labelled as such. Secondly, it is safe to say that as the twenty-first century progresses, every scrap of wood that can be produced will be needed. It is vital that tree planters and tree managers retain the goodwill of the public. The impact of tree felling can be lessened if replacements can begin some time before the cutting does. The newly established trees can disguise the effects of felling the older ones. In doing so they provide continuous tree cover instead of unsightly bare ground and obtrusive visible stumps.

Diversity

Diversity of species is a kind of tree-planting insurance policy. If we plant every sort of tree that we know will grow in Britain, some of them are certain to be a great success and still here in 100 years' time. Greater site diversity is another factor that will increasingly

NEXT PAGE: *A redwood which tumbled to earth in 1936, at Leighton in Powys, has now thrown up a sturdy new row of trees from its long prone trunk. A sight that is something of a rarity in Britain, this is typical of the dense interiors of many North American forests, and typifies the kind of natural regeneration of trees that stretches back for millions of years.*

Alternative Species

The following is a list of some alternative species of trees and situations where they could be used to good effect:

Production of home-grown quality timber in lowland Britain:
- Cherry birch (*Betula lenta*)
- Box (*Buxus sempervirens*)
- Turkish hazel (*Corylus colurna*)
- Holly (*Ilex aquifolium*)
- Green ash (*Fraxinus pennsylvanica*)
- White ash (*Fraxinus americana*)

Quality timber combined with high amenity value:
- Tulip tree (*Liriodendron tulipifera*)
- Laburnum (*Laburnum* spp.)
- Empress tree (*Paulownia tomentosa*)
- Fruitwood – apple (*Malus*) and pear (*Pyrus communis*)

Tropical hardwood look-alike timber producers:
- Black locust (*Robinia pseudoacacia*)
- London plane (*Platanus* x *hispanica*)
- Black walnut (*Juglans nigra*)
- Yew (*Taxus baccata*)

Quality softwoods for moist western forest sites.
- Japanese cedar (*Cryptomeria japonica*)
- Coast redwood (*Sequoia sepervirens*)
- Bosnian pine (*Pinus peuce*)
- Low's fir (*Abies concolor var. lowiana*)
- Leyland cypress (X *Cupressocyparis leylandii*)

Community woodland enrichment species:
- Turners oak (*Quercus* x *turneri*)

Southern beech (*Nothofagus nervosa* and *N. obliqua*)
Tree of heaven (*Ailanthus altissima*)
Indian horse chestnut (*Aesculus indica*)

Decorative timber trees in hot dry conditions:
- Amur cork tree (*Phellodendron amurense*)
- Kentucky coffee tree (*Gymnocladus dioica*)
- Tree privet (*Ligustrum lucidum*)
- Smooth Arizona cypress (*Cupressus glabra*)
- Monterey pine (*Pinus radiata*)

Safe trees in very cold places:
- Swedish whitebeam (*Sorbus intermedia*)
- Noble fir (*Abies procera*)
- Sycamore (*Acer pseudoplatanus*)

Trees for wet places:
- River birch (*Betula nigra*)
- Pin oak (*Quercus palustris*)
- Hybrid American poplar (*Populus* x *generosa*)
- Bitter nut (*Carya cordiformis*)

Seaside trees:
- Evergreen oak (*Quercus ilex*)
- Bishop pine (*Pinus muricata*)
- Phillyrea (*Phillyrea latifolia*)
- Aspen (*Populus tremula*)

Particularly windfirm trees:
- Grey sallow (*Salix cinerea subsp. cinerea*)
- Mountain pine (*Pinus mugo*)
- Wellingtonia (*Sequoiadendron giganteum*)

Magnolia (*Magnolia* spp.)
Environmentally useful trees:
- Goat willow (*Salix caprea*)
- Field maple (*Acer campestre*)
- Common buckthorn (*Rhamnus cathartica*)
- Crab apple (*Malus sylvestris*)

Trees for built-up areas where space is limited:
- Yellow wood (*Cladrastis lutea*)
- Mulberry (*Morus nigra* and *M. alba*)
- Lenga (*Nothofagus pumilio*)

Pest and vandal-resistant trees:
- Hybrid wingnut (*Pterocarya* x *rehderana*)
- Shagbark hickory (*Carya ovata*)
- Silver maple (*Acer saccharinum*)
- Syrian juniper (*Juniperus drupacea*)

Unusually high-quality amenity and shade trees:
- Canary Island oak (*Quercus canariensis*)
- Hop hornbeam (*Ostrya carpinifolia*)
- Maximowicz birch (*Betula maximowicziana*)
- Japanese rowan (*Sorbus commixta*)
- Snow gum (*Eucalyptus pauciflora* subsp. *niphophila*)

Trees for ceremonial planting:
- Indian horse chestnut cv. (*Aesculus indica* 'Sydney Pearce')
- Italian alder (*Alnus cordata*)
- Columnar hornbeam (*Carpinus betulus* 'Fastigiata')
- Wild service tree (*Sorbus torminalis*)

Trees for Timber

Very little top-quality decorative timber has been deliberately grown in Britain since imported tropical hardwoods became so widely available. Even British oak, so vital to shipwrights until the late nineteenth century, has largely been superseded by foreign oak, teak and mahogany. Oak will only produce solid, crack-free timber on 'good oak ground' in Britain, and because it has so often been planted on marginal sites in the past, home-grown oak has acquired a poor reputation in the trade. Unfortunately, some growers appear to have convinced themselves that most other British hardwoods are not worth growing either. This is not true – ash, beech and cherry wood grown in Britain is equal to any produced elsewhere in the world. British-grown sycamore, when its arch enemy the grey squirrel is deterred, is also very fine. Cricket bat willow, of course, cannot be grown anywhere but in eastern England if 'genuine' cricket bats are required.

There are a number of other species, either not immediately thought of as timber producers, or hardly ever planted, which should have good production and commercial prospects in Britain. Cherry birch, for instance, claimed to be the best marketable birchwood in eastern North America, where it grows naturally, is at present limited to a scattering of specimen trees, and one tiny wood in Kent. A good looking tree, the

ABOVE: *Cherry birch in autumn.*

be available to us as farming practices change. It will enable new species to be tried out in a range of different conditions. In addition the opportunities to strengthen non-tree benefits such as recreation and environmental enhancement fit in very well alongside increased species diversity.

BELOW: *Cricket bat willows in a small Essex plantation.*

cherry birch has rough purplish-brown 'cherry like' bark and paper-thin fresh-green, sharply toothed and pointed leaves. These turn to rich golden yellow in autumn before falling. The twigs are peculiarly aromatic, smelling strongly of wintergreen ointment when scratched. It is a tough, windfirm 65-foot woodland tree that roots deeply for a birch, and seldom blows down. In America it tolerates winter cold to -40°F and endures hot, dry summers, so it thrives very well in the more moderate British climate. Above all, it prefers slightly acid, sandy soils. It has been growing well here in limited numbers since 1759 but has hardly ever been used commercially.

Box and holly are two small native trees in Britain which produce world-class high-density wood, but their value is now forgotten by everyone except for a few craft workers and antique-furniture specialists. These common enough trees and shrubs are familiar as ornamentals but not often as wood producers. Boxwood, including the root, was once used to produce drawing instruments, bobbins, printing blocks and musical-instrument parts. Holly is similar except that it is pure white to pale green and not straw yellow. It was formerly used as a substitute for box and, when dyed black, for ebony. In its own right it was used for high-quality joinery, turnery and white veneers. Box and holly are, of course, also useful and decorative as living plants. They can be clipped and trained into topiary, hedges and ornamental shelter. Box will thrive in the shade of many other large tree species. Holly is most at home under the dappled shade of oaks. They can both be coppiced and, once established, they will remain in production almost indefinitely. A box stem 10 inches in diameter may have taken 200 years to grow. Trees over 33 feet tall are rare. Holly grows faster and larger, reaching up to 65 feet with stems about 20 inches thick in about 100 years.

Another producer of fine timber is the exotic species Turkish hazel. Its wood has great strength and stability. It has occasionally been used in high-quality furniture and as a veneer, often cut from the root. Pinkish-brown in colour, it exhibits a pale, lustrous sheen when cut on a radial line through or near the centre of the tree. Albania and Georgia are its native strongholds, where pure natural forests extend nearly a mile up into the mountains. It grows very well in Britain, sometimes producing annual shoots up to a yard long. Trees over 80 feet tall with stems 27 inches thick are not unusual. It is seldom planted here, but it is grown commercially in central and southern mainland Europe. Except for seed provenances collected at low elevation this tree will withstand the coldest British winters. It also tolerates hot continental summers. Town conditions do not stunt its growth or spoil its shapely appearance – there are fine trees in London, Cardiff, Derby and elsewhere.

Finally, American ash should not be overlooked as a quality-timber tree. Several species including green ash and white ash will produce fine wood equal to the best European ash when grown in Britain. They are totally hardy, unless collected from southern seed origins, and grow quickly. The main advantage, however, is that unlike European ash the seed does not require special treatment to break dormancy. A good crop of seedlings can be relied upon in the first year after sowing. Trees in Britain 100 feet tall are known, and a stem nearly 31 inches thick has been measured in Scotland. Except for developing slightly rougher bark on the trunk, American ashes look very much like their European cousins.

Quality Timber Plus High Amenity

There are several trees that combine the potential to produce high-quality timber with good amenity. One of the best of these is the North American tulip tree. It is related to magnolia but will grow larger and faster. American tulip trees were introduced to Britain – Fulham Palace garden in London, to be precise – shortly before 1688. Specimens up to 120 feet tall are now known, and a huge trunk in Surrey measures 9½ feet across. The pale grey bark develops an intricate lattice pattern of vertical ridges. The leaves are unique in shape having three lobes with the top one truncated, as if it had been cut squarely across with scissors. They are pea-green in summer and dusty-gold in the autumn. Flowers like tulip heads occur on side branches in early summer. They are palest green with an irregular orange-yellow band on the outside. The seeds float on water so river valleys and flood plains are where tulip tree forests occur naturally in America, from Florida to North Carolina. The wood is slightly scented and creamy-buff with a silky sheen. It is durable, resistant to splitting, and relatively easy to work and polish.

Laburnum is a well-known garden tree, but is less familiar to most people as a wood producer. The heartwood is dark coffee-brown with a thin layer of bright yellow sapwood. In former times it replaced

BELOW: *Tulip tree with its distinctive flowers, leaves and bark.*

BOTTOM: *The small, delicate, and seldom appreciated, flowers of the box tree set off by its deep green glossy foliage.*

ebony as an inlay and was widely used as a veneer in its own right. It has distinct rays making it suitable for 'oyster' figure work used to decorate fine furniture. This involves cutting oblique cross-sections of small-diameter wood, often highlighting the contrast between the dark brown and pale yellow. Laburnum is poisonous, especially the unripe seeds, so developing infertile hybrids for more extensive planting would be an advantage. There are two main European species and a multitude of variable hybrids between them. They all develop and mature quickly, often producing flowers in around seven years, heartwood in 10–15 years, and they reach maturity (as wood producers) in 90 years. Hollow trunks well in excess of three feet across survive in Scotland and Wales.

The Empress tree from central China is usually grown as an ornament in Western countries, notably in Italy and France. It has glorious large, fragrant, mauve and cream clusters of flowers like great bunches of foxgloves. They occur before the huge leaves in spring, so unfortunately in Britain they are often spoiled by frost. In the Orient the wood, known in Japan as Kiri, is highly valued. It is very light and pale-coloured with lustrous radial surfaces. In musical-instrument manufacture it has unique properties that improve resonance. In Japan there is a long history of superior cabinet work and turnery using Kiri. In China it is the *t'ung* tree of the ancient classics. In modern China it is used in farm woodlands and along field edges, where the clammy leaves act like fly papers to control aphids and other insect pests of cereals.

Fruitwood is a resource that is now sadly neglected. Many fruit trees are grown on dwarfing rootstocks so new apple and pear trees as such are a rarity. There are tiny remnants of a native pear and crab-apple population left, and feral roadside apple trees are still common. However, they are often abused by hedge cutters, and stunted by traffic pollution. There are few other trees besides apple, pear, cherry and plums that provide superb blossom for us to delight in and bees to forage, fruit and drink to savour, and then beautiful wood to cherish when they have gone. Apple, pear, and even damson and plum woods are tough, durable and brightly coloured with streaks of pink, gold and warm brown. There are often traces of purplish-red and yellow reminiscent of a glorious winter sunset. Country furniture, cog-wheels, floor blocks and turned items were commonly made of fruitwood in former times, while the offcuts made superior firewood.

The Tropical Hardwood Look-alikes

Black locust, often called false acacia, is perhaps the closest thing we can grow in Britain to many of the imported tropical timbers. It is tough, hard, durable and straight-grained. The best material is used for joinery and cabinet-making. The finest wood-producing trees grow on the poorest of soils. Trees on good fertile sites often become brittle and break up prematurely. They also tend to rot in the centre of the trunk. Black locust was noted by Columbus in its native Appalachian Mountains soon after 1492, but it was not introduced to Europe until after 1600. William Cobbett admired it very much and sold over a million plants in the 1820s. It is still a common tree in British parks and gardens. The fragrant white summer pea flowers occur in hanging clusters with the delicate pinnate leaves. Each leaf consists of around 21 small, pointed, oval leaflets. New plants continually come up close to old ones because of the tendency for this species to produce numerous sucker shoots from its wide-spreading surface roots. Cut or storm-damaged individuals do this at a furious rate, often producing a dense thicket of spiny new stems in a very short time.

London plane, known in the timber trade as lacewood, is a very common large city tree. The wood cut through, or near, the centre exhibits an intricate pattern of golden-brown flecks and lacy rays on a paler background. It is usually seen as a highly polished veneer and is also used for marquetry. Although suitable for cabinet-making, joinery and light construction work, the wood is not durable, and the sapwood is attacked by the common furniture beetle. The origin of London plane remains in dispute. Some botanists regard it as a hybrid between the American and Asiatic planes. The earliest herbarium evidence is from a plant grown at Oxford in the late seventeenth century.

Juglan/10 Black walnut, from central North America, produces rich dark brown to purplish-black wood. The heartwood is completely durable and very tough. In America it is used for virtually every high, quality woodworking purpose imaginable – from gun stocks to clock cases. Curly-grained veneers are cut

OPPOSITE: Generally considered to be the oldest London planes in London, these mighty specimens towering high above Berkeley Square were planted in 1789, and have weathered many a gale, not to mention two World Wars, to continue providing shade and tranquillity for hot and bothered city folk.

ABOVE: Transverse section of a laburnum stem showing the dark heartwood.

LEFT: Impressive laburnum specimen at Bodnant Gardens.

RIGHT: *This black locust, or false acacia, is one of the oldest trees in Kew Gardens, being one of the few remaining trees from the original botanic garden founded in 1759 by Princess Augusta, the mother of King George III. William Cobbett was probably the greatest advocate of the locust tree, encouraging landowners to plant copious quantities and claiming that its commercial potential was far better than most of the native hardwoods. In this assumption he was slightly misguided, as although the wood was ideal for small components such as pins or trenails for shipwrights, the larger-scale sections of timber were never a match for oak. Truth to tell, Cobbett had a vested interest in promoting the tree as he owned many acres of nurseries full of locust saplings, and earned a substantial income from their sale.*

from burrs, branch crotches and roots. The living tree is stately and wide-spreading with slightly aromatic foliage and fruits. The roughly fissured trunk is reminiscent of an old oak tree. The long, pale green, pinnate leaves turn bright yellow in the autumn. It was introduced to Britain before 1656, and the best examples in Britain are mostly from New England. Unfortunately, this tree is fairly difficult to transplant, and mice, rabbits and hares love to chew the young bark of saplings.

The yew tree is also a quality timber producer, although the wood is temperamental to work with. It will shake and split if seasoned badly. This familiar, sombre, native evergreen tree is dark, mysterious and poisonous. The timber, by contrast, is brightly coloured. The heart of the tree is coppery-gold, with black, red and purple streaks, while the narrow sapwood is pale cream. The wood is very hard and oily and consequently it takes an exceptionally fine polish. Clusters of tiny 'birds eye' pin knots add interest. Both straight- and curly-grained wood occurs, and often cavities and pockets of enclosed bark are found inside the stem. Veneers were traditionally stored in water to produce a deeper purple colour.

Quality Softwoods

In order to meet a constant demand for wood and paper, it is essential to continue to plant sustainable forests of conifers in Britain. There are several very diverse species, such as Japanese cedar, Coast redwood, Low's fir, Island cypress and Bosnian pine, that are seldom used on a large scale at the moment. They not only produce high-quality wood but also show considerable visual diversity. Furthermore, many of them coexist nicely, so monocultures might easily be avoided if their use is combined with careful planning and sensitive landscaping.

In sheltered parts of upland western Britain, Japanese cedar seems to be an obvious alternative choice. It has tough, spiky evergreen foliage and coppery-red bark. Forests resemble redwood groves but the bark is stringy and hard, not soft and spongy. Straight trees up to 148 feet tall are known in southern Japan. The species was introduced to Britain from China in 1842, from Java in 1853, and then from Japan in 1861 and 1879. However, they all appear to have come originally from Japan, where the natural range is restricted to the Yukushima Island region. Under

RIGHT: *This is one of the greatest London planes in Berkeley Square. Can a slightly haughty demeanour be detected?*

RIGHT: *Evening sunlight on this ancient yew tree hints at the warm colours in the wood.*

the commercial name '*sugi*', this is one of the most valuable softwood timbers in the world. It is fine-grained, smooth, scented and incredibly light when dry. Its colour ranges from pale brown to pinkish-buff. Sugi has great strength and is suitable for all kinds of construction work; in Japan it is even used for boat building and cooperage. British-grown timber is of equally good quality and it grows well here.

A tall straight tree, Coast redwood grows in California to over 320 feet and is the tallest tree in the world. It also thrives in Britain on sites where Douglas fir is usually planted now. Fossil ancestors of redwoods from the Cretaceous period occur in Europe, so when it was introduced in 1843 via Russia, it was simply being returned to its old Pre-glacial haunts. Existing British plantations live up to their media image in America, promoted so enthusiastically by the tourist authorities. The timber is red-brown and durable. There are a wide range of uses for it. As the trees are shade-tolerant, forests will stand at close spacing and are easy to manage. Cut stumps sprout freely to produce spontaneous evergreen cover in which wildlife thrives, and from which a new crop of trees will eventually be produced.

The Bosnian pine is a relative newcomer to British forestry. So far, only a limited number of experimental forest plots and a few nineteenth-century arboretum specimens, including original 1863 trees, exist. This species is particularly interesting because it thrives on damp acid sites that are generally reserved in Britain for spruces and not pines. Visually it is a fresher green than most of the more familiar present-day upland conifers. The stems are straight and the timber is strong but resinous. A remarkable tree at Stourhead in Wiltshire reached a height of 111 feet and produced a huge stem 52 inches thick in about 120 years.

Low's fir is an alternative species which is superficially similar to its many silver fir relatives. It is a variety of the tough Colorado white fir growing in southern Oregon where it intergrades with stands of grand fir. As a potential forest tree in Britain it does have some advantages. It appears to be less prone to stem damage caused by drought or frost, which in grand fir can send gaping cracks spiralling vertically up the whole length of the tree. Low's fir will also tolerate drier sites than grand fir. Productivity is less, but wood quality is about the same or better. Although visually this variety is just another fir, it does come in smooth-barked and rugged-barked forms. There are already small plantations throughout Britain, and new provenance trials have been established to test this variety since the 1960s.

Leyland cypress, that ubiquitous and self-willed suburban garden hedge tree, is actually a producer of good durable timber similar to the western red cedar of garden shed fame. Leyland cypress will grow enormously fast and will rapidly produce a thick trunk. The first specimens, hybrids between North American Monterey cypress and Nootka cypress, were produced unintentionally in Wales in 1888 and 1911. Strangely, the cross has never occurred spontaneously in America where the wild populations of the parents coincide. As a plantation Leyland cypress produces an environmental desert. There is no undergrowth, no light and virtually no wildlife. As fertile seed is not available, plantations are always clonal. Every tree is likely to be identical, and as they tolerate deep shade, they grow closely packed together. Not a pleasing environmental prospect, but almost certainly a commercially viable one.

Alternative Community

The community woodland is a relatively new concept that brings trees into close proximity to towns and cities. To survive, the tree species used must be tough, quick to establish and attractive to look at, but not so attractive that the trees will be stolen. Very often, rapid growth and quick screening are required, especially at eye level. At the end of the tree rotation, some wood product is likely to be an advantage, to demonstrate purpose for the woodland and help to finance the continuation of the project.

There are many tree contenders for a place in community woodlands, of which a newly rediscovered oak is of particular interest. Turner's oak was originally produced artificially in Essex before 1785 but was

BELOW: *Bosnian pine in an exposed mountain top location.*

hardly ever used. A hybrid between evergreen oak and common oak, it looks like a small compact traditional oak tree with acorns and lobed leaves, but in winter the leaves mostly remain green and stay on the tree. Turner's oak is an ideal semi-evergreen camouflage tree which thrives on lime-rich soils and in polluted air. To be true to type it has to be grown from cuttings, which is not easy; however, large numbers have been produced recently from long-established parent trees in London and Edinburgh.

For a relatively quick deciduous woodland effect, southern beech is worth considering in mild areas. There are two potential species that are suitable for this, *Nothofagus nervosa* (previously called *procera*), which is fast and straight but is slightly tender, and *Nothofagus obligua*, which is less straight but more robust. Provenance trials are showing significant seed origin differences when these South American trees are grown in Britain. They will produce a shady woodland big enough to walk in after only 15–20 years. They have fresh green leaves in early spring and good yellow and orange autumn foliage colour. Oak and other native species will survive quite well amongst them and eventually take prominence if required.

Even faster-growing than southern beech in sheltered fertile areas is the tree of heaven, imported from China in 1751 to the Chelsea Physic Garden in London. This giant of a tree tolerates air pollution and is virtually free of any damaging pests and diseases. The tall stem is sinuous and silvery-grey. It is called tree of heaven because of the speed with which it grows towards the sky. The 16-inch-long leaves are pinnate with around 30 leaflets. Some of these have basal glands which give off a musty, slightly spicy, smell on hot summer days. This is a tree that is hard to confine, and harder still to get rid of, so care is needed in choosing a site for it. The whole widely spreading root system can produce sucker shoots at any time. They usually appear singly but over a period of years they can produce a vast forest around the place where only one tree was originally planted.

There is a choice of several flowering trees that might grace our community forests. One that stands out is Indian horse chestnut, from the Himalayas. In many respects it is like common horse chestnut, but it flowers later, when few other trees are out, and does not produce conkers until early winter. It is more resistant to spring frost damage than common horse chestnut. Drought and air pollution seldom cause it any difficulties. Although not widely planted until 1900, it appears to thrive on a wide range of soil types, and is possibly at its best when planted in mixture with several different species of trees.

Decorative Timber Trees for Hot and Dry Conditions

Trees that can withstand hot, dry conditions are becoming increasingly sought after as the frequency of hot, dry summers in southern Britain increases. Within this category are species which are primarily grown for amenity but also produce interesting timber, and other species which are grown especially for their timber.

The amur cork tree from China and Manchuria is highly decorative with aromatic deciduous foliage which in autumn turns pale yellow and contrasts well with the clusters of small black fruits. The leaves are pinnate, each with five to 13 leaflets, and the fruit, a ½-inch black berry, smells of turpentine. The most distinctive feature of the amur cork tree, however, is its softly fissured, pale yellowish-grey corky bark. The best specimens in Britain are around 50 feet tall, mostly with a neat, rounded head of branches.

The Kentucky coffee tree from North America, now naturalised and cultivated over much of the eastern United States, is another good amenity tree. Native Americans used to roast the seeds for a beverage. The tree grows rapidly on moist ground in sheltered places to around 50 feet tall. It produces huge, bi-pinnate pea-green leaves, up to three feet long, which are quite remarkable. Up to 150 individual

OPPOSITE: Turner's oak – whole tree with green leaves in wintry view.

LEFT: Amur Cork tree with its distinctive bark.

LEFT: Kentucky coffee tree – a popular amenity tree in parts of South Wales.

ABOVE: *Tree privet in flower – Battersea Park, London.*

ABOVE: *Downy underside of a Swedish whitebeam leaf.*

ABOVE: *The deeply lobed leaves of a pin oak.*

ABOVE: *River birch.*

OPPOSITE: *Massive Monterey pine on the Devon coast.*

leaflets have been counted on a single leaf. Flowers, when they occur, are nectar-rich and good for bees.

A close relative of the olive, tree privet, has deep evergreen glossy leaves reminiscent of a rain forest tree. Casting deep shade but individually sparkling and wet-looking in the sun, tree privet grows to around 52 feet tall. Town parks suit it well: Battersea Park in London boasts several, one of which has a massive stem over three feet thick. Panicles of ivory-white, scented flowers stand up above the rounded outline of the branches in summer. These are succeeded by black berries but they are soon taken by birds. Sir Joseph Banks introduced the first specimen to Britain in 1794, from seed collected during Captain Cook's expedition in the preceding years to the Far East. The pale yellow-brown wood is hard and heavy, like the wood of the olive tree.

Two representatives of a small number of potential forest trees for hot dry conditions are recommended here. Smooth Arizona cypress is a desert tree with wispy sage-grey foliage. It will grow to more than 65 feet tall in Britain, which is larger than in its native Arizona. The stems are straight and heavy with thin, scaly, greenish-brown, red-brown, and purplish tinted bark. It survives burning heat with almost no summer rainfall and withstands winter cold to at least 5°F. Exactly when it first arrived in Britain is uncertain. Some authorities suggest it could have been as recently as 1900, but there has always been some confusion about the true identity of the tree here.

Monterey pine has exercised the minds of forest researchers in Britain for many years. For geographical reasons this species was unable to migrate far enough northwards in California following the last ice age to find ideal climatic conditions in which to grow, and it is confined to a few hot, dry areas along the Monterey Pacific coast in the sea-fog belt. It was discovered there languishing in the summer heat in 1831. The first British trees arrived at Chiswick in 1833, but Britain was generally too dull and cold for them. However, it subsequently found its way to various warmer countries, where it has been enormously successful and important. In Chile and New Zealand, for instance, it has become a major component of the productive forest estate. It has been subjected to rigorous provenance testing and selection in the British Isles, where second generation, locally grown stock appears to survive best. In southern England and South Wales reasonably close to the sea it will produce good pine timber faster than any other species. It also appears to grow well in cooler western coastal areas of Wales and Scotland.

Trees for Very Cold Places

There are many alpine species that withstand the cold, but in Britain the coldest places sometimes experience freak periods of hot summer weather too. Selecting trees for cold places would be simple if during their long lives, often in excess of 80 years, hot periods were unlikely to occur and kill them: only a few trees can withstand both sustained cold and occasional warmth. These very special species provide essential tree cover in north Britain. The first one is surprising for those of us who know it as a familiar broadleaved street tree in southern towns, but its name, Swedish whitebeam, holds a clue to its cold origins. This tough little tree also occurs naturally in Norway, Denmark and along the Baltic coast. It withstands severe exposure, salt-laden wind and atmospheric pollution. Its seed produces copiously and germinates true to type. Although generally small – less than 39 feet tall – it is indispensable as an amenity fringe, a windbreak and a source of food for birds and other wildlife.

The best 'all weather' conifer for really cold, exposed places in Britain is perhaps the North American noble fir. The health and vitality of noble fir progressively improves towards the wetter west and cooler north of Britain; many west Scottish trees exceed 40 feet in height. In the 1950s a single experimental row of noble fir was planted at 6-foot spacing straight up a bare mountainside in North Wales to an elevation of 1,800 feet. It survived for years almost without any losses. Although progressively shorter, the topmost trees were not deformed or deflected from the vertical by the wind. In its native Cascade Mountain forests in western North America it grows at an elevation of up to 4,900 feet.

The sycamore fills another niche in this cold-tolerance category to which it is well suited. How sycamore with its big leaves and thin bark has managed to contend with northeasterly gales and exposure to the North Sea is a mystery, but it has done so, often in total isolation and on poor moorland soils. Its natural origin, between the Pyrenees, the Alps and the Carpathian Mountains, contains a wide range of climatic conditions, and native trees may experience very hot and also viciously cold conditions there. Sycamore has few physical limits in Britain, but unfortunately it is no match for the American grey squirrel. This alien rodent has succeeded in almost destroying this alien tree in recent years, to a greater extent than the North Sea and the north wind ever managed to do since its introduction in the thirteenth century.

Trees for Wet Places

Many tree planting sites are inherently wet even when they don't appear to be. Soil covered by paving, for example, retains a lot of moisture because it is partially insulated from the dry atmosphere. Many urban sites are also heavily compacted and have impaired vertical drainage. Trees that grow naturally on flood plains are perfectly adapted to growing where there is a surplus of ground water, and some of them take very well to being planted in such artificially wet places. River birch, introduced from the eastern United States in 1726, is one. It looks attractive with its persistent frilly peeling bark and

TOP: *Bishop pine in the Pinetum at Kew Gardens.*

ABOVE: *Bitternut from the American continent displaying its distinctive leaf.*

small diamond-shaped leaves. It withstands greater summer heat than is usually experienced in Britain, but more importantly, it will grow in soggy, sporadically flooded ground. It will coppice freely and its multiple stem regrowth is most attractive.

Also from the same part of America is the pin oak, introduced to England in 1880. An elegant, fairly large tree with smooth, grey bark, it has typical American oak leaves with deeply cut, angular, pointed lobes. Much of Britain is scarcely warm enough for it to produce acorns, and the tree thrives best only in the hottest parts of the country. Pin oak is very easy to establish, even when surprisingly large trees are transplanted. Its tolerance of wet soil and hot town conditions is well known, and it is already used to great advantage by planners in shopping precincts and paved open spaces.

Another wetland tree is the very much larger hybrid American poplar. It is not actually American at all, but it does have American parents. The best clones have recently been bred in Belgium. The parents are the eastern and the western cottonwoods: as their respective natural ranges do not overlap, the hybrid cannot occur naturally. The young tree is easily the fastest growing poplar in Europe, and the newest clones are also apparently disease-free. The first cross was made in 1912 and tree breeding improvements, using various selected already cultivated parents, have been made ever since. In the 1960s a cultivar called 'Oxford' produced veneer logs at the incredibly young age of 11 years in southwest England. Several clones reached heights in excess of 100 feet in 16 years. Most of these original trees, however, eventually succumbed to bacterial canker. Modern clones such as 'Beaupre' and 'Boelare' have so far avoided this debilitating disease. Fertile damp lowland valley bottoms and alluvial flood plains are ideal planting sites for this hybrid, which requires a large amount of space and good light. Trees should not be positioned closer than 26 feet apart,

and even at that spacing the canopy will soon fill in.

Yet another American tree, the bitternut, also withstands a great deal of water around its roots. As a wild tree it ranges from northern Florida to eastern Canada. It is a hickory and therefore produces remarkable timber of great strength and elasticity. The nuts are not edible, but they are seldom produced in Britain anyway. Although introduced in 1766, in Britain it has never been widely planted here because it is sensitive to root damage during transplanting. To be successful it has to be carefully placed in the ground in late spring, preferably without any root disturbance at all. Established trees can be coppiced, thus providing an endless sustainable supply of wood. Trials in America suggest the timber from coppice stems is of even better quality than from maiden trees. In Britain a specimen on the Thames gravels at Kew has already grown to a massive 92 feet tall, with a stem 25 inches thick.

Seaside Trees

Facing the sea on a winter's day is very different at Bridlington than at Bournemouth. The four species selected here will all tolerate salt-laden wind, but not all of them will survive Arctic gales. In most circumstances seaside trees are supported by wind-resistant shrubs, at least on the littoral edge. In the south this might be tamarisk, *Phillyrea latifolia* or kangaroo apple, while in the colder north it might be blackthorn and shrubby willow. Areas of shifting sand dunes pose additional problems for trees and tree planters, and in this situation grasses usually play an important intermediate role in stabilising sand.

A good all-year-round seaside windbreak in reasonably mild districts can be created using evergreen oak, originally from southern Europe. Although not fast-growing, evergreen oaks are very salt-tolerant and strong. The microclimate under and behind their crooked stems and tangled branches is often attractive to other shrubby species, and trees such as grey poplar and maritime pine are likely to thrive there. Together these trees provide a good range of contrasting colours and textures.

Bishop pine from California is another species resistant to salt spray and wind. It will grow in dune slacks and out to the back of sandy beaches. Trees closest to the sea may be no more than stemless mushrooms of spreading branches, but they do provide shelter and a foothold for other plants – other bishop pines even. By about the tenth plant in from the seaward edge of a bishop pine plantation, trees should have recognisable trunks and slightly lighter horizontal branches. There are blue and green foliage forms. The most northerly blue form is certainly the best for British conditions. Of particular interest are the clusters of spiny 3½ woody cones, which are held on the branches for many years.

Phillyrea latifolia is usually regarded as a shrub, but seaside trees of it up to 23 feet tall are known from north Kent to southwest England. This plant looks like a small, shiny-leaved version of evergreen oak. It is especially useful at the back of beaches where some shelter is required but a view of the sea over the treetops is desired.

Finally there is aspen, which is considerably hardier than the others, but because it is slender and deciduous a lot more are needed to create year-round shelter. It will grow on sand-dune slacks, along estuaries and on rocky coasts. In a constant sea wind the stems become very pale grey in winter resembling dry bones. Aspen grows freely from suckers, and given a space it will soon fill it: a single tree on the Severn flood plain near Chepstow was reckoned in 1990 to have 1,000 individual stems over four inches thick. This British native tree is environmentally very useful. Its foliage and dead wood provide a food source for numerous insects. Like other trees in the poplar family, it responds to abuse by redoubling its growth rate. Storm-damaged aspens usually grow more vigorously than trees that escape the tempest.

Windfirm Trees

It is a sad fact that trees sometimes blow down. Roots often have a surprisingly limited outward extension, and they will only go down as far as the first impermeable layer in the soil: this may be bedrock, the permanent water-table or a layer of anaerobic soil. A large proportion of trees in Britain have roots no more than a few feet deep. If you consider the height and weight of a tree above ground as a ratio of its root depth and bulk, it is quite amazing how trees ever stand up at all, especially when the wind persistently rocks them and tugs at the branches. Many trees that blow down show symptoms of root reduction through disease, which aggravates an already poor root-to-shoot ratio. Sudden changes in physical conditions also exacerbate the risks of windthrow. The most obvious are when roots are cut through or excavated, or when a specimen is suddenly isolated from the support it has grown up with.

Two small trees that are outstandingly windfirm are grey sallow and mountain pine. The more remarkable is the sallow, which is native to eastern Britain, especially exposed fenlands, and to mainland Europe. It produces a massive bulk of fibrous roots very quickly. In undisturbed conditions they would be instrumental in the gradual conversion of boggy fens into carr, an intermediate environment in which scrub and sedges eventually build up the level of the ground to create dry woodland. Other willows and alder play a part in this process, but none of them has a root system comparable in size to grey sallow. Although the tree itself is tough and fairly short, seldom over 33 feet tall, with flexible shoots, it is the strength of grip the roots have on the ground that holds it in place come what may. The multi-stemmed European mountain pine has a different strategy. It is an upland

and sub-Alpine tree-line species with a spreading but unremarkable root; but it produces only a small amount of new vertical top growth each year. The main stems and branches are sinuous, and the short, needle-clad shoots are pliable enough to bend away from the wind and deflect most of it up and over the tree. In this way mountain pine clings on to its precarious elevated position in the most severely exposed places where few other trees could survive.

Two other apparently windfirm trees in Britain have come to light since the great storms of the 1980s and 1990s in southern England. It is hard to explain exactly why Wellingtonia and magnolia stood up in areas where virtually every other exotic tree was demolished. Wellingtonia has flexible foliage which might absorb the energy of strong gusts of wind, but the tree as a whole is often so vast that its root-to-shoot ratio is usually very poor. In one instance trees were uprooted that only had 2-feet-deep roots supporting 115 feet of top growth. Magnolia has more credible, strong, spreading roots, though most have big, heavy branches and most large, wind-resisting leaves. Cultivated plants near houses in the hurricane belt of south-eastern North America help protect them from the blast. Branches are even pinned down to root round the main stems, which further improves their stability.

BELOW: *The downy seeds of grey sallow are cast to the fenland breeze.*

BOTTOM: *A huge magnolia in Kew Gardens shows its amazing resilience – it was one of the few massive specimens to remain totally unmoved by the severe 1990 storms.*

Environmentally Useful Trees

All trees are environmentally useful in some respects, and a few species are worth encouraging primarily for this reason. Perhaps the most environmentally important tree in the British native flora is goat willow. As a food source alone it is almost equal to common oak in Britain. Pure goat willow is actually something of a rarity. The sallow willow group, of which there are at least nine, interbreed with complete freedom, so most specimens of what appears to be goat willow are actually of hybrid origin. Cultivated goat willows are seldom true to type either, because authentic plants do not grow as easily from winter cuttings as the hybrid imposters do, so raising plants is more difficult. However, hybrid or true, it makes little difference environmentally, because all sallows are extremely valuable to wildlife.

The willows and sallows collectively are hosts to more kinds of wildlife than any other group of British trees. Another British native species, the field maple, also supports a large number of insect species. It provides a tangle of twiggy cover for birds and lots of rich decomposing leaf litter on the ground which benefits numerous other inhabitants of the countryside. From a human point of view, this tree also looks good in the landscape. It is never excessively large or obtrusive and it colours well in the autumn. Planting field maple in Britain is ecologically acceptable because there are no apparent genetically distinctive regional variations. It can be planted on a wide scale anywhere with no risk of causing the genetic pollution of an individual wild population.

Two other native trees, buckthorn and crab apple, are worth encouraging purely for the benefit of the wildlife they support. They should not be widely planted though, except from authentic local seed stocks in most circumstances. The true crab apple is a very special tree. Although apple trees in the countryside, either cultivated or escaped, are useful for insects – bees in particular – the authentic wild species is considerably better than the rest. It is part of an ancient woodland environment, which is almost lost and deserves special protection in its own right, as well as for all the organisms that depend on it. Common buckthorn is not a rare plant but it probably still has an unspoilt regional distribution which is worth preserving. Several insects are specific to it, including the well-known brimstone butterfly.

Trees for built-up areas where space is limited

The ideal town tree has to be small with roots that are unlikely to cause damage. A small tree, however, needs to be identifiable as a tree and not confused with a shrub. It needs to grow reasonably rapidly to a predictable and manageable size, and then not exceed it by much more for a very long time. Although old town trees may develop widely spreading roots, this depends very much more on the soil type and the location of moisture than on the actual species of tree. Slow-growing elderly trees seem to have less aggressive roots than fast-growing pioneer species. A rare and beautiful species that meets all the requirements of an urban tree is yellow wood from the central United States. It is cultivated widely there in gardens, and although introduced to Britain in 1812, it has never been used much. It is hardy but thrives best on sheltered lowland sites, especially in town parks in the south. White flowers appear in June, but this tree is primarily grown for its foliage. The pinnate leaves are rich shady green in summer and butter yellow in the autumn. The wood, if it is ever exposed, is bright yellow.

The mulberries are quite familiar as town trees, especially the deliciously fruiting black mulberry. It is never very large – specimens over 26 feet tall are rare – but it does tend to recline on its side in old age. This should not present a safety or amenity problem. The tree withstands severe lopping so if it leans it can be drastically cut back to a new shape. With careful cutting, a reclining specimen can be made to fit exactly into the space it took up previously when it was standing up straight. It is usually possible to utilise some stout lower branches as living props. With this support, ancient trees usually remain healthy and stable, and often become an uncommonly picturesque feature. The species arrived in Britain from western Asia in the sixteenth century, and trees over 200 years old are quite common. The white mulberry, which produces leaves for silkworms but only inferior pale-coloured fruit, is a rugged little tree. It arrived in Britain from China in 1596. One possible advantage of white mulberry over the black in a garden is that it does not make such a mess when the over-ripe, uneaten fruit falls to the ground. However, many people would consider this a high price to pay for a clean path or lawn. Both mulberries have rough nettle-like leaves

BELOW: *Goat willow male catkins.*

BOTTOM: *Fruit, or keys, of the field maple.*

ABOVE: *Hybrid wingnut growing at Westonbirt arboretum. Highly prone to reproducing by suckers as shown in this picture.*

BELOW: *The great range of colours of silver maple leaves in autumn.*

del Fuego are used. Young tree trunks are reminiscent of an alder or a dark-stemmed birch, and the young winter twigs are very birch-like, with prominent buff lenticels and reddish buds. It is completely hardy in Britain, even in elevated areas, but southern provenances may suffer from excessive summer heat. Although apparently not introduced until the 1950s, it has already been widely grown in botanic gardens and in experimental forest plots. It clearly has unrealised potential as a tough small-leaved hardy amenity species.

Pest and Vandal-resistant Trees

Trees have various defences against organisms that set out to attack them. Most rely on their ability to regenerate if the damage is purely mechanical, but vascular damage, internal poisoning or infection is less effectively resisted. A few trees have ingenious methods of preventing serious damage from occurring in the first place. The hybrid wingnut, for instance, produces a whole forest of shoots, and eventually new trunks, from root suckers. This is a particularly good defence against casual browsing or mindless vandalism, where the antagonists are not usually systematic enough to destroy the whole plant and soon move on to something else. The black locust, described already for its timber qualities, adopts the same strategy and in addition arms its young growths with particularly vicious spines.

American grey squirrels are at present one of the most seriously damaging pests that trees have to contend with in Britain, and two trees show interesting mechanisms for surviving their attacks. The shagbark hickory develops rigid peeling bark that becomes so rough and projects so far outwards that, it is said, squirrels will not climb it. It is not clear how the trees survive long enough to produce the rough bark. The silver maple also somehow resists squirrel damage to a greater extent than the very susceptible red maple, sugar maple, Norway maple and even beech. Exactly how or why this should be is not known. Adjacent forest plots of these species consistently show silver maple as the least palatable to squirrels, though the species is not entirely immune to damage.

Syrian juniper is possibly the most defensive tree that it is possible to grow in Britain. In its native western Asia, it has learned to be goat-proof. The foliage consists entirely of needle-sharp spines, which point in every possible direction on a multitude of stiff vertical stems, and will draw blood if touched with an unprotected hand. Their sharpness is far greater than holly prickles, and they are considerably more numerous than the sharply pointed scales of the monkey puzzle. Even young plants, like monkey puzzle trees, are needle-sharp. However, suppliers are few and far between, planting is a difficult and potentially painful business and taking cuttings is almost out of the question.

and warm pinkish-brown bark. They provide good shade from spring until late summer, and a focal point all year round.

Lenga from southern Argentina and Chile is not at all well known in Britain, and it is not yet widely used as a town tree here. It has great potential, though, and meets all of the standard requirements of an urban tree. The deciduous leaves are small – only 3/4 – 1 1/4 inches long, and round to broadly oval. Their deep green colour and lustrous appearance give the impression of an evergreen, which this tree is not. Lenga is small, or very small if provenances from Tierra

Unusual Amenity and Shade Trees

These trees have been chosen purely for their exceptional merit as living plants. The value of their timber is either negligible or not known in Britain. They are simply some of the most decorative and treasured examples of amenity trees that are available to us. Each one has special characteristics that set it apart from even its close relatives. The Canary Island or Algerian oak is a typical and prodigious oak tree, often 80 feet tall and wide, with a rough-barked mossy trunk over 3 feet thick. It has unmistakable oak leaves, except that they are a bit larger than common oak. They are bright, fresh green and not at all like most sombre, dark evergreen leaves. However, in the autumn they do not fall, and in mid-winter this tree is still as fresh and verdant as it was in the summer – a rare delight. The old leaves do eventually fall as the twigs thicken in spring. At the same time a new flush of foliage emerges to cover their departure. Although Canary Island oak was introduced in 1845, it is seldom seen in cultivation. For a long time the delights of this tree appear to have been available only to the landed gentry. Most of the prime specimens are still to be found on large private estates and usually close to noble houses.

The hop hornbeam is an unusual tree, often spreading wider than it is tall in old age. A member of the birch family, it is deciduous and has rough scaly bark. The early spring clusters of catkins are a distinctive feature, but less so than the fruits. These are reminiscent of hops, numerous hanging 2-inch clusters of small nutlets each individually enclosed in a tiny pale greenish-brown papery bag. The leaves are deep green in summer and yellow in the autumn, when they are complimented by the distinctive bunches of 'hops'.

Birch trees are nearly always picturesque. One of the rarest, Maximowicz birch, is perhaps the most interesting of them all, not for dazzling whiteness of stem, but because it gives the distinct impression that it can grow faster than any other birch tree. It is said to be the largest of all the Japanese birches, but in Britain it has not yet exceeded 80 feet in height. The ample, evenly rounded stem is usually the most striking feature. Often in excess of 20 inches thick, it is white with grey bands of only slightly peeling bark that seem to accentuate its relentlessly increasing girth. Shoot growth of a few feet a year regularly occurs for a time on favourable moist sites. The leaves, which are up to 6 inches long, on glossy brown twigs, turn gold in the autumn. The late winter to early spring catkins, in small pendulous bunches, may be up to 5 inches long. Like most birches, this species is prone to hybridisation. To ensure that pure-bred trees, and not hybrids, are obtained, only imported wild-source Japanese seed should be sown.

Arguably one of the finest autumn-colour feature trees in British collections is the Japanese rowan. It is small, seldom ever reaching 45 feet in height, but this does not detract from its great beauty. In late autumn the delicate pinnate leaves turn deep claret-red with flecks of orange and a haze of purple as the late autumn season progresses. Added to this brilliant collage are numerous bunches of small, bright orange-red berries. The most effective trees are ones with branches almost to the ground, or previously coppiced

TOP: *The monarch birch is one of the most splendid of the tribe, and boasts the largest leaves of all, having been introduced from Japan in the late nineteenth century.*

ABOVE: *A large spreading hop hornbeam tree.*

RIGHT: *One of the most distinctive pest- and vandal-resisitant trees is the Shagbark Hickory.*

FAR RIGHT: *Snow gum with its distinctive striated bark.*

OPPOSITE: *An unusual occurrence of a wild service tree in a hedgerow, since it is more commonly found in woodland. It may be that this hedge is the old boundary of a defunct wood.*

plants with three to five individual stems and a thick mantle of foliage. There are named cultivars such as 'Embley' and 'Jermans' that have been selected for their outstanding autumn tints.

In total contrast is the snow gum from Tasmania and the highlands of southeast Australia. The subspecies *Eucalyptus niphophila* commonly cultivated in Britain has unreal flecks of pure white and mussel-shell blue bark, with patches of pale brown and pink. The young shoots are bloomed white. It makes a sinuous and twisted, seldom straight, little specimen. Often long weeping shoots develop and the whole tree ends up reclining almost on its side. The narrow, grey-green, aromatic, waxy leaves are up to 4 inches long. They hang on the twigs and very sensibly turn edgeways to the sun on hot days in order to stay cool. The creamy-white flowers, occurring at unpredictable times of year in the northern hemisphere, are very decorative

RIGHT: *European Mountain pine in precarious situation*

and provide nectar for bees. This is one of the few eucalyptus species that develop a good fibrous root system. It can therefore be transplanted reasonably safely as a large specimen. Coppiced plants thinned out to three to five stems give the most stunning display of bark colour.

Ceremonial and Commemorative Species

Four very safe and undemanding alternative trees are suggested here to meet various different requirements of flowers, shape, colour, size and place of origin. The Indian horse chestnut cultivar 'Sydney Pearce' is an outstanding average-sized symmetrical garden tree that produces a spectacular annual display of flowers in early summer. A faster-growing, larger but less wide-crowned tree with glossy leaves, but insignificant flowers, is Italian alder from southern Europe. It has a dark-grey stem which can exceed 12 inches in diameter in only 15 years on a favourable site. It is an altogether less flamboyant tree than 'Sydney Pearce', but very reliable even on dryish ground, and produces a substantial tree within a single human generation. Narrower still, but very strong and storm-resistant, is fastigiate hornbeam. This is mostly grown for its elegant columnar shape and upswept branch pattern. It is a foliage rather than a flowering plant. Finally, a standard tree that has historical, environmental and amenity interest is the wild service tree. This is a British native species with maple-like deciduous leaves and good golden-brown autumn colour. Specimens are usually shapely and neat at first but tend to have a more natural tousled look after 30 years or so. They live for a very long time, and produce small, apple-like fruits, which were eaten in former times and were cultivated for that purpose.

TREE MEASURING

In Britain there has been a long tradition and fascination for measuring things and collecting numbers. Trees are an obvious target for ardent measurers: they just stand still and quietly expand. Accurate measuring of them, and importantly, re-measuring, is possible because, unlike people or animals, trees grow consistently larger for as long as they live. They do not change their outward appearance much with adulthood, as a child or a puppy does, they simply add a new layer of wood over what is already there. Size is never reduced except through damage caused by some outside agency. If wounded, trees do not heal up as an animal might – they simply 'wall off' damaged tissue and grow on round it. Thus, every subsequent girth measurement of the stem of a particular tree will be greater than the last, if no structural damage has occurred, the tree is still alive and the measurement is in the same place on the stem. A national register of notable trees has been compiled in the British Isles which consists of a monumental list of about 120,000 measurements. This has resulted from the painstaking work of various dendrologists and tree enthusiasts over a period of almost 100 years. It is a unique and priceless record of heights to the topmost branch and stem diameters at five feet from the ground. In addition, the location of each tree and the dates of measurement have been included. If the tree has been measured more than once, its past dimensions and dates are listed too, often with the initials of the measurer who did it. For people who need to know more about the potential of trees on a particular kind of site, all of this information is extremely valuable, especially if the actual planting date of the recorded specimen is known.

The History of Tree Measuring

In the early twentieth century the renowned dendrologists and landowners Henry John Elwes and Augustine Henry set off on an odyssey to find notable trees throughout the British Isles. They travelled thousands of miles, and, it is said, wore out six motor cars by the time they had finished. They visited at least 600 estates and corresponded with numerous landowners and foresters about the origins and planting dates of over 500 species and varieties of tree. Wherever they found a tree of a size that they deemed to justify the attention, they recorded its measurements; they also often took high-quality photographs, so that specimens could be recognised again with certainty

years later. Elwes and Henry published this work privately between 1906 and 1913 in the epic seven volume *Trees of Great Britain and Ireland*. Their record marked the beginning of systematic tree measuring in the British Isles and is arguably the cornerstone of the modern tree register. Authors such as Evelyn and Loudon had published tree measurements many years earlier, but they were less comprehensive than those of Elwes and Henry, and many of the measured trees have been lost, died or confused with others. A few of the trees painstakingly illustrated and recorded by Jacob Strutt in the nineteenth century can also still be identified today.

Tree measuring became a popular pastime in the eighteenth century when newly introduced exotic trees from around the world became established in Britain. This was especially so with landowners, who were usually motivated by fierce competition between neighbouring estates to have the best collection of tallest and fattest trees on their land. Enthusiasm for this has been known to influence the figures on occasions, depending on whether the owner or the rival was holding the tape measure. Tree heights have always been notoriously difficult to measure accurately, so there was plenty of room for argument: who was to say whether a top branch had fallen off or died back when a later measurement proved, as it often did, to be less than the previous one?

Many old estate planting books and forest maps have contained pencilled-in annotations plotting the progress of this or that tree. By the 1960s a plethora of individual record books had surfaced, but each one was an independent entity. They told us useful but different things in varying degrees of accuracy. In an attempt to systematise the data, the Forestry Commission authorised their Research Dendrologist, who at the time was the geneticist Alan Mitchell, to set about co-ordinating this jumble of information. A card index was begun, and over the next 20 years it was expanded vastly, pulling together estate records from all over Britain and Ireland. Eighty thousand measurements were eventually added to what was then called the National Tree Register. Many of the old Elwes and Henry trees, and even some of Loudon's specimens, were found and re-measured in the process. In Ireland corresponding measurements by James Fortescue and Samuel Hayes were also picked up, as well as numerous trees recorded later on by H. M. Fitzpatrick and the Hon. Maynard Greville in the 1950s.

In the forest industry the scientifically accepted point on the stem where diameter is measured is 'breast height' nominally 4 feet 3 inches from the ground. The traditional height for girthing, however, is 5 feet and this is retained in Britain's National Tree Register because to do otherwise would make nonsense of all the older repeat measurements taken at 5 feet. Trials have shown that measuring at 4 feet 3 inches instead of 5 feet can make a significant difference on small and medium-sized trees. It is less crucial on most ancient trees. The register does not conform either with the forest industry practice for recording irregular stems with deformities at 4 feet 3 inches by measuring equidistantly above and below the distortion and calculating an average. Tree register specimens that are deformed or in some way non-standard at 5 feet are measured at the nearest uniform point above or below the deformity and the actual height of this is noted. Deformed and badly forked trees that cannot be measured at 5 feet are treated with caution when champion lists are compiled or productivity is being assessed. As a rule, trees with burrs or swellings, or a tendency to spread directly from the base, are measured at the narrowest point on the stem. Multi-stem trees are often listed as two or more separate measurements. Allowance should be made for ivy on tree trunks if the tape cannot be squeezed under it.

OPPOSITE: *A mature beech in autumn colours stands majestic over a country lane in Shropshire.*

In 1988, Alan Mitchell co-founded a charitable trust, the Tree Register of the British Isles (TROBI), to administer the development of the National Tree Register. A nucleus of dedicated volunteer measurers has gradually been expanded to become a regional network of people who contribute new information to a central register. They provide accurate measurements and precise location details, including National Grid references to locations in Britain. In addition to measuring skills, these people have invaluable local knowledge about the trees in their own regions. The Forestry Commission continued to provide tree measurements for the register after 1988, but concentrated its efforts more towards researching and developing computerised systems for recording tree data. Various like-minded people in Australia, North America and elsewhere have been inspired to do the same.

Tree Measuring Today

The emphasis now is to measure with a purpose and not simply for the sake of it. The objective must be to identify significant trees, record where they are, and determine their dimensions precisely. Surprisingly new, big or rare trees are still being regularly discovered in forgotten and neglected places. The accolade of 'Champion', the largest-known specimen of each species, is constantly moving from tree to tree as new specimens are found or old ones are lost. The

detective work undertaken by a modern-day tree measurer is never dull and can often be quite exciting. It has an element of discovery and history about it that few other outdoor pursuits can boast.

Why Measure Trees?

The most important reason to measure trees, especially old ones, is to keep a historical record. Trees have a horrible knack of suddenly vanishing. Although a tree may have stood for 50 years, or 500 years, and look very permanent, its future can never be guaranteed for another day. The slightest environmental change, lightning strike, thoughtless fire, increase in water abstraction, overeager cultivation, chemical fertiliser enrichment or building development can cause a tree to be instantly less permanent than the written record of its measurements. The short time taken to measure and photograph a tree is certainly time well spent if a day, or a week, or a year later the tree is gone. A comprehensive register of tree measurements has numerous other uses and its value increases with every new addition.

The unique information the register contains can in some circumstances obviate the need to initiate new experimental work to test the suitability of a species for a certain purpose in a particular region or type of site. It can provide a ready-made shortlist of potentially good, safe tree species to plant within a given area and,

a corresponding list of unsuitable, failed or previously slow-growing species which should probably be avoided. It is well known that the safest choice of species for any site is the one based upon the indisputable evidence of what is growing well there already, but how do we know what has grown well there before? Only an inventory such as the tree register can say.

Distribution maps of measured trees can now be produced from the data. These might show, for example, all the specimens of a single species over a certain size in the British Isles. The range and limits of tender species show up particularly well because there are often extensive areas where they do not occur at all. Although at present, where data is somewhat limited, such maps have a tendency to show clusters of choice trees in the vicinity of the most active measurers' homes, their potential value to future growers and planners is unquestionable. Conservationists, designers, foresters, local-authority tree officers, and timber companies alike can benefit by consulting such extracts from the register and a wide range of other information resulting from its analysis.

In addition to its value as a means of choosing species for future planting, the register yields over time a tangible indication of expected growth rates. They will vary according to species and site, as well as the condition and age of the tree, and can only be accurately quantified by the analysis of detailed and repeated measurements. To be of maximum value, these need to have continued periodically for a long time, perhaps several hundred years. How else can the growth rate and life span of an individual tree be accurately calculated without cutting it down or harming it by boring holes in the trunk to count the annual rings? For conservation managers, repeat measurements from the register can also give an indication of potential longevity of a species. This same information provides commercial growers with an indication of annual timber production, and for arboriculturists it suggests what safety procedures may eventually be necessary and the likely time-scale. Designers can be provided with some positive evidence with which to make critical and long-lasting decisions about landscape plans. It is also possible that climatologists will find useful evidence about the effects of climate change on the growth rates of trees from this ever lengthening historical record.

Which Trees to Measure

In Britain alone it is estimated that around 200,000 trees are of dendrological significance, out of probably 200 million non-forest trees of measurable size (over 8 inches in girth) at 'breast height' or higher. It would be impossible to measure every one of these, so a priority system has been devised for the guidance of measurers and landowners. Variation from region to region is of interest. The ultimate size and age for silver birch in western Ireland, for example, is not

necessarily the same as for silver birch in East Anglia. So both records are justified, as are records for silver birches in every other distinctively different climatic and geographical region of the British Isles. Species of botanical scarcity or horticultural interest are another special category, worthy of measurement especially if few records already exist. Very often little is known about the potential of such trees, particularly the rate of growth and ultimate size.

Priority candidates for measurement are trees of known planting date. They are greatly valued for the information they provide, and they are included in the register automatically even if they are still young or small. The planting dates for ancient native trees are generally fairly speculative, but most historical evidence should probably be given the benefit of the doubt. Occasionally there are references to individual trees on old parish or estate records and maps that prove to be valuable. After some fairly well-documented introductions of trees by the Romans, around 1,500 years appear to have elapsed before reliable records were kept again. The British have clearly not always been as keen on writing down numbers as they are now. William Turner (1547) was the first 'recent' systematic recorder of trees. The earliest precise introduction and planting date for an exotic species, common laburnum, is 1560. The next were Turkish hazel in 1582 and alpine laburnum in 1596. After that a plethora of introduction dates are available, and dated plantings associated with historical events are common.

Even if the actual planting date is not known, repeat measurements are very important in the register. There is usually a set interval between measurements stipulated for the tree register, but trees not seen for over 10 years are likely to need checking, if only to see whether they are still there. Some remarkable specimens have been measured at intervals for almost two centuries: in such

OPPOSITE: *The Whitty pear in the middle of the Wyre forest, Worcestershire*

BELOW: *Ancient pollard beech in Burnham Beeches, Buckinghamshire.*

solid long-term research the tradition of the Victorian naturalists lives on to this day. Second and subsequent measurements indicate trends in growth rates, and only from such data can accurate information about size relative to age be obtained. Incremental tables for similar species on similar sites can gradually be compiled and checked against ring counts and measurements of adjacent cut stumps. When several measurements have accumulated it is sometimes possible cautiously to extrapolate patterns of growth beyond the period that has actually been assessed. Mean annual ring widths can be calculated for short intervals between measurements – though, after the initial formative growth period of the tree the widths tend to decrease as the stem diameter increases. They cannot be regarded as being equal for long periods of years, so a ring width in the outer part of a mature tree can on no account be taken to represent the ring width towards the centre of the stem or even a mean for the whole tree. The current annual increment (the area of new wood contained within each annual ring), on the other hand, may be expected to remain more or less constant after mature size has been reached. The area (and volume) of new wood produced each year depends upon the healthy leaf area exposed to the sun, so while the crown of the tree is intact this remains more or less the same. Of course, equal areas of new wood produced each year, spread over an ever increasing stem circumference, automatically result in diminishing ring widths. Current annual increment of new wood in the stem is a useful basis on which to estimate the age of a tree from its diameter. It is also possible to calculate the probable stem size at some predetermined time in the future. In calculating the age of a tree, the central core of rapidly increasing formative growth must be considered separately. So too must the period of decline when major limbs are shed, and photosynthesis is reduced due to foliage loss. As additional measurements are taken, tables of age according to species, size and site type can be compiled and perfected.

Other definite candidates for measurement are trees exhibiting one or more of a range of special attributes. They may show a particularly good rate of growth, outstanding genetic vigour, good straight form or other desirable commercial features. Of equal value to the size and species of the tree are its provenance details if they are known. The provenance of a tree is its exact origin within its natural range. High elevation, latitude and exposure are particularly important because seedlings retain their parents capacity to tolerate adverse conditions regardless of where they are planted. It is sometimes vital to future success for planters to know exactly which provenance of a particular species is doing well on a site similar to their own. Trees that make a contribution to a historic landscape, including remnants of native vegetation, are also important in this measurement category.

Although priorities for measurement have been

suggested, it is usually appropriate to measure as many notable trees as possible on an estate or in one particular location during a single visit. Although this may be time-consuming, it is generally easy to do and very much in the Elwes and Henry tradition. It gives an overall picture of the range and diversity at a particular place and time. A repeat visit some years later can, by comparison of the inventories, indicate subtle changes to the treescape that might otherwise be overlooked. Third and subsequent visits build up a picture of individual tree survival, species performance and patterns of growth.

Measuring a Tree

The process of measuring a tree consists of running a tape measure round the stem to obtain girth, and using some geometry to ascertain the height from the ground. The latter requires the use of a hypsometer, an optical device. In addition, a notebook, map, pencil and tree identification book will be needed. If previous measurements are available, these may also be taken along.

Measurements in the British Isles have traditionally been in feet and inches. Height was always in feet to the nearest foot, and girth was measured in feet and inches at a point five feet above ground level. In the modern register the height measurement is converted to metres by multiplying the number of feet by 0.3048. New measurements are made directly in metres, which is the internationally accepted practice. Using units of less than one metre is acceptable if confidence allows.

Height is notoriously difficult to measure, even with a precision instrument. It is not always easy to see the actual top of the tree. Leaning stems present additional problems. Preferably, sightings should be made from two or more directions and the results compared. In the absence of any specialist equipment the simplest way to measure the approximate height of a tree is to use a straightish stick, which is cut or broken off to exactly the same length as the distance from your eye to the tip of your thumb when your arm is fully stretched out in front of you. Hold the stick vertically near its centre between your thumb and fingers, keeping your arm at full stretch. Move away from the tree to a place where you can see the top and the bottom at the same time. Walk backwards or forwards until, using one eye to look at the stick and the tree simultaneously, the top and the bottom of the tree appear to align with the top and the bottom of the stick. If you have done everything properly, the height of the tree will be about the same as the horizontal distance that you are away from it.

The second measurement to obtain is the diameter of the stem. Girth is still regarded as the best and easiest measurement to record in the field. It can be done very accurately with a long tape measure and is therefore of greater scientific merit than height. Measuring the circumference of an irregular cross-section, as most tree trunks are, automatically evens out any deviations

from a true circle. To conform with modern international standards in tree measuring, girth is expressed in the register as diameter (girth divided by 3.14159, or pi), giving a mean of an infinite number of variable diameters. Diameter is usually recorded in centimetres, but there is a strong case for changing this to metres (to two decimal places), so that height and diameter on the register are in the same unit of measurement. Commercially produced tapes are available which are actually calibrated in units of diameter. In the field, however, it is probably easier for an already overloaded measurer to carry just one ordinary 30-metre tape and use it for girthing and measuring out points from where height sightings can be taken.

Keeping Records

Modern computerised data management means that information can be stored, sorted and presented in many sophisticated ways. For years the product of the tree register was a single publication, a 'champion' tree list. The list arranged tallest and fattest trees alphabetically by species and variety followed by height, diameter, date of measurement and location. Production of this list continues but now it can be updated and printed out at the touch of a button. In addition the range of subsidiary files that can be taken from the register is considerable. A single species or a single locality listing, for instance, can be produced and supported and enhanced by graphics if necessary. As an increasing number of accessions include National Grid references, distribution maps will become more comprehensive. The register as it stands is already extensive and unique, and it expands on a daily basis. Compatibility with other computerised inventories using similar data with some additional fields is now technically feasible. Ancient-tree recorders in Britain, for example, tag veteran trees using a menu-driven veteran-tree database employing paper data collection forms and palm-top portable-computer methods. The eventual aim is to achieve a national database of sites and ancient trees. Likewise, virtually all botanic gardens and arboretums now use similar kinds of database to catalogue plants. This ensures compatibility, often on a world scale. In local-authority offices similar systems are an indispensable part of municipal tree management. They are compatible to a very large extent because of the similarities between the available equipment and programs they use. With so much unanimity of purpose and consistency of procedure, it seems likely to be only a matter of time before all the various computerised registers of trees become a single functional facility, a 'super register' for everyone who works with trees to refer to.

TREE PLANTING

About the easiest thing associated with planting a tree is actually planting it – that is, digging the hole and popping it in. There are much harder things to do, and fascinating decisions to make, before any planting takes place, and sustained effort is also required afterwards. Great care is needed to establish a plant successfully and nurture it until it becomes a mature and healthy tree that is able to fend for itself. Even then, protection will be essential for many more years to keep it safe and vigorous.

The key to lasting success is that the tree and the site should be perfectly matched long before the spade is taken out of the tool shed. We should not be deceived by the apparent adaptability of many tree species. They may appear to succeed in less than perfect circumstances, but trees and woodlands that are merely surviving and not thriving will be under constant stress, making them vulnerable to disease and predation. The choices associated with proper selection of species are not the same for plantations or community woodlands as for gardens and urban open-space planting.

Choice of Species for Woodlands

Satisfactory choice of species for a woodland or plantation requires a detailed ecological and physical analysis of the site. The approach to amenity considerations may be more relaxed, since, after all, only the edge trees of a woodland, unless it is on a steep hillside, are open to public view. Particular notice must be taken of the soil, and any changes in its type, and of drainage, or lack of it, right across the area to be planted. The degree of exposure may also not be uniform over the whole site because of variable topography. Hills and exposed edges may require different species to the valleys and sheltered hollows. The effects of all this tend to show up in most established monoculture plantations. Even where the choice of species is generally good for the site there will be patches that vary. Trees will have grown faster or straighter in some parts than in others, or may even have failed altogether. Pest attacks and windthrow tend to be prevalent where the species choice was least appropriate. In upland and marginal areas, cultivation is often employed to bring a site up to an acceptable overall standard. Even this is not always completely effective.

In an ideal silvicultural situation different species would be used in different conditions across the area. Windfirm edge trees might be employed to provide shelter and to filter gale-force winds. Strong-rooted pioneers might be used to protect more valuable but perhaps less stable species. Wetland trees would be used to soak up water from wet hollows. In some instances mixtures of species using nurses – hardy but less valuable trees – might be interplanted throughout the area to provide quick temporary cover and encourage uniform upward growth. Alder and larch, for example, are good nurses for oak and beech in some circumstances. However, it is not always practical or economic to fill a wood with so many different sorts of trees. There may be a legal or conservation requirement to use only British native species, or only common broadleaved trees. Funding and grant aid considerations might also favour a much less complicated planting proposal. Invariably, some form of compromise has to be reached, though, where choice of species is concerned, it inevitably reduces the quality of the woodland to some extent in years to come.

Choice of Species for the City

A municipal park, an urban street or a private garden each requires a careful choice of tree species. The amenity aspect is especially important here because every tree will be seen and enjoyed by a great many people. The native tree versus non-native tree debate is less of an issue in town tree planting, so a wide range of species might be considered from the start. There are about 2,500 kinds of tree that will grow somewhere in the

OPPOSITE: *Italian hybrid and Lombardy black poplars, Herefordshire.*

BELOW: *A 'conifer caterpillar' roams the hills north of Lydham, Shropshire.*

Above: *Coral bark maple* (Acer palmatum 'senkaki').

Opposite: *Norway maple in full flower, Pychard, Herefordshire.*

Below: *Autumn leaves of the full moon maple in striking contrast against the blue sky.*

British Isles. Add to this the 30 or so ecological, physical and social decisions that need to be made about the site and the complexity of the task becomes clear.

Most species are soon rejected: poisonous, tender, very large, very small, invasive or odorous plants stand little chance of being chosen. Poisonous plants such as yew or laburnum should not be used in public places. Tender trees may be tempting but they will almost certainly encounter damaging winter cold at some time in their expected lifespan. Excessively large trees, such as most poplars, soon dominate the landscape and could damage the infrastructure. Dwarf conifers and other small trees are slow-growing and vulnerable to theft. Invasive black locust and wingnut trees, amongst others, sucker freely over a wide area and do not respect boundaries. Several species – female maidenhair trees and male trees of heaven, for example – smell horrid in the summer.

Once non-starters are eliminated from the list, the site should come in for some detailed scrutiny. If the geology is shrinkable clay and the site is of limited size in a residential area, only very small trees should be chosen, or the whole idea of planting at all should perhaps be reconsidered. In addition to looking at soils, many of which in towns will be artificially reformed, and climatic limitations, pay particular attention to space: is there enough to support a tree or group of trees? How will it look in 10 and 50 years' time with trees growing in it? Checking on the health, size and age of existing specimens in the neighbourhood will provide an idea of the answer. How much space do they each appear to take up? Are there any signs of disease or stress? Is there an epidemic waiting to happen because of over-planting of one or just a few species? Note what is doing well in the locality and make a list of the potentially good families or genera. Most importantly, list the names of trees you particularly like the look of which fit nicely into the local landscape.

Garden sites require a lot of detailed preliminary consideration, especially in heavily built-up areas. Note the proximity of walls, paths, roads, buildings and neighbours' windows. These should not be obstructed or shaded out by your tree. Look out for ponds, statuary, steps, obelisks and other hard features that a new tree might physically damage or obscure from view. Underground services are more tricky to locate, but in most urban areas there will always be cables, pipes and drains somewhere nearby. The assistance of specialists from the various utilities will probably be needed to find them all. Overhead wires are more obvious but calculating whether the tree will ultimately reach them or not is an inexact science. Failure to get this right could eventually be dramatic, inconvenient and costly. Television reception may also have to be considered because large trees can cause interference. Slippery dead leaves on pathways in winter are another hazard that it would be prudent not to overlook. The number of leaves a mature tree will generate in its lifetime is not something that springs automatically to mind when a young leafless sapling is being planted on a fine spring day.

A tree will ultimately either enhance or obstruct a designed landscape or view. Trees that block out an existing view themselves become the backdrop to a new view and should therefore be interesting and attractive. The effectiveness of a design also depends, of course, upon the viewpoint. If the tree is most often seen from one direction, lighting will play an important part. White-flowered or pale-leaved specimens should never be backlit or the detail is lost against the sky. Backlighting, however, will enhance reds, purples and golds, such as a copper beech or a good autumn colour maple. They glow when the sun shines through their leaves from the back but can look rather flat when lit from the front. The main problem about designing with trees is that the effect you wish to create may take

RIGHT: *'Removal of a tree by Barron's machine'.*
London Illustrated News, *1855*

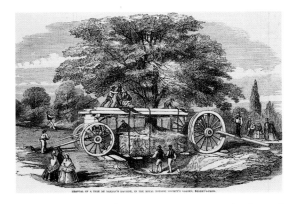

20 years or more to become apparent. Newly planted saplings seldom give an observer the impression that they are part of a carefully landscaped plan, and in the early years a well-designed treescape often looks very similar to any other more haphazard planting.

Commemorative trees are often the most difficult to establish. The choice of tree species and site has probably already been made in light of personal preferences, which may not be compatible with the considered opinions of an arborist or the biological requirements of the tree. It is extremely disappointing, especially for the donors or the sponsors, when a commemorative or ceremonially planted tree fails to grow. Unfortunately, it happens all too frequently and demonstrates in an emotive way the importance of thorough preliminary planning. An obvious solution to this problem if you are intending to donate a tree is to take advice from a tree specialist before deciding what species to choose. There is almost certainly a suitable tree for every possible kind of site, and if the right one can be found, it will give great pleasure and satisfaction to all. A small selection of unusual commemorative trees are described in Chapter 11.

The Technique of Planting Trees

Preparing a site and planting a tree is usually a once-

BELOW: *Strange topiary at Cross, Somerset.*

in-a-lifetime opportunity to cultivate or improve the soil in which it will grow. If the tree stands for 100 years the half an hour or so taken to plant it can surely be regarded as time well spent. The season and the weather must be right. It is not worth being impatient – just keep that 100-year time-scale in your mind. A dull, damp autumn or spring day is best. The plant should be dormant, that is, if it is deciduous, the leaves should be off. If it is evergreen it should be planted just before the new growth begins. This is so that the evergreen leaves are not left thrashing about in the wind and cold all winter without any support from their unestablished roots.

The planting hole should be larger than the rootball of the tree so that any nasty diseased roots and rubbish in the soil can be taken out and disposed of. This also ensures that the medium around the plant is thoroughly cultivated and well mixed. If a stake is to be used it must be pressure-treated with a fungicidal preservative and it should be driven into the bottom of the hole now, and not forced through the tree's root system later on.

The tree, with its roots still in their packing or pot and kept in the shade until the very last minute, can now be prepared. Cut off any damaged or diseased shoots cleanly with secateurs. Unpack the rootball, pull any winding roots out straight, and, if they won't have it, cut them back. Cut off any diseased, dead or damaged root ends cleanly, but do not let the plant dry out while you search for them. Fill the hole until the tree, sitting on a small mound of soil in the bottom, has its root collar level with the original ground level. Then fill the hole as quickly as possible, pressing the soil down gently at intervals and occasionally shaking the tree until the job is done. Finally, firm up the surface to expel any air pockets. Be careful not to skin the stem or roots with your feet in the process. The area immediately round the tree should then be treated or covered so that weeds or grass do not encroach for at least five years. Mulching, using either a manufactured plastic or fibrous sheet or some organic material such as composted bark chips, is good for this. Organic mulching should not be overdone, however, because it can reduce the oxygen supply available to the roots through oxidation as it decomposes, and it also locks up soil nitrogen for a time in the process. Fertilisers should be avoided at this stage unless the soil is deficient. They encourage excessive top growth and upset the fine and already stressed balance between root and shoot. This results in unsightly shoot-tip death on young trees until the balance is redressed.

Care of Young Trees

The amount of maintenance required to establish a tree depends very much on how good the original choice of species was. A happy tree well suited to its planting site will grow quickly and easily without much assistance. Weed control however is always essential because any competition for moisture or nutrients will set the tree back. Mown grass is probably the tree's worst enemy in this respect, however pretty it may look. Weeds should be pulled out or killed with a herbicide; hoeing is not advisable due to the possible surface root damage it can cause. Weeds should never be cut, as their regrowth is usually more harmful than the original plant was. Strimming is also risky near young trees, even if you do it yourself. If watering is required, it should never be superficial. Give the tree a good soaking or do not water at all. Heavily chlorinated or lime-enriched tap water in large quantities may be harmful to some species. Remember that once watering is started, it should be kept up until the end of the summer. If feeding is needed, yellowing foliage or other mineral-deficiency symptoms may begin to show up; a balanced, slow-acting general fertiliser is best. Trees should never be overfed and should not be fed at all if there is no need for it.

Pruning

A well-chosen tree will not generally need much pruning, except to produce a particular shape or a stem without forks. Normally only broken, diseased or dead twigs have to be cut off. When pruning becomes lopping – the removal of large branches – specialist tree surgeons with ladders, ropes, professional equipment and good insurance cover are usually brought in. Wound painting afterwards is not recommended unless there is a risk of infection locally from airborne diseases, such as fireblight or silverleaf. The paint used should always contain a chemical or biological ingredient to kill bacteria or fungal spores. Cosmetic wound painting with other oil-based paints can cause increased damage by sealing in living spores that cause infection.

Pests and diseases

It cannot be emphasised enough that the key to success with growing trees is correct choice of species, which is of paramount importance when it comes to damage and disease prevention. A multitude of different predators attack trees and this varies from one area to another. The best defence is avoidance of susceptible subjects. Some of the most difficult browsers to contend with now in Britain are muntjac deer and grey squirrels, but even they seldom touch certain tree species (see Chapter Two). There are also trees that withstand excessive air pollution, salt spray and gales. Many can cope with a warming climate and even reduced summer water availability. There are trees, and ways of planting trees, that quite remarkably provide some resistance to vandalism and accidental damage, such as that caused by motor vehicles. There is hardly a situation, natural or artificial, that precludes tree planting altogether.

NEXT PAGE: a beautiful example of a mature hornbeam in open parkland, Devon.

BELOW: An atlantic cedar dominates the horizon of a residential street. Always give careful consideration to the final size and shape of a tree before planting!

ARBORETA: THE TOP 20 TREE COLLECTIONS

BATH BOTANIC GARDEN AND ROYAL VICTORIA PARK, BATH.

Bath City Council, founded 1881, open all year.

One mile from the centre of Bath, North of A431. ST7365. Free adjacent parking. WC. Rail one mile.

BATSFORD ARBORETUM, MORETON-IN-MARSH, GLOUCESTERSHIRE.

Batsford Estates Company, open April–October.

Five miles north of Stow-on-the-Wold, A429, turn west at the northern end of Moreton-in-Marsh (signposted). SP1833. Admission charge, adjacent parking. WC, tea-room and shop. Rail two miles.

BICTON GARDENS, DEVON.

Privately owned, pre-1820, open March–October.

Three miles north of Budleigh Salterton on A376. SY0786. Admission charge, parking and all amenities.

BUTE PARK ARBORETUM, CARDIFF. OPEN ALL YEAR.

North of city centre.

Admission free, no car park. WC. Rail ½ mile.

EDINBURGH BOTANIC GARDEN, EDINBURGH.

Open all year.

North of city. NT2475. Bus route. Admission charge, no car park. WC. Rail 1½ miles.

HELIGAN, MEVAGISSEY, CORNWALL.

Open all year.

B3273, Pentewan, three miles south of St. Austell (signposted). Admission charge, free parking. WC, shop and cafe.

INVEREWE GARDEN, POOLEWE, HIGHLAND REGION.

NT (Scotland), founded 1862; open April–October.

Six miles northeast of Gairloch on A832 by Poolewe. NG8582. Admission charge, parking. WC, shop and cafe, summer only.

KEW ROYAL BOTANIC GARDENS, LONDON.

Open all year.

On A307, south of Kew Bridge, TC1876. Admission charge, limited parking. WC, shop, cafe and museums.

KILMUN FOREST GARDEN, DUNOON, ARGYLL.

Forestry Commission, open all year.

A880, six miles north of Dunoon (ferry), NS1872. Limited free adjacent parking. Steep hill, strong footware advised.

LYNFORD ARBORETUM, MUNDFORD, NORFOLK.

Forestry Commission.

East of A1065, ½ mile north of Mundford roundabout (signposted). Free adjacent parking.

NATIONAL PINETUM, BEDGEBURY, KENT.

Forestry Commission, founded 1925, open all year.

East of A21 at Flimwell (signposted). TQ7233. Admission charge, parking. WC, refreshments and information (summer). Rail seven miles at Tunbridge Wells.

NESS GARDENS, WIRRAL, CHESHIRE.

University of Liverpool, open March–October.

A540, ten miles north-west of Chester. SJ3076. Admission charge, parking.

SIR HAROLD HILLIER GARDENS AND ARBORETUM, ROMSEY, HANTS.

Hampshire County Council, open all year.

North of A31, two miles east of Romsey (signposted). SU3824. Admission charge, adjacent parking. WC, shop, cafe and exhibitions. Rail two miles.

SPEECH HOUSE ARBORETUM, FOREST OF DEAN.

Forestry Commission, open all year.

South side of B4226. 2½ miles west of Cinderford. SO6212. Free adjacent parking, hotel.

THORP PERROW ARBORETUM, BEDALE, YORKSHIRE.

Privately owned, open March–November.

Ten miles north west of Ripon, A6108 and B6267 (signposted), SE2685. Admission charge, adjacent parking. WC.

TRESCO ABBEY GARDENS, ISLES OF SCILLY.

Privately owned, open all year.

SV8914. Ferry one mile, heliport nearby. Admission charge. WC, cafe, and shop.

WAKEHURST PLACE GARDEN, SUSSEX.

Royal Botanic Gardens (Kew), open all year.

Four miles north of Haywards Heath on B2028. TQ3331. Admission charge, parking, cafe.

WESTONBIRT ARBORETUM, TETBURY, GLOUCESTERSHIRE.

Forestry Commission, founded 1829, open all year.

A433, 3½ miles south–west of Tetbury. ST8590. Admission charge, parking. WC, shop, visitor centre and cafe.

WISLEY GARDEN, SURREY.

Royal Horticultural Society, founded 1870, open all year.

Off A3, seven miles east of Guildford (signposted). TQ0558 Admission charge, parking. WC, shop and cafe.

YOUNGER BOTANIC GARDEN, DUNOON, ARGYLL.

Royal Botanic Garden (Edinburgh), founded 1928, but planted from 1865, open April–October.

A815, seven miles north of Dunoon (ferry). NS1385. Admission charge, parking. WC and cafe.

OPPOSITE: *Silk wood, Westonbirt arboretum*

Tree Organisations

INSTITUTE OF CHARTERED FORESTERS

7A, St Colme St., Edinburgh EH3 6AA

Founded in 1925 as the Society of Foresters of Great Britain, and becoming the Institute in 1973. The Institute's objectives include maintaining and improving practice and understanding of forestry in all its aspects, promoting the professional status of British foresters, and standing as the representative body for the profession. The Institute is deeply committed to education with its own examinations structure for admission, linked to several university and college courses in forestry.

ASSOCIATION OF PROFESSIONAL FORESTERS

7/9, West St., Belford, Northumberland NE70 7QA

Aimed primarily at those who derive their livelihood from forestry, whether it be at the practical forefront of the profession, as a woodland owner or consultant, or even a supply industry with close forestry ties. The Association has a comprehensive network of services providing practical support to its members, including: insurance cover, educational and provident funds, training schemes, along with regular technological, legislative and political updates through conferences, exhibitions, visits and the Association's own quarterly journal.

WOODLAND HERITAGE

P.O. Box 2950, Epping, CM16 7DG

A registered charity originally formed by a group of traditional cabinet makers keen to improve British forestry practice, and quickly joined by numerous like-minded landowners, tree nurserymen, consultants and contractors. The core aim is to vigorously promote the importance of multi-purpose sustainable forestry, with particular emphasis on management excellence in harvesting and replanting, which will provide longterm timber resources of all types for Britain's future needs. In achieving this goal due respect is paid to structural and biological diversity using variable term management systems. A regular update of information and advice keeps tree growers up to the mark with the very latest strategies and technical data for successful woodland development. Woodland Heritage also makes a determined effort to work with schools and young people, thus nurturing the knowledge, skill and respect for trees of future generations.

THE NATIONAL FOREST

Stanleigh House, Chapel St., Donisthorpe, Swadlincote, Derbyshire DE12 7PS

Originally set up under the guidance of a Countryside Commission Development Team, who spent three years planning the National Forest, it was established as an independent public company in 1995, under joint membership of the D.O.E. and MAFF. The site chosen covers some 194 square miles across the counties of Staffordshire, Derbyshire and Leicestershire and the aim has been to convert an area in which a mere six percent of land was given over to woodland into one where 70 percent becomes woodland (on a 60:40 ration in favour of broadleaves). The whole forest is being planned with a total integration strategy, to encompass communities, industry, agriculture, tourism, recreation and nature conservation incorporating sensitive building development policies and progressive restoration programmes for derelict land.

COMMUNITY FORESTS

The Community Forest Unit, The Countryside Agency, John Dower House, Crescent Place, Cheltenham Glos. GL50 3RA

There are twelve Community Forests in England, varying in size from 38 to 300 square miles, but covering about 1,800 square miles in total. The main criterion for the establishment of these forests is that they will surround and link urban areas with the aim of improving the environment and revitalising the communities they encompass and, with this aim in view, the intention is to increase the tree cover of these regions from about seven percent to 30 percent over the next forty years. A partnership between the Countryside Agency, the Forestry Commission and all the relevant local authorities is implementing policies which will enhance the landscape; work sympathetically with wildlife conservation; encourage diverse uses for agricultural land; develop new opportunities in sports, arts and recreation; remain sensitive to historical, archaeological landscape heritage; create a better environment in which to live and work.

BROGDALE HORTICULTURAL TRUST
(NATIONAL FRUIT COLLECTIONS)

Brogdale road, Faversham, Kent, ME13 8XZ

A non-profit making charitable organisation providing research and development into fruit varieties, and hence of great value to the industry and equally interesting to the consumer and amateur fruit grower. Holds a massive collection of over 2,300 apple varieties, 550 pears, 350 plums and 220 cherries. Offers lectures and workshops in orchard culture, guided tours of the orchards, as well as an identification service.

COED CYMRU

23 Frolic St., Newtown, Powys SY16 1AP

An advisory body established to provide information on the management and conservation of the native woodlands of Wales. There are officers in each of the Welsh County Councils and National Parks who can provide free advice to farmers and landowners.

THE GREEN WOOD TRUST

Station Road, Coalbrookdale, Telford, Shropshire TF8 7DR

Promotes woodmanship through information and training in order to stimulate revitalisation of sustainable woodland management, linked to the greater understanding and development of coppice wood products. A large range of courses are always available.

NATIONAL SMALL WOODS ASSOCIATION

Hall Farm House, Preston Capes, Northants. NN11 6TA

The association promotes management and conservation strategies for small or neglected woodlands, linked to commercial revitalisation. Networks of specialists aim to make the trees and woodlands in their hands commercially viable and ecologically sustainable.

TREE REGISTER OF THE BRITISH ISLES

77a, Hall End, Wootton, Bedford MK43 9HP

Founded by the late Alan Mitchell, the Register was established to identify and record the location and dimensions of exceptional trees in Britain. Measurements are taken at regular intervals which sets the statistics in context with the trees' location and growth conditions, thus making a valuable data bank for future silviculture. In recent years the list of champion trees has grabbed the public imagination. (See Appendix A for more details).

OPPOSITE: An ancient oak in Moccas deer park, Herefordshire.

THE TREE COUNCIL

51, Catherine Place, London SW1E 6DY

Stated aims are: the improvement of town and countryside environments through the promotion of the planting and conservation of trees and woods in the UK; dissemination of knowledge and management techniques of trees; to act as a forum for all manner of organisations associated with trees and to foster cooperation between such bodies; provision of grant aid, especially to schools, for the planting and maintainance of trees. The Tree Council is responsible for promoting and coordinating the Tree Warden Scheme, a nation wide network of volunteers who gather information, offer advice, and take a general interest in the planting and conservation of trees within their own locality. Two annual events raise the profile of trees and woodland, working in conjunction with the Esso Living Tree Campaign. There is the Esso Walk in the Woods from early May to the end of July, which promotes the enjoyment and understanding of the wide diversity of woodlands, including many special events such as guided walks, talks and demonstrations. Esso National Tree Week, in late November, is a campaign aimed at encouraging people everywhere to actively go out and plant trees. The Esso Living Tree Campaign has subsequentely given rise to:

TREES OF TIME AND PLACE

c/o Esso UK plc, Mailpoint 08, Esso House, Ermyn Way, Leatherhead, Surrey KT22 8UX

The principal aim is to raise awareness of the full range of benefits which trees bring to everyone. This is realised through celebration of the emotional, cultural, historic and environmental contributions of trees. A strong emphasis is placed upon the actual gathering of seed, sowing and growing seedlings, and planting and tending personal new trees; with a focus on the cumulative effect of many individual efforts towards the quality of life for everyone in the future.

FORESTRY COMMISSION

231, Corstorphine Road, Edinburgh EH12 7AT

Established in 1919, this is the body most readily identified as the largest producer of commercial timber woodland in Britain. They currently own around 40 percent of British woodland, most of which comprises spruce, pine and larch, although recent years have seen an increase in hardwood planting.

FOREST ENTERPRISE, designated an executive agency of the Commission, will be the name most familiar to the public, and it's objectives combine the efficient production of timber; with due regard to education, management and financial affairs, alongside a policy of public access, recreational benefit, and wildlife and landscape conservation.

OPPOSITE: *Sheep browsed hawthorns struggle for survival on the Limestone pavement at Twistleton Scar, Ingleton, North Yorkshire.*

FORESTRY AUTHORITY, is principally the regulatory body of the Commission, and also largely responsible for grant aid advice and allocation for the growing of commercial forestry.

FORESTRY COMMISSION RESEARCH AGENCY

Alice Holt Lodge, Wrecclesham, Farnham, Surrey GU10 4LH

An arm of the Forestry Commission with a specific remit to advance the theory and practice of productive tree and forest management, particularly in respect to larger areas of woodland.

TREE HELPLINE (ARBORICULTURAL ADVISORY AND INFORMATION SERVICE)

Alice Holt Lodge, Wrecclesham, Farnham, Surrey GU10 4LH

Until 1993 the AAIS was part of the Research Branch of the Forestry Commission; at which time the Service was privatised and placed under the management of the Tree Advice Trust, since when it has remained independent of the Commission although a tenant at the same address. AAIS is one of the best authorities from whom to seek information and advice on the care and management of trees, most especially those in urban and peri urban situations. They specialise in dealing with problems of pests and diseases, as well as general management advice and traditionally difficult areas such as legal disputes involving trees.

ANCIENT TREE FORUM

c/o Treework Services Ltd., Cheston Combe, Church Farm, Backwell, nr. Bristol BS19 3JQ

Describe themselves as "a network of experts and managers who exchange hints, tips and useful information" on the subject of ancient trees. The aims of the group are to raise awareness off ancient trees, their significance in the landscape and as wildlife habitats.

TREE AID

28, Hobbs Lane, Bristol BS1 5ED

Tree Aid is a charity working to improve conditions for villagers in some of the most arid regions of Africa, currently raising funds for projects in the Sahel, Mali and Zimbabwe, in particular where, "trees are a lifeline providing food, wood for shelter, fuelwood, medicines, animal fodder and many other important needs as well as protecting the precious top soil so people can grow food." In many of these areas adverse climatic conditions coupled with poor understanding of land and resource management have led to massive loss of trees followed by desert incursion. Tree Aid aims to fund tree planting, particularly around settlements, with all the necessary education and information to enable the indigenous people to grow their own trees from seed and successfully nurture them to maturity, thus creating sustainable tree cover.

THE ABORITCULTURAL ASSOCIATION

Ampfield House, Romsey, Hants, SO51 9PA

The association for professional tree consultants and contractors, providing technical information and publishing its own Directory of 'Registered Arboricultural Consultants and Approved Tree Contractors'.

BRITISH TRUST FOR CONSERVATION VOLUNTEERS

36, St, Mary's Street, Wallingford Oxon. OX10 0EU

'Aims to harness people's energies, talents and enthusiasm to promote and improve the environment by practical action.' Working in partnership with communities, businesses, local authorities and landowners BTCV's thousands of volunteers across the country give their time and practical support to numerous countryside projects, many of which focus on woodland sites. BTCV provides transport, training and equipment.

COUNCIL FOR THE PROTECTION OF RURAL ENGLAND

Warwick House, 25, Buckingham Palace Road, London SW1W OPP

(With related bodies for Wales and Scotland). 'Campaigning for the Countryside'. Exists to promote the protection of the countryside in the face of civil, industrial and agricultural developments. Trees, hedgerows and woodland are prominent amongst their concerns.

INTERNATIONAL TREE FOUNDATION
(ORIGINALLY – MEN OF THE TREES)

Sandy Lane, Crawley Down, W. Sussex RH10 4HS

The Society was founded by Richard St. Barbe Baker in 1922 while he was staying in Kenya. Having both experienced spells of practical forestry and studied the subject in some detail he became increasingly aware of the ecological importance of trees worldwide. His first major project in Kenya was to restore forest in the face of desert incursion. He was able to encourage and influence whole tribes to plant and tend trees (these were the original Men of the Trees). He subsequently established reafforestation programmes in Nigeria, Palestine and the Sahara, as well as helping to protect the Redwoods of North America. Although the founder died at the grand old age of 93 in 1982, he was still actively involved in the promotion of his cause, and even today the society continues with its global remit to foster tree planting in some of the most ravaged environments. The society also contributes to programmes in Britain; notably replanting after the 1987 hurricane and administering the Family Tree Scheme as a member of the Tree Council. This Scheme makes it possible for people to dedicate newly planted trees to family or friends in return for a donation to the Foundation.

NATIONAL TRUST

*36, Queen Anne's Gate,
London SW1H 9AS*

'Founded in 1895 for the preservation of places of historic interest or natural beauty.' One of the largest owners of woodland in Britain, as well as many historic homes and parks well endowed with specimen trees from all over the world.

PLANTLIFE (THE NATURAL HISTORY MUSUEM)

Cromwell Road, London SW7 5BD

Main aim to conserve plants and their habitats, with particular attention to endangered species.

THE ROYAL FORESTRY SOCIETY OF ENGLAND, WALES & NORTHERN IRELAND

102, High Street, Tring, Herts. HP23 4AF

An independent educational charity, the RFS was founded in 1882, and is now the largest forestry association in Britain. The Society is broad-based, and includes many professionals involved in different aspects of forestry, as well as members of the general public. The RFS promotes conservation and expansion of tree resources through good forestry management, with particular regard to wildlife, landscape, recreational and socio-economic values.

THE WOODLAND TRUST

Autumn Park, Dysart Road, Grantham, Lincs. NG31 6LL

The United Kingdom's leading charity dedicated solely to the protection of native woodland heritage. Currently their aims include: curtailing the loss of ancient woodland; restoration and improvement of the biodiversity of woods; increasing the area of new native woodland; generally increasing the public's awareness and enjoyment of woodland. To achieve this they combine woodland and site acquisition with effective planting and management strategies, according to their woodland management principles, while adopting wider advocacy on matters of policy and practice which will improve the future of the UK's native woodland.

COMMON GROUND

PO Box 25309, London NW5 1ZA

Promotes the importance of local distinctiveness enmeshed in the nation's common culture; including vernacular architecture, regional plants and animals and threatened habitats such as old meadows and orchards by, 'exploring and developing links with the arts and encouraging local conservation action emphasising the cultural, aesthetic and emotional bonds that we develop with the places in which we live.'

THE COUNTRYSIDE AGENCY

John Dower House, Crescent Place, Cheltenham, Glos, GL50 3RA

An amalgamation of the Countryside Commission and the Royal Development Commission. The agency advises the government on all matters of landscape conservation, access and recreation, and is responsible for policy strategy and management implementation of National Parks, AONBs, Community Forests and many Long Distance Paths. It also provides advice and financial assistance for certain conservation projects.

ENGLISH NATURE

Northminster House, Northminster Road, Peterborough PE1 1UA

Originally established as the Nature Conservancy Council in 1973, English Nature is the principal scientific advisory body reporting to government on all matters of nature conservation. It owns, leases or manages the National Nature Reserves, which are often ecologically rare or special habitats, and is instrumental in the designation and monitoring of SSSIs. It has also compiled the definitive national inventory of ancient woodland. Its Welsh and Scottish counterparts are the Countryside Council for Wales and Scottish Natural Heritage, respectively.

THE VETERAN TREE INITIATIVE, is part of English Nature, and is currently compiling a register of important sites where ancient trees exist and providing practical advice on management issues. They have also been running numerous training and awareness courses around the country to raise the historical and cultural profile of Veteran Trees.

THE LONDON TREE FORUM

PO Box 15146, London WC2B 6SJ

The Forum's principal aims are: to bring about concerted action to sustainably manage London's trees; raise public awareness; investigate funding for planting and maintenance. It is estimated that London has more than six million trees and 65,000 woodlands and small stands of trees. The Forum seeks to link up with individuals, communities, councils and industry to forge strategies for planning, planting, maintaining and protecting trees in the capital.

NATIONAL URBAN FORESTRY UNIT

Red House, Hill Lane, Great Barr, W. Midlands B43 6LZ

Established to, 'develop, expand and promote the concept and practice of urban forestry.' To achieve this the Unit aims to increase public awareness and support, encourage development strategies, and promote technical and commercial excellence of tree and woodland cover in towns.

THE CONSERVATION FOUNDATION

1 Kensington Gore, London SW7 2AR

The Foundation tends to adopt specialised channels of interest and then actively seeks commercial sponsership to fund project development. One of the most high profile tree related projects has been the ongoing Yew Tree Campaign, in association with *Country Living*
magazine. Initiated by Allen Meredith and Professor David Bellamy OBE in 1988 the aim has been to raise public awareness of Britain's most ancient yews, to map their distribution, accrue data in order to better understand their morphology and assess their antiquity and, moreover, to engender local and national pride in some of the oldest living plants in Britain. The Foundation is also actively involved with Elms Across Europe, a project developing disease-resistant elms in a bid to replace the thousands of elms lost to Dutch Elm Disease.

TREES FOR LIFE

The Park, Findhorn Bay, Forres IV36 OTZ

This is a charity dedicated to the regeneration of the Caledonian Forest in the Scottish Highlands. Established in 1981 as part of the Findhorn Foundation, the name Trees for Life was derived from the recognition that global tree cover is fundamental in creating a sustainable future. The once massive Caledonian Forest, estimated to have stretched coast to coast across some 1.5 million hectares of the Highlands, is currently reduced to about one percent of its original size. Trees for Life aims are to encourage natural regeneration by fencing off existing stands of Scots pine so that seedlings will not be predated by red deer and replanting open and long denuded stretches of upland with native species such as rowan, birch, aspen as well as the pines. In some areas the strategy stretches to the removal of non-native trees, where they have been planted for commercial reasons in the past, but currently impede natural regeneration. More controversially, as part of the total ecological restoration project there are plans to reintroduce bears and wolves.

THE ROYAL SOCIETY FOR NATURAL CONSERVATION, ENCOMPASSING THE WILDLIFE TRUSTS

The Green, Witham Park, Waterside South, Lincoln LN5 7JR

'The Wildlife Trusts are a nation-wide network of local trusts which work to protect wildlife in town and country. Through their care of 2,000 nature reserves, The Wildlife Trusts are dedicated to the achievement of a United Kingdom richer in wildlife, managed on sustainable principles. Sharing this goal and making a vital contribution to its attainment are the junior branch, Wildlife Watch, and the urban wildlife groups.' *Natural World* magazine, the journal of The Wildlife Trusts is published three times a year. Issues concerning trees, woodland and hedgerows are perennially at the forefront of their concerns.

OPPOSITE: *Crack willow on the river Windrush at Burford, Gloucestershire.*

NEXT PAGE: *Morning sunlight shines through a plantation of Scots pine (Pinus sylvestris).*

BIBLIOGRAPHY

CHAPTERS 1,2,3,4,5,6,7 AND 9.

Anon, *English Forests and Forest Trees* (Ingram, Cooke & Co. 1853)

Anon., *The Fruiterer's Secrets* (1604)

Heather Angel, *The Natural History of Britain and Ireland* (Book Club Associates 1981)

Lyn Armstrong, *Woodcolliers and Charcoal Burning* (Coach Publishing House and Weald and Downland Open Air Museum 1978)

James Arnold, *The Shell Book of Country Crafts* (John Baker 1977)

Ralph Austen, *Treatise of Fruit Trees* (Second Edition. 1657)

H. E. Bates, *Through the Woods* (Victor Gollancz 1936)

G. S. Boulger, *Familiar Trees* (Cassell & Co. 1886)

James Brown, *The Forester* (1894)

R.W. Brunskill, *Traditional Buildings of Britain* (Victor Gollancz 1985)

William Cobbett, *Rural Rides* (Cambridge University Press 1922)

Common Ground, *Orchards, A Guide to Local Conservation* (Common Ground 1989)

Nicholas Culpeper, *The Complete Herbal* (London 1653)

D. F. Cutler and I. B. K. Richardson, *Tree Roots and Buildings.* (Longham Science and Technical, Harlow, Essex 1989)

W. Dallimore, *Holly, Yew & Box* (John Lane, The Bodley Head 1908)

H. L. Edlin, *Woodland Crafts in Britain* (Batsford 1949)

H. L. Edlin, *Trees, Woods and Man* (Collins 1970)

Elwes and Henry, *The Trees of Great Britain and Ireland* (Edinburgh 1907–13)

John Evelyn, *Silva or, A Discourse of Forest-trees* (First Hunter edition 1776)

Frederick Faulkner, contributing to *The Journal of the Royal Agricultural Society of England* (1843)

John Fowles & Frank Horvat, *The Tree* (Aurum 1979)

John Gerard, *Herball* (1597)

William Gilpin, *Essay on Picturesque Beauty* (London 1792)

Robert Graves, *The White Goddess* (Faber & Faber 1948)

Geoffrey Grigson, *The Englishman's Flora* (Phoenix House, J. M. Dent 1987)

Miles Hadfield, *Landscape with Trees* (Country Life 1967)

Thomas Hardy, *Selected Poems* (Macmillan 1940)

Margaret F. Harker, *Henry Peach Robinson* (Blackwell 1988)

Norman E. Hickin, *The Natural History of an English Forest* (Country Book Club 1972)

Thomas Hinde, *Forests of Britain* (Victor Gollancz 1985)

Robert Hogg, *Apples and Pears as Vintage Fruit* (London 1886)

Robert Hogg, *British Pomology* (1851)

Hogg and Bull, *Herefordshire Pomona* (London 1876–85)

Christina Hole, *English Customs and Usage* (Batsford 1950)

Christina Hole, *British Folk Customs* (Hutchinson 1976)

Homer, *The Odyssey*

W. H. Hudson, *Nature in Downland* (Dent 1923)

W. H. Hudson, *Hampshire Days* (Duckworth 1928)

N. D. G. James, *A Book of Trees* (The Royal Forestry Soc. 1973)

Rev. C. A. Johns, *The Forest Trees of Britain* (SPCK 1899)

Gertrude Jekyll, *Wood and Garden* (Longmans, Green & Co. 1899)

Richard Jefferies, *Wildlife in a Southern County* (Lutterworth 1949)

Hugh Johnson, *The International Book of Trees* (Mitchell Beazley 1973)

Rev. Francis Kilvert, *Diaries Vol. III* (Jonathan Cape 1940)

Angela King and Susan Clifford, *Trees Be Company* (Common Ground 1989)

Thomas Andrew Knight, *Pomona Herefordiensis* (1811)

Adrien Le Corbeau, *The Forest Giant* (Jonathan Cape 1935)

Clare Leighton, *Country Matters* (Victor Gollancz 1937)

Nathaniel Lloyd, *Garden Craftsmanship in Yew and Box* (Ernest Benn 1925)

J. C. Loudon, *The Hardy Trees and Shrubs of Britain* (Warne & Co. 1883)

Antoine Lorgnier, *Forests AGEP* (1992)S. Lousada, *A Year of Fruits* (Grange Books 1993)

Gareth Lovett Jones and Richard Mabey, *The Wildwood* (Aurum 1993)

Luckwill and Pollard, *Perry Pears* (University of Bristol 1963)

Richard Mabey, *Flora Britannica* (Sinclair–Stevenson 1996)

Richard Mabey, *The Common Ground* (J. M. Dent 1993)

Richard Mabey and Tony Evans, *The Flowering of Britain* (Hutchinson 1980)

E. Maple, *The Secret Love of Plants & Flowers* (Robert Hale Ltd. 1980)

Peter Marren, *The Wild Woods* (David & Charles 1992)

M. Mességué, *Health Secrets of Plants & Herbs* (William Collins & Sons 1979)

J. Edward Milner, *The Tree Book* (Collins & Brown 1992)

Andrew Morton, *The Trees of Shropshire* (Airlife 1986)

Thomas Packenham, *Meetings with Remarkable Trees* (Weidenfeld & Nicolson 1996)

R. Philipps, *Trees in Britain, Europe & N. America* (Pan Books 1978)

E. Pollard, M. D. Hooper, N. W. Moore, *Hedges* (Collins 1979)

Oliver Rackham, *The Illustrated History of the Countryside* (Weidenfeld & Nicolson 1994)

Oliver Rackham, *Trees and Woodland in the British Landscape* (Weidenfld & Nicolson 1995)

C. W. Radcliffe Cooke, *A Book About Cider and Perry* (London 1898)

Reader's Digest, *The Living Countryside* (1986)

R. H. Richens, *Elm* (Cambridge University Press 1983)

F. A. Roach, *Cultivated Fruits of Britain* (Blackwell 1985)

Simon Schama, *Landscape and Memory* (Harper Collins 1995)

P. Schanenberg and F. Paris, *Guide to Medicinal Plants* (Lutterworth Press 1977)

Prideaux John Selby, *A History of British Forest Trees* (John Van Voorst 1842)

Brian Shuel, *The National Trust Guide to Traditional Customs of Britain* (Webb & Bower 1985)

E. Soothill and M. J. Thomas, *Nature's Wild Harvest* (New Orchard Editions 1983)

J. G. Strutt, *Sylva Britannica* (Longman, Rees, Orme, Brown & Green 1830)

Homer Sykes, *Once a Year* (Gordon Fraser 1977)

H. V. Taylor, *The Apples of England* (London 1948)

T. F. Thiselton Dyer, *The Folklore of Plants* (London 1889)

Edward Thomas, *The Heart of England* (J. M. Dent 1932)

Edward Thomas, *The Woodland Life* (Wm. Blackwood & Sons 1897)

Jean Valnet, *Heal Yourself with Vegetables, Fruits and Grains* (Cornerstone Library 1976)

R. Vickery, *A Dictionary of Plant Lore* (Oxford University Press 1995)

Charles Watkins, Sonia Holland, Stephanie Thomson, *A Survey of Black Poplars in Herefordshire* (University of Nottingham 1997)

Gilbert White, *The Natural History of Selborne* (Penguin Classics 1977)

John White, *Forest and Woodland Trees in Britain* (Oxford University Press 1995)

J. E. J. White, *The History of Introduced Trees in Britain.* (Proc. Native and Non-native in British Forestry, Institute of Chartered Foresters, Edinburgh 1995)

Ralph Whitlock, *A Calendar of Country Customs* (Batsford 1978)

Gerald Wilkinson, *Epitaph for the Elm* (Book Club Associates 1978)

Gerald Wilkinson, *Trees in the Wild* (Stephen Hope 1973)

Richard Williamson, *The Great Yew Forest* (Readers Union 1978)

M. (Ed.) Woodward, *Gerard's Herbal* (Senate 1994)

Marcus Woodward, *The New Book of Trees* (A.M. Philpot 1920s)

John Worlidge, *Vinetum Britannica, or A Treatise of Cider* (1676)

WWF, *The Vital Wealth of Plants.* (World Wide Fund for Nature, Gland, Switzerland 1993)

John Wyatt, *Reflections on the Lakes* (W. H. Allen 1980)

CHAPTER 8

Y. Aburrow, *The Enchanted Forest,* (Capall Bann 1993)

A. Clapham, *The Oxford Book of Trees* (Oxford University Press 1975)

Culpeper, *Culpeper's Complete Herbal* (Wordsworth Editions 1995)

R. Elliot and C. de Paoli *Kitchen Pharmacy* (Tiger Books 1994)

W. T. Fernie, *Herbal Simples* (Wright & Sons Ltd. c.1914)

Sir J. G. Frazer, *The Illustrated Golden Bough* (George Rainbird Ltd. 1978)

D. Furnell, *Health from the Hedgerow* (Batsford 1985)

H. A. Guerber, *The Norsemen* (Senate, 1994)

L. Gordon, *Trees* (Grange Books 1993)

Mrs M. Grieve, *A Modern Herbal* (Penguin Peregrine Books 1976)

A. Koehn, *A Gift of Japanese Flowers* (Tuttle 1992)

C. Hole, and E. Horsley,
The Encylopedia of Superstitions
(Hutchinson & Co., 1980)

S. Lousada, *A Year of Fruits*
(Grange Books 1993)

E. Maple, *The Secret Love of
Plants & Flowers*
(Robert Hale Ltd. 1980)

M. Mességué, *Health Secrets of
Plants & Herbs*
(William Collins & Sons. 1979)

R. Philipps, *Trees in Britain, Europe & N.
America* (Pan Books. 1978)

P. Schanenberg and F. Paris,
Guide to Medicinal Plants
(Lutterworth Press 1977)

E. Soothill and M. J. Thomas,
Nature's Wild Harvest
(New Orchard Editions 1983)

M. (Ed.) Woodward, *Gerard's Herbal*
(Senate 1994)

Jean Valnet, *Heal Yourself With Vegetables,
Fruits and Grains*
(Cornerstone Library 1976)

R. Vickery, *A Dictionary of Plant Lore*
(Oxford University Press 1995)

CHAPTER 10

The Arts Council,
Landscape in Britain 1850-1950
(The Arts Council, London 1983)

John Blakemore, *Inscape*
(Zelda Cheatle Press 1991)

Roger Cardinal, *The Landscape Vision of
John Nash* (Reaktion, London,1989)

R. A. Clarke, *The Blasted Oak*
(Herbert Art Gallery, Coventry 1987)

Rex Vicat Cole, *The Artistic Anatomy of
Trees* (Seeley Service London, nd.)

Common Ground, *Pulp!*
(Common Ground, London 1989)

Denis Cosgrove and Stephen Daniels,
(Eds), *The Iconography of Landscape*
(Cambridge University Press 1988)

Kaherine Crouan, *John Linnell*
(Cambridge University Press 1982)

Stephen Daniels, *Fields of Vision*
(Cambridge, Polity Press 1992)

Stephen Daniels and Charles Watkins
(Eds), *The Picturesque Landscape*
(Nottingham University 1994)

H. L. Edlin, *Trees, Woods and Man*
(London, Collins 1956)

Brent Eliott, *Victorian Gardens*
(London, Batsford 1986)

James George Frazer,
The Golden Bough (1890)
(Oxford University Press 1994)

Andy Goldsworthy, *Hand to Earth*
(London, Harry N Abrams 1990)

William Gilpin, *Remarks on Forest
Scenery* (London, R. Blamire, 1791)

John Hayes, *Thomas Gainsborough*
(London, Tate Gallery,1980)

N. D. G. James, *A History of English
Forestry* (Oxford, Blackwell 1981)

Ian Jeffrey, *The British Landscape
1920-1950* (London, Thames and
Hudson, 1984)

Edward Kennion, *An Essay on Trees in
Landscape* (London 1815)

Angela King and Sue Clifford,
Trees Be Company
(Bristol, Classical Press 1989)

Raymond Lister,
The Paintings of Samuel Palmer
(Cambridge University Press 1985)

Raymond Lister,
Samuel Palmer, His Life and Art
(Cambridge University Press 1987)

Helen Luckett, *The Tree of Life*
(London, South Bank Centre 1989)

Richard Mabey (Ed.) *Second Nature*
(London, Jonathon Cape 1984)

Christopher Martin, *The Ruralists*
(London, Academy Editions, 1991)

David Mellor, *A Paradise Lost*
(London, Lund Humphries 1987)

Roger Miles, *Forestry in the English
Landscape* (London Faber, 1967)

Mary Russell Mitford,
*Our Village (1825-32),
Illustrate edition*
(London, Bracken Books 1992)

David Morris, *Thomas Hearne
and His Landscape*
(London, Reaktion 1989)

Leslie Parris and Ian Fleming-Williams,
Constable
(London, Tate Gallery 1991)

J. Ernest Phythian, *Trees in Nature, Myth
and Art* (London, Methuen 1907)

Oliver Rackham, *Trees and Woodland in
the British Landscape* (London 1993)

R.H. Richens, *Elm*
(Cambridge University Press 1983)

Michael Rosenthal, *British Landscpae
Painting* (Oxford, Phaidon 1982)

John Ruskin, *Modern Painters
Vols. 3 & 5 (1856)*
(London, Everyman edition 1904)

John Ruskin, *Praeterita (1885-89)*
(Oxford University Press 1949)

Simon Schama, *Landscape and Memory*
(London, Harper Collins 1995)

Alan Staley, *The Pre-Raphaelite
Landscape* (Oxford, Clarendon 1973)

Jacob George Strutt, *Sylva Britannica*
(London 1822)

Kim Taplin, *The English Path*
(Ipswich, Boydell Press, 1979)

Kim Taplin, *Tongues in Trees, Studies in
Literature and Ecology*
(Bideford, Devon, Green Books 1989)

Tate Gallery, *The Pre-Raphaelites*
(London, Tate Gallery, 1984)

Keith Thomas, *Man and the Natural
World* (London, Allen Lane, 1983)

Alfrew Watkins, *The Old Straight Track*
(London, Methuen, 1925)

Charles Watkins (Ed.) *Rights of Way*
(London, Pinter, 1996)

Ralph Whitlock, *The Oak*
(London, Allen and Unwin, 1985)

J. H. Wilks, *Trees of the British Isles in
History and Legend* (London,
Frederick Muller, 1972)

Gerald Wilkinson, *Epitaph for the Elm*
(London, Hutchinson, 1978)

Yale Center for British Art, *John Hubbard*
(New Haven, Yale Center
for British Art, 1986)

CHAPTER 11

M. G. R. Cannell, J. Grace and A. Booth,
*Possible Impacts of Climatic Warming
on Trees and Forests in the United
Kingdom* (Forestry, Vol. 62, No.4,
Institute of Chartered Foresters,
Edinburgh 1989)

C. Elliot, *Sustainable Tropical Forest
Management by 1995*
(World Wide Fund for Nature,
Gland, Switzerland 1990)

H. L. Elwes and A. H. Henry,
The Trees of Great Britain and Ireland.
(SR Publishers Ltd, Edinburgh 1913)

H. Godwin,
The History of the British Flora
(Cambridge University Press 1956)

B. Gruffydd, *Tree Form, Size and Colour*
(E. & F.N. Spon, London 1987)

L. Hill, *Pruning Simplified*
(Rodale Press, Storey
Communications, Inc. Pownal,
Vermont, USA 1986)

JCLI. *Trees and Shrubs for Landscape
Planting. Joint Council for Landscape
Industries* (The Landscape Institute,
London 1989)

J. Kelly and J. Hillier,
*The Hillier Gardeners Guide to Trees
and Shrubs* (David and Charles,
Newton Abbot, Devon 1995)

Land Use Consultants *Trees in Towns*
(DoE., HMSO, London 1993)

Land Use Consultants.
Urban Tree Strategies
(DoE., HMSO, London 1994)

S. Leathart, *Whence Our Trees*
(Foulsham, London 1991)

W. Linnard, *Welsh Woods and Forests,
History and Utilization* (Amgueddfa
Genedlaethol Cymru, Cardiff 1982)

J. S. Maini, *Sustainable Development
of Forests* (Unasylva Vol. 43,
No.169, FAO, Rome 1992)

A. F. Mitchell, V. E. Schilling and J. E. J.
White, *Champion of Trees in the British
Isles. Technical Paper 7*
(Forestry Commission,
Edinburgh 1994)

O. Rackham, *Trees and Woodland in
the British Landscape*
(J.M.Dent and Sons Ltd.,
London 1990)

A. R. Perry and R. G. Ellis (Eds.),
*The Common Ground of Wild and
Cultivated Plants.* (Botanical Society
of the British Isles Conf. Rep. No.22.,
Amgueddfa Genedlaethol Cymru,
Cardiff 1994)

L. D. Pryor, *Biology of Eucalyptus*
(Studies in Biology 61, Edward
Arnold, London 1976)

O. Rackham, *The History of The
Countryside* (J. M. Dent & Sons Ltd.,
London 1986)

H. J. Read, (Ed.) *Pollard and Veteran Tree
Management II*
(Corporation of London 1993)

J. S. Rodwell (Ed.),
*British Plant Communities Vol.1.
Woodlands and Scrub*
(Cambridge University Press 1991)

G. D. Springthorpe and N. G.
Myhill, *Wildlife Rangers Handbook*
(Forestry Commission Handbook 10,
HMSO, London 1994)

R. G. Strouts and T. G. Winter,
Diagnosis of Ill-Health in Trees.
(DoE. HMSO, London 1994)

P. Thoday and J. Wilson,
Landscape Plants
(Proc. Inst. Hort. Conf. Plants for
Landscape Sites. Institute of
Horticulture, London 1996)

G. W. Watson (Ed.), *Selecting and
Planting Trees* (Morton Arboretum,
Lisle, Illinois 1990)

J. E. J. White, *Trees for Community
Woodland and Ornamental Plantings.*
(See Thoday and Wilson (1996) Proc.
Inst. Hort. Conf. 1996)

J. E. J. White, *New Tree Species in a
Changing World* (Arboricultural Jnl.
Vol. 18, A B Academic Publishers,
London 1994)

J. E. J. White, *Forest and Woodland
Trees in Britain*
(Oxford University Press 1995)

J. E. J. White, *Estimating the Age of
Large Trees in Britain.*
(Research Information note 250.
Forestry Commission, Edinburgh
1994, Revised 1998)

J. E. J. White, *Black Poplar, the Most
Endangered Native Timber Tree in
Britain* (Research Information
Note 239, Forestry Commission,
Edinburgh 1993)

J. E. J. White, *The Possible Effects of
Cultivated Introductions on Native
Willows (Salix) in Britain.*
(In Botanical Society of the British
Isles Conf. Rep. No.22 [Perry and
Ellis] 1994)

J. E. J. White, *What is a Veteran Tree and
Where Are They All ?*
(Quartely Jnl. of Forestry Vol. No.3
Royal Forestry Society, London 1997)

P. S. Wyse Jackon and J. R. Akeroyd,
*Guidelines to be Followed in the Design
of Plant Conservation or Recovery
Plans.* (Council of Europe Press,
Strasbourg 1994)

P. S. Wyse Jackson, *Convention on
International Trade in Endangered
Species (CITES)* (Botanic Gardens
Conservation international, Kew,
London 1994)

H. Young, (translator) Arnoldo
Mondadori Editore, Milan, 1977,
The Macdonald Encyclopedia of Trees
(MacDonald & Co,. London. 1982)

INDEX

ACKNOWLEDGMENTS

Principal photography Archie Miles

20 (Illustrations of maps) Ian McKinnell, 40 Clare Leighton, 65 E. A. Janes, 87 Ardea, 91 E. A. Janes, 94-5 Bob Gibbons/Ardea, 99 Raymond Blythe/Oxford Scientific, 110 C. Dillion McGurk, 111 Bob Gibbons/Ardea, 117G. A. McClean/Oxford Scientific, 120-2 Deni Brown/Oxford Scientific, 122 Geoff Kiff/Oxford Scientific, 124-5 Julie Meech/NHPA, 126 Brian Shoel/Collections, 127 David Dixon/Ardea, 128 Ardea, 154/155 Niall, Benvie/Oxford Scientific, 159 Heather Angel, 159 Bob Gibbons/Holt Studios, 159 Bob Gibbons/Holt Studios, 162 Heather Angel, 162 Local Studies Dept., Newcastle City Library, 164-5 Martin Garwood/NHPA, 167 C. Dillion McGurk, 173-4 Bob Gibbons/Ardea, 176 WH Eldridge, 178-9 Niall Benvie/Oxford Scientific, 180 Terry Heathcote/Oxford Scientific, 181 Michael W Richards/Oxford Scientific, 182 Ardea, 184 Niall Benvie/Oxford Scientific, 185 C. Dillion McGurk, 141 John Mason/Ardea, 148 Jeff Pick, 189 Ardea, 191 C. Dillion McGurk, 204 Niall Benvie/Oxford Scientific, 204 Laurie Campbell/NHPA, 206 John Clegg/Ardea, 215 Stephen Turner/Common Ground, 219 Brian Shuel/Collections, 219 H. P. Robinson/Royal Photographic Society Picture Library, 220 Jarrold Publishing/Collections, 221 Davis Woodfall/NHPA, 225 Weaver/Ardea, 227 Garden and Wildlife Matters Photographic Library, 229 Bob Gibons/Ardea, 232 John Lindau/Ardea, 240 Harold Taylor/Oxford Scientific, 243 Ardea, 246-7 John Daniels/Ardea, 249 G. I. Bernard/Oxford Scientific, 252 Brian Shuel/Collections, 259 P. Peacock/Holt Studios International, 259 Wildlife Matters, 261 Laurie Campbell/NHPA, 269 Ardea, 250 Geoff Kidd/Oxford Scientific, 273 Rural History Centre, 276 Brian Shuel/Collections, 280 Andy Hibbert/Collections, 284 Hans Reinhard/Oxford Scientific, 285 Brecknock Museum, 286 Malcom Crowthers/Collections, 286 Tate Gallery Publications, 287 Rural History Centre, 294 Bob Gibbons/Ardea, 295 Rural History Centre, 297 Miriam McGregor An Apple Tree (from artist), 298 The Bridgeman Art Library, 299 V&A Museum/The Bridgeman Art Library, 300 The Bridgeman Art Library, 301 John Ruskin, 302 The Bridgeman Art Library, 303 The Bridgeman Art Library, 304-5 The Bridgeman Art Library, 306 John Webb/Tate Gallery Publications, 307 Hearn 'Oak Tree'/British Museum , 308 The Bridgeman Art Library, 309 The Bridgeman Art Library, 309 Tate Gallery Publications, 310 National Gallery Picture Library, 311 John Hubbard/Private Collection, 312 The Bridgeman Art Library, 313 National Gallery of Canada, Ottowa, 313 The Bridgeman Art Library, 314-5 Tate Gallery Publications, 316 John Farleigh In My End is My Beginning /kind permission of Mrs. Barbara Kane, 317 Paul Nash Black Poplar Pond /Tate Gallery, London, 318 Thomas Bewick Robin Hood, 318 Eric Ravilious Birdsnesting /© Estate of Eric Ravilious, 1999 All Rights Reserved, DACS, 319 Peter Smith Fallen Tree /from artist, 319 Betty Pennell Bird Garden /from artist, 320 Sarah Van Niekerk Otmoor Oak /from artist, 320 Monica Poole Hollow Tree /from artist, 321 Agnes Miller ParkerWoodcutters /Courtesy of Mrs. Anne Quickenden, 323 Edwina Ellis Parterre, Llanerchaeron /from artist, 323 George Mackley Broken Willow /Felix Dennis Collection, 324 Simon Brett King Willow /from artist, 324 Gwen Raverat Elms by a Pond /© Estate of Gwen Raverat, 1999, All Rights Reserved, DACS, Howard Phipps Wistman's Wood (from artist), 364 Judith Wilson/Oxford Scientific, 367 David Woodfall/NHPA, 380-1 Geoff Kidd/Oxford Scientific, 390-1 Liz Bomford/Ardea, 394 Tom Ang/NHPA

Every effort has been made to trace the copyright holders and we apologise in advance for any unintentional omissions. We would be pleased to insert the appropriate acknowledgements in any subsequent edition of this publication.